AF380483

Integrated Electronic Circuits

Jad G. Atallah • Mohammed Ismail

Integrated Electronic Circuits

 Springer

Jad G. Atallah
Notre Dame University – Louaize
Zouk Mosbeh, Lebanon

Mohammed Ismail
Wayne State University
Detroit, MI, USA

ISBN 978-3-031-62706-4   ISBN 978-3-031-62707-1   (eBook)
https://doi.org/10.1007/978-3-031-62707-1

This Springer imprint is published by the registered company Springer Nature Switzerland AG
The registered company address is: Gewerbestrasse 11, 6330 Cham, Switzerland

If disposing of this product, please recycle the paper.

To Nora, Liam, Lilia
Jad G. Atallah
To the memory of my parents and to my family
Mohammed Ismail

Preface

A major challenge that has been facing education in basic electronic circuits is the preparation of students to continue their higher education or pursue a career in electronics given the accelerating pace of expansion of the field and the constant, sometimes even shrinking, time allocated for electronics in academic programs.

A key to meeting this challenge is for academia to incorporate Electronic Design Automation (EDA) tools as a core entity in its programs, blending it seamlessly with its classical tools. This requires that course books (textbooks) strike a delicate balance between the required breadth and depth of the material to be covered, all while naturally paving the way for the usage of EDA tools. As such, all the results in this book originate from mathematical models where derivations are shown in details as well as from integrated circuits simulations.

The circuits are meticulously analyzed theoretically in order to show the students that in electronic circuits formulas only apply to fundamental relations with everything else being a result of analysis. The circuits are also simulated using BSIM models partly in order to corroborate the theoretical analysis, but more importantly to show the difference between the theoretical results and more accurate results, mainly due to the differences in the models and the approximations done in the theoretical part.

The students need to hone both theoretical and simulation skills and choose when to use each appropriately. Theoretical analysis is appropriate to understand the dynamics of the circuits in order to improve them or invent new ones. This is why they should be done symbolically as a first step. Also they are appropriate to provide the simulator with an initial circuit to start with. Afterward, simulations take over, and the higher the competence the students have in this, the faster the specifications can be met.

With these observations in mind, this book constitutes a unique offering as a first formal contact with electronic circuits whereby the students are exposed early-on to integrated circuit technology and EDA tools in order to make them ready for all the advancements that are taking place in the field today and in the foreseeable future.

The pre-requisites needed for the students to follow this book consist of knowledge and skills in basic electrical circuits, which are usually covered in basic circuits courses. These include basic components including resistors, capacitors, inductors, and sources in addition to mesh analysis (KVL), nodal analysis (KCL), Thevenin and Norton models, input and output impedances, and frequency response including transfer functions and bode plots. Also, basic knowledge in logic design is needed, which is also covered in a separate course.

This book is the fruit of several years of research and education experience in the field of integrated circuits. It is built from the ground up and follows a novel hands-on approach to fundamental electronics education that combines a unique pedagogy along with the right amount of theory and practice. The information for example is presented visually not only for the circuits but also for the signals involved. It aims to be easily read and understood by the students, which gives the instructors ample time to concentrate on the important points.

The book is accompanied by free online resources such as the integrated circuit libraries whose simulation results are at the basis of the graphs found in this book. As such, it constitutes an integral element within the multitude of resources that students have at their disposal.

The book is divided into four parts:

Part I	Signals and Basic Applications: Chaps. 1 and 2
Part II	Devices: Chaps. 3 and 4
Part III	Integrated Circuit Technology Overview: Chaps. 5 and 6
Part IV	Integrated Circuits: Chaps. 7, 8, 9, 10, 11, and 12

Chapter 1 starts with sample applications of electronic circuits. It then presents fundamental aspects of circuits such as information-related signals in the voltage and current domains. This is followed by best practices to read and analyze electronic circuits in addition to unit conversions and logarithmic scales. Finally, feedback basics are presented along with Miller's theorem.

Chapter 2 presents three of the most basic and important applications of electronic circuits: signal amplification, signal filtering, and logic computation. This chapter starts by looking into the two main variations of amplifiers, regular and operational. This is followed by continuous-time filtering. Finally, logic gates for computation are briefly discussed along with their non-idealities.

Chapter 3 presents several ways to model a diode. The method to devise models and how to choose between them given a certain context applies not only to diodes, but to all electric and electronic circuits. As a result, these are important skills to acquire and hone in the long run. This chapter also presents some important applications of diodes such as AC-to-DC conversion, lighting, and photodetection.

Chapter 4 discusses the general behavior of a transistor. This is followed by the behavior of technology-specific transistors, namely FETs and BJTs, including their characterization. Afterward, basic circuit applications are presented.

Chapter 5 covers the physical aspects of technology starting at the system level down to the device level. The chapter discusses the materials used, the fabrication process, and the limitations including process variations, parasitics, and electromigration. Additionally, reliability and aging are also discussed along with heat management. This chapter concludes with an overview of past, present, and future trends in technology advancement.

Chapter 6 presents a custom IC design flow with all the major steps that need to be taken. This includes the theoretical analysis, schematic entry, simulations, layout, and post-layout simulations. This design flow is meant to be applicable to any IC design suite since it targets the fundamentals without going into vendor-specific features.

Chapter 7 discusses single-ended amplifiers using MOSFETs. The approach is generic, which means it can be generalized to any other types of transistors.

Chapter 8 describes circuits for logic applications. It focuses particularly on MOSFET-based implementations. It starts by categorizing the different logic circuit families. This is followed by detailed discussions regarding each logic circuit family, starting with static logic and finishing with dynamic logic.

Chapter 9 describes differential structures for analog and digital applications. It starts by introducing the concept of electrical noise. Afterward, the main idea behind differential circuits is presented before detailing analog followed by digital differential structures.

Chapter 10 provides an overview of the general power distribution scheme in integrated circuits, followed by a discussion about regulators and converters, and reference and bias circuits.

Chapter 11 focuses on output stages designed to drive loads with small input impedances.

Chapter 12 introduces the concept of mixed-signal circuits and some of their applications. It looks into comparators, switched-capacitor circuits, sample-and-hold circuits, digital-to-analog converters, analog-to-digital converters, as well as oscillators and phase-locked loops.

This book can be covered in two single-semester courses, amounting to about 90 h in total. The first course can cover parts I, II, and III, giving the students a very good overview of the field and a solid foundation in the basics. This course, particularly parts I and II, can also be given to students in other fields such as communications and mechanical engineering since it covers the basics of electronics across various applications. The second course can cover Part IV. Here the students are immersed into integrated electronic circuits and all their practical aspects.

Since electronics is an applied field, it is expected that at least the same amount of time allocated to lectures be allocated to hands-on sessions where the students can use EDA tools to experiment with the given IC design libraries and to design and implement their own circuits.

We would like to thank all the current and former students of the basic electronic courses and the ndu ecas lab at the Notre Dame University—Louaize (NDU) for their indirect contribution to this work by providing valuable feedback regarding the efficacy of the educational methods across the years. We would also like to extend our gratitude to the Wayne State University for its support.

Zouk Mosbeh, Lebanon   Jad G. Atallah

Abbreviations

ADC	Analog-to-Digital Converters
APF	All-Pass Filter
BEOL	Back End Of Line
BPF	Band-Pass Filter
BSIM	Berkeley Short channel IGFET Model
CMC	Compact Model Coalition
CMFB	Common-Mode Feedback
CNTFET	Carbon Nanotube FET
COB	Chip-On-Board
CP	Charge Pump
CTAT	Complementary-To-Absolute-Temperature
CTM	Capacitor-Top-Metal
CVD	Chemical Vapor Deposition
C-VD	Constant-Voltage Drop
DAC	Digital-to-Analog Converters
dB	Decibel
D-FF	D-type Flip Flop
DRC	Design Rule Checking
DUT	Design Under Test
EDA	Electronic Design Automation
EDS	Energy Dispersive Spectroscopy
EKV	Enz, Krummenacher, Vittoz
EM	ElectroMigration
ESD	Electrostatic Discharge
FDSOI	Fully Depleted Silicon on Insulator
FEOL	Front End Of Line
FET	Field-Effect Transistor
FinFET	Fin FET
FoM	Figure-of-Merit
GAAFET	Gate-all-around FET

GaN	Gallium Nitride
GDSII	Graphic Design System II
GENF	General Filter
HCI	Hot Carrier Injection
HEMT	High-Electron-Mobility Transistor
HFET	Heterostructure FET
HiSIM	Hiroshima-University STARC IGFET Model
HPF	High-Pass Filter
IC	Integrated Circuit
IGBT	Insulated-Gate Bipolar Transistor
IPF	Intrinsic Parameter Fluctuations
JFET	Junction FET
LDO	Low DropOut
LED	Light-Emitting Diode
LER	Line Edge Roughness
LNA	Low-Noise Amplifiers
LNB	Low-Noise Block downconverters
LPF	Low-Pass Filter
LTI	Linear-Time-Invariant
LVS	Layout Versus Schematic
MBCFET	Multi-Bridge Channel FET
MESFET	Metal-Semiconductor FET
MGG	Metal Grain Granularity
MIM	Metal-Insulator-Metal
MOSFET	Metal-Oxide-Semiconductor FET
MPW	Multi-Project Wafer
NBTI	Negative Bias Temperature Instability
OBM	Original Brand Manufacturer
ODM	Original Device Manufacturer
OEM	Original Equipment Manufacturer
Op Amp	Operational Amplifier
OTA	Operational Transconductance Amplifier
OTF	Oxide Thickness Fluctuations
PA	Power Amplifiers
PBTI	Positive Bias Temperature Instability
PCB	Printed Circuit Board
PDK	Process Design Kit
PDN	Pull-Down Network
PEX	Parasitic EXtraction
PFD	Phase-Frequency Detector
PIV	Peak Inverse Voltage
PLL	Phase-Locked Loop
PRD	Process Rule Deck
PSG	Polysilicon Granularity

PSU	Power Supply Unit
PTAT	Proportional-To-Absolute-Temperature
PUN	Pull-Up Network
PVD	Physical Vapor Deposition
PVT	Process Voltage Temperature
PWM	Pulse Width Modulation
RDF	Random Dopant Fluctuations
RF	Radio Frequency
SEM	Scanning Electron Microscope
SHE	Self-Heating Effects
Si2	Silicon Integration Initiative
SiC	Silicon Carbide
SiGe	Silicon Germanium
SiP	System in Package
SISO	Single-Input-Single-Output
SMPS	Switched-Mode Power Supply
SNR	Signal-to-Noise Ratio
SoC	System-on-Chip
SOI	Silicon-On-Insulator
SPICE	Simulation Program with Integrated Circuit Emphasis
TEM	Transmission Electron Microscope
THD	Total Harmonic Distortion
TIA	Transimpedance Amplifier
TSV	Through-Silicon Via
TVS	Transient-Voltage-Suppression
USB	Universal Serial Bus
VCO	Voltage-Controlled Oscillator
V-VD	Variable-Voltage Drop

Contents

Chapter 1
Circuits and Signals

Fundamental aspects of circuits and signals are discussed in this chapter. These aspects include sample applications of electronic circuits such as embedded systems with sensors and actuators, power supply units, particularly AC to DC converters, and communication systems. Additionally, basic information related to signals and their domains is discussed. This is followed by best practices to read and analyze electronic circuits in addition to unit conversions and logarithmic scales. Finally, feedback basics are presented along with Miller's theorem.

1.1 Sample Applications

The importance of circuits stems from the importance of the signals that they process. Therefore, we need to keep in mind that eventually, it is the signal that matters, not the circuit.

These signals are created to either carry power, information, or both. Power-carrying signals are found in power electronics applications. Information-carrying signals are found in information technology applications.

Electronic circuits have become so widespread that we do not notice their presence anymore. In fact, these circuits have become an integral part of our infrastructure and are continuously expanding in that direction, even for applications that were once electronics-free.

1.1.1 Embedded System with Sensors and Actuators

A very common application of electronic circuits is sensing, computing, and controlling. This is the case for all embedded electronic systems found in household

© The Editor(s) (if applicable) and The Author(s), under exclusive license to
Springer Nature Switzerland AG 2024
J. G. Atallah, M. Ismail, *Integrated Electronic Circuits*,
https://doi.org/10.1007/978-3-031-62707-1_1

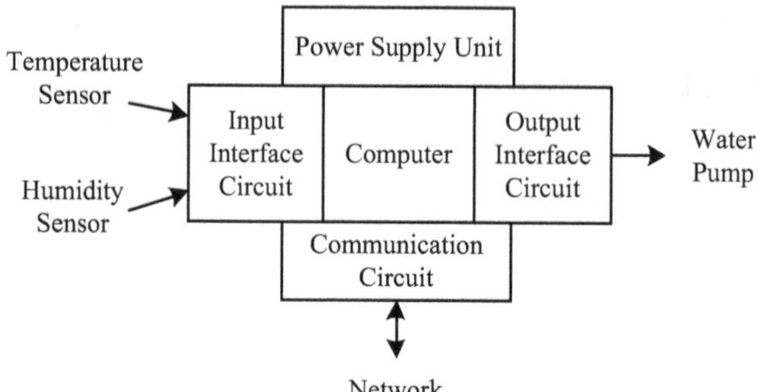

Fig. 1.1 Monitoring device for agricultural applications

Fig. 1.2 Sensor interface circuit

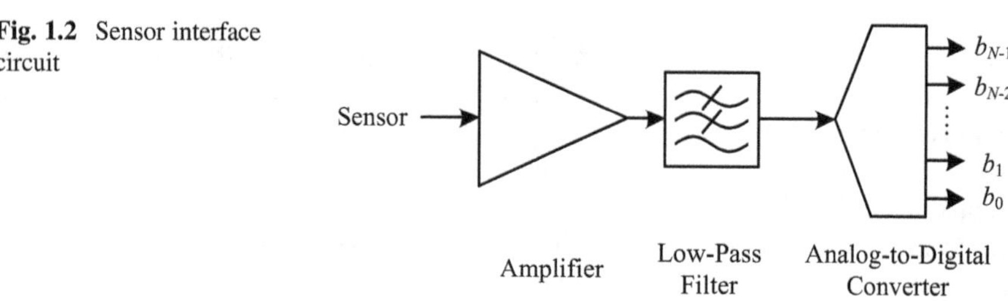

appliances, industrial machines, and cars, for example. The center of such a system is a computer connected on one side to an input coming from a set of sensors and on the other side to an output going into actuators, displays, etc.

For example, a monitoring device for agricultural applications is shown in Fig. 1.1.

The central unit of the device is the computer operating in the digital domain. The circuits surrounding it are there either to provide power or to interface it with the outside world. These are typically radio frequency circuits, mixed-signal circuits, and/or analog circuits. The input interface might contain an amplifier, a low-pass filter, and an analog-to-digital converter as in Fig. 1.2.

An example of the output interface is shown in Fig. 1.3. It consists of a transistor and a relay that drives the motor of the water pump.

1.1.2 Power Supply: AC-DC Conversion

All electronic devices require DC power to operate. Often, this power comes from an AC source, so an AC-DC converter is needed. This converter is typically part of the power supply unit (PSU), which is present in all AC-powered electronic devices. Sometimes, the main PSU is inside the device such as for a desktop computer and at other times it is outside the device, such as for a display screen.

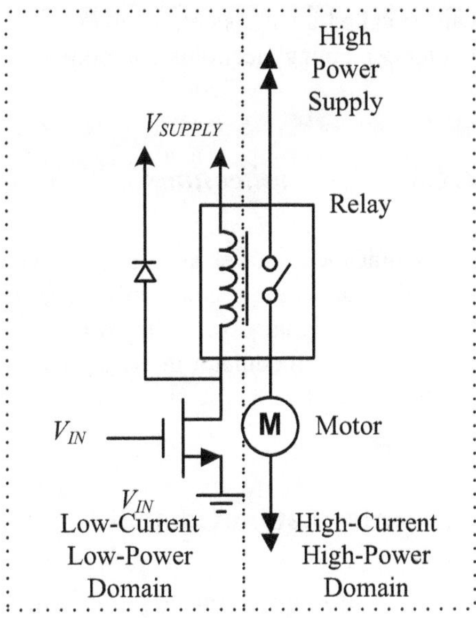

Fig. 1.3 Output circuit to control the motor of a water pump

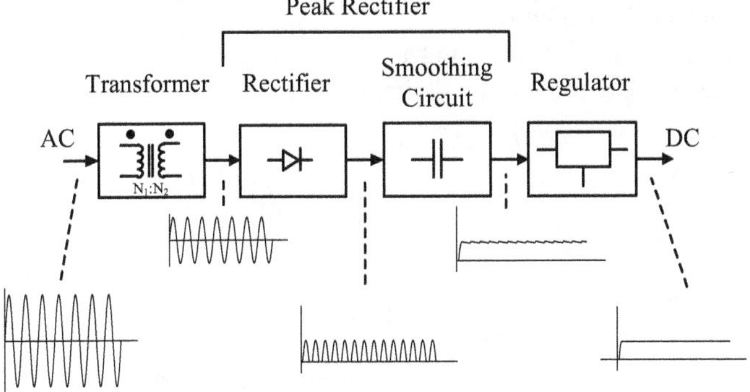

Fig. 1.4 AC-DC converter block diagram

In many applications, several PSUs exist in series where most of them are inside the device, taking a certain DC-level power and converting it using a DC-DC converter to another DC-level power compatible with the internals of the device.

The device that the PSU powers might contain a battery and the PSU will have special circuitry to charge that battery in an optimal manner. For example, this is the case for what is commonly known as the mobile phone power adaptor which connects to the AC power outlet at the input side and to a Universal Serial Bus (USB) port on the output side.

A typical AC-DC converter is shown in Fig. 1.4. If we call the complete electronic device, such as the mobile phone, a system, then the AC-DC converter can be called a subsystem, whereas the rectifier can be called a block, which is a circuit containing electronic components.

1.1.3 Communications

Communications is yet another prominent application of electronic circuits. For example, for wireless communications a receiver needs to receive the radio frequency (RF) signal, down-convert it to lower frequencies, convert it to digital, then pass it to the demodulator in the digital domain. A very simplified receiver chain is shown in Fig. 1.5.

1.2 Information Content in Signals

There are several methods to put information in a signal. These methods affect the signal shape resulting in the following general signal types:

- continuous signal, also known as an analog signal,
- discrete-time signal, also known as a mixed signal,
- discrete signal, also known as a digital signal.

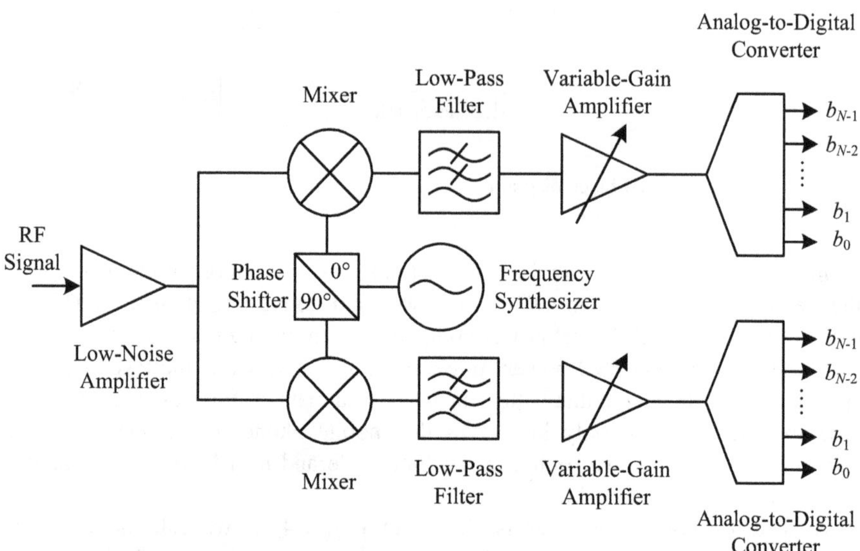

Fig. 1.5 RF receiver signal processing chain

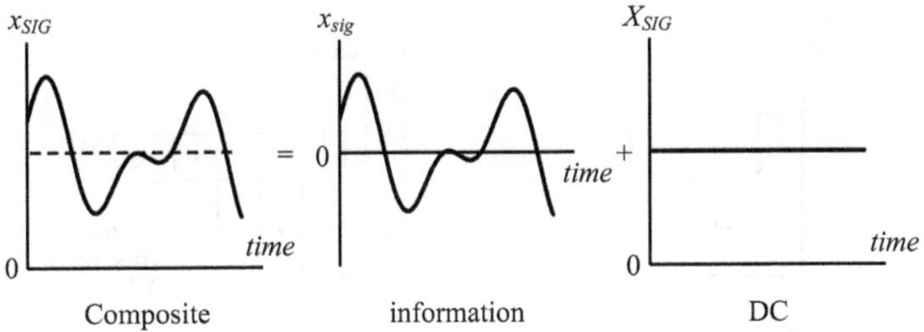

Fig. 1.6   A continuous signal and its component categories

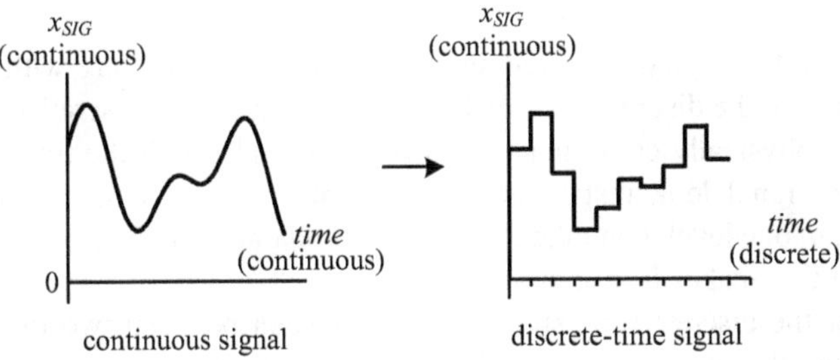

Fig. 1.7   A discrete-time signal generated from a continuous signal

In this discussion, we will concentrate on baseband signals, which are signals that do not have a high-frequency carrier. Baseband signals have most of their power at low frequencies and are typical in most applications.

The best way to differentiate between these types of signals is to plot them versus time.

A continuous signal is continuous both in time and amplitude. A typical continuous signal is shown in Fig. 1.6.

The signal is denoted using the general variable x. The signal can be a voltage signal, current signal, or power signal. The complete signal is called composite, and is denoted using small letters followed by capital letter subscripts. The composite signal is made of two component categories:

1. The information category, which consists of all the frequencies, except for DC. This is where the information resides and is denoted using small letters followed by small letter subscripts.
2. The DC component, which carries no information, but is frequently needed for the correct operation of the circuit or for power delivery. It is denoted using capital letters followed by capital letter subscripts.

A discrete-time signal is discrete in time, but continuous in amplitude. A typical discrete-time signal is shown in Fig. 1.7.

If the generating continuous signal contains a DC component, so will the resulting discrete-time signal.

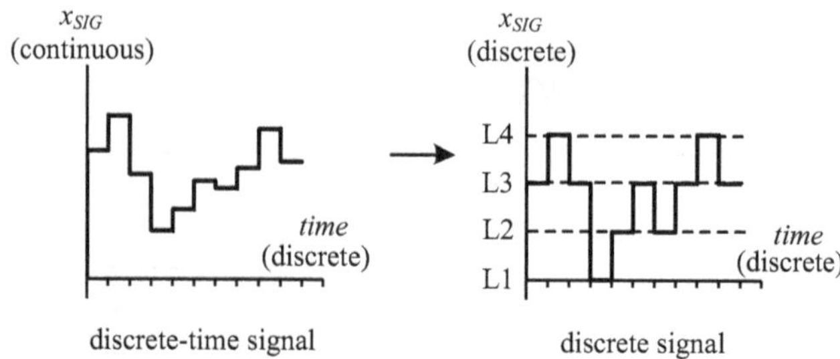

Fig. 1.8 A discrete signal generated from a discrete-time signal

To generate this signal, a mixed-signal circuit is needed along with a reference clock resulting in the discrete time scale. Even-though time is discretized, the whole signal is still physically continuous. This is because although the transitions in the discrete-time signal look instantaneous, in reality the signal needs some time to transition from one level to another. Therefore, fundamentally, it is our interpretation of the signal that makes it discrete-time.

Note how the discrete-time signal is held constant between two time instances. This is because the amplitude is sampled at the beginning of the time period and then held constant using a sample-and-hold circuit. Therefore, no changes in the amplitude occur between the sampling time instances.

A discrete signal is discrete in both amplitude and time. A typical discrete signal is shown in Fig. 1.8. In this case, we say that the amplitude is quantized.

In this example, four levels are used for the amplitude. This creates a difference in the amplitudes between the generating discrete-time signal and the discrete signal. This difference is known as quantization noise. Of course, the more levels used, the less is the quantization noise.

1.3 Voltage and Current Domains

Although circuits contain voltages and currents, it is up to the designer to choose whether, in a certain section of the circuit, the information should be in the voltage, the current, or the power domains.

Most of the circuits these days have the information in the voltage of the signal. This is because it is easy to store voltages using very tiny capacitors or devices having capacitances in them. Additionally, logic gates operate primarily in the voltage domain, which is best suited for computation.

Less frequently, the information might be in the current of the signal. The current is more difficult to store since it requires inductors which are physically large. However, given a certain technology where the allowable voltage range is limited, the current can be increased, thus overcoming the noise that couples to the circuit.

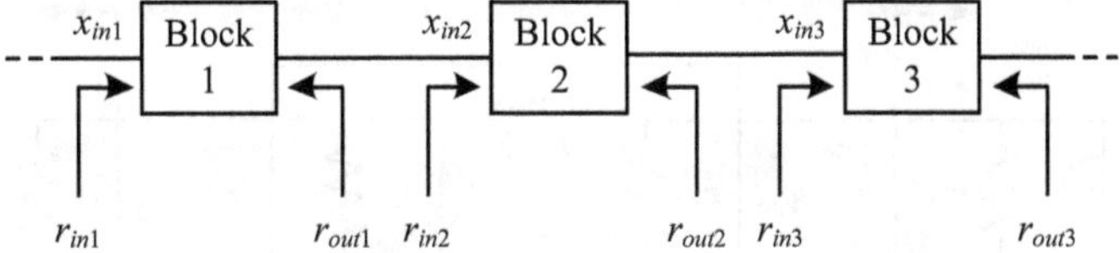

Fig. 1.9 A three-stage cascade of circuit blocks

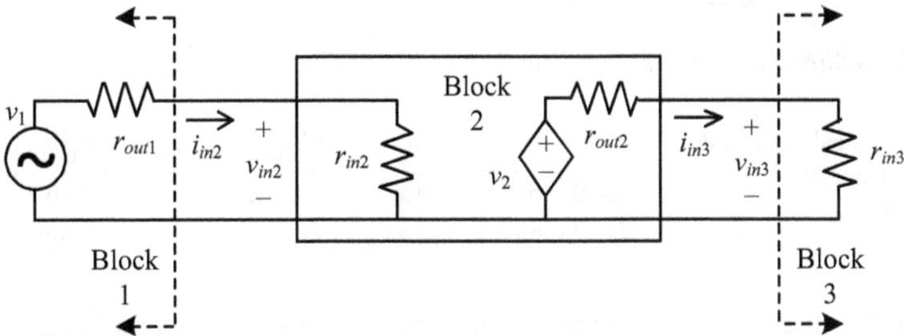

Fig. 1.10 A three-stage cascade of voltage-domain circuit blocks

In some applications such as the ones that deal with power generation and distribution or that make use of radio frequency (RF) circuits, it is the power that is important. RF circuits in particular deal with electromagnetic waves at their interfaces, and these waves carry information in their power.

Among other things, the choice of the information-carrying parameter affects the choice of input and output resistances and in general impedances of the blocks in a circuit.

Consider a three-stage cascade of circuit blocks as shown in Fig. 1.9.

If the circuit is operating in the voltage domain, that is the information is in the voltage of the signal, not its current, then Block 2 and its neighbors can be modeled as in Fig. 1.10.

Since the information is in the voltage of the signal, and in spite of having a current also, Thevenin-equivalent models are used. For Block 2, v_2 is a function of v_{in2}, irrespective of the functionality of the block.

In order to keep the information, we should be able to communicate the voltage efficiently. Looking at the interface between Block 1 and Block 2, the best-case scenario would be to have v_{in2} equal to v_1. This can be achieved by ideally having either r_{out1} zero, that is short, or r_{in2} infinite, that is open, or both. In particular, having an infinite r_{in2} will result in a zero i_{in2}, in which case no current flows from Block 1 to Block 2. In this situation, we say that Block 2 does not load Block 1 since there is no power transfer from Block 1 to Block 2, only voltage transfer.

The same discussion can be made at the interface between Block 2 and Block 3 resulting in an ideal situation where r_{out2} is zero and r_{in3} is infinite.

As a result, an ideal voltage source has a zero output resistance and an ideal voltage load has an infinite input resistance. In practice we try to get as close as possible to this situation.

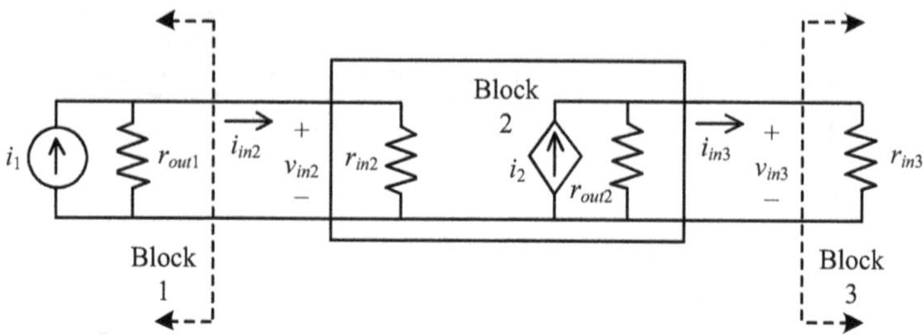

Fig. 1.11 A three-stage cascade of current-domain circuit blocks

Table 1.1 Ideal input and output resistances

	Voltage domain	Current domain
Ideal input resistance	Infinite	Zero
Ideal output resistance	Zero	Infinite

A similar discussion can be made regarding current-mode circuits. If the circuit is operating in the current domain, that is the information is in the current of the signal, not its voltage, then Block 2 and its neighbors can be modeled as in Fig. 1.11.

Since the information is in the current of the signal, and in spite of having a voltage also, Norton-equivalent models are used. For Block 2, i_2 is a function of i_{in2}, irrespective of the functionality of the block.

In order to keep the information, we should be able to communicate the current efficiently. Looking at the interface between Block 1 and Block 2, the best-case scenario would be to have i_{in2} equal to i_1. This can be achieved by ideally having either r_{out1} infinite, that is open, or r_{in2} zero, that is short, or both. In particular having a zero r_{in2} will result in a zero v_{in2}, in which case no voltage is communicated between Block 1 and Block 2. In this situation, we say that Block 2 does not load Block 1 since there is no power transfer from Block 1 to Block 2, only current transfer.

The same discussion can be made at the interface between Block 2 and Block 3 resulting in an ideal situation where r_{out2} is infinite and r_{in3} is zero.

As a result, an ideal current source has an infinite output resistance and an ideal current load has a zero input resistance. In practice we try to get as close as possible to this situation.

A summary of the results above is shown in Table 1.1.

Some circuits are a mixture of both modes. For example, the information can be in the voltage domain at the input and in the current domain at the output. The same conclusions above apply, that is, from Table 1.1, the ideal input resistance would be infinite and the ideal output resistance would also be infinite.

1.4 Vertical Circuits

It is natural to think of circuits as closed loops, since in reality this is what they are. The current from the power supply goes out of the positive terminal and tries to find the easiest way to reach the negative terminal. By easiest, here we mean the path with least resistance. We take advantage of this tendency and make sure the current passes through a load before reaching the negative terminal of the supply, thus achieving a desired functionality.

Electronic circuits can have either one or two power supplies. These two configurations are shown in Fig. 1.12.

A reference node is chosen in a circuit so that all the voltages are reported with respect to that node instead of being reported between any two arbitrary nodes.

In a single power supply circuit as in Fig. 1.12a, the reference node is usually taken at the negative terminal of the power supply itself, and is shared by all the signals in the system. This is a typical situation in most information-processing circuits for example.

In a dual power supply circuit as in Fig. 1.12b, two power supplies, usually having the same voltages, are used. The reference node is taken between the two power supplies. This setup can be found in power-processing circuits for example.

In the single power supply case, the total voltage range across the circuit is V_{SUPPLY} whereby in the dual supply case when the two supplies are equal, the total voltage range across the circuit is $2V_{SUPPLY}$. This is important since in either case, except for a few special cases, the circuit will only accept signals and output signals that are within this range.

When we draw electronic circuit schematics, we do so in such a way that they become more readable, just like a sentence. In this context, what we are reading are voltages, currents, impedances, etc. Now, we will take every circuit in Fig. 1.12 and redraw it accordingly.

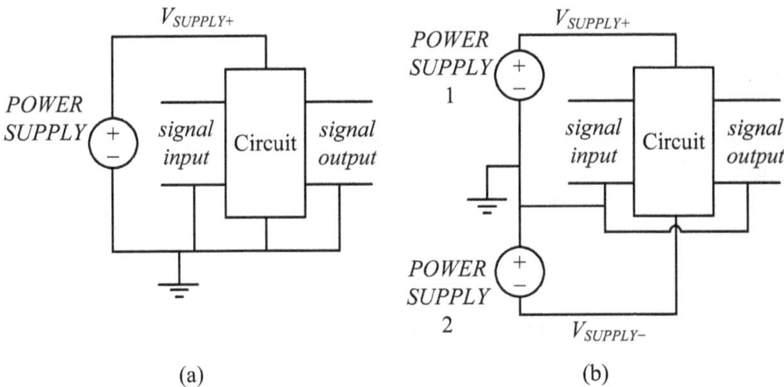

Fig. 1.12 Single power supply (**a**) and dual power supply (**b**) circuit

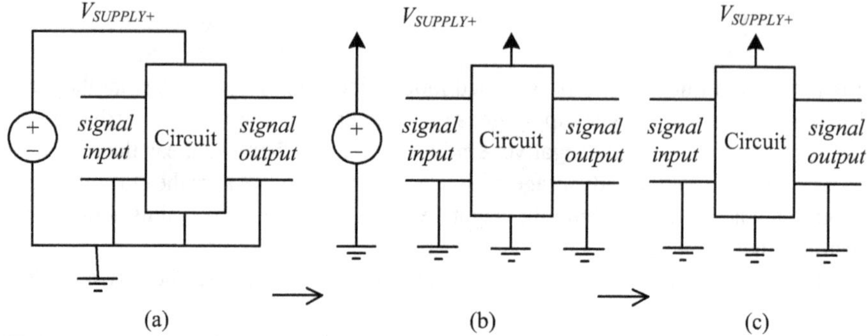

Fig. 1.13 Visual transformation of a single power supply circuit from circular to vertical

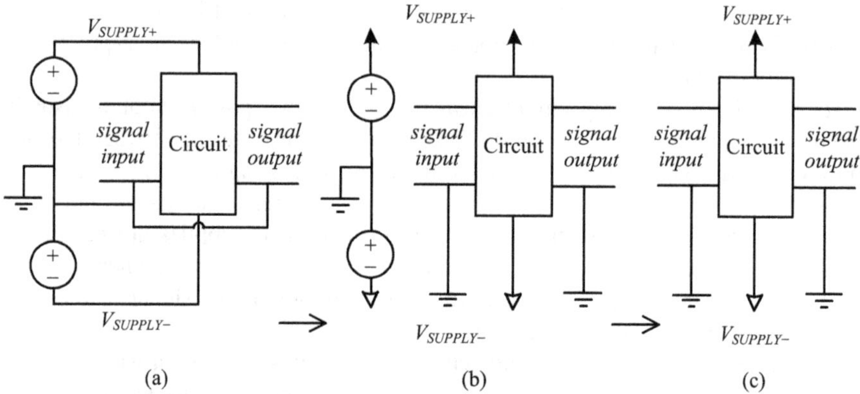

Fig. 1.14 Visual transformation of a dual power supply circuit from circular to vertical

Just like sentences are read in English from left to right, so are electronic circuit schematics. Therefore, the input is on the left and the output is on the right. The high voltage of the supply is at the top and the low voltage is at the bottom, as can be seen in Fig. 1.12.

Considering the circuit in Fig. 1.12a, the visual transformation is shown in Fig. 1.13.

In Fig. 1.13b, $V_{SUPPLY+}$ is given a certain symbol, distinct from that of the reference node and the circuit is broken visually. It is understood that the two nodes at the top having the same symbol are actually connected. The same visual disconnection is done for the reference node at the bottom. In Fig. 1.13c, the power supply is visually eliminated altogether, since it is understood that there is one whose positive terminal is at $V_{SUPPLY+}$ and its negative terminal is at the reference node.

Since the resulting circuit looks vertical now compared to the original circular one, we will call it the vertical version.

Now, consider the circuit in Fig. 1.12b, the visual transformation is shown in Fig. 1.14.

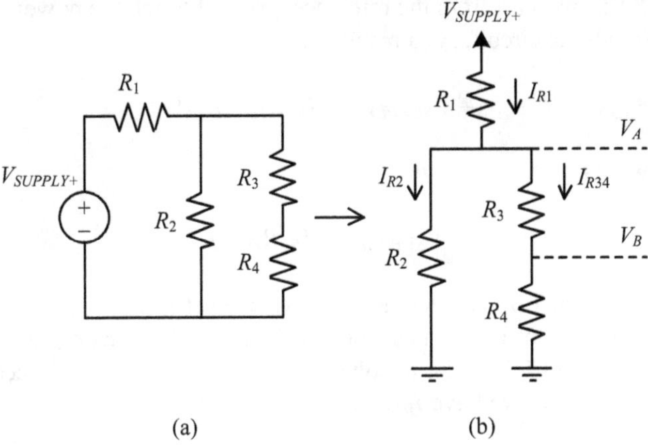

Fig. 1.15 Circular circuit schematic (a) and its vertical version (b)

In Fig. 1.14b, $V_{SUPPLY+}$ and $V_{SUPPLY-}$ are given distinct symbols, both different from that of the reference node, and the circuit is broken visually. In Fig. 1.14c, the power supply is visually eliminated altogether since it is understood that there are two, one whose positive terminal is at $V_{SUPPLY+}$ and whose negative terminal is at the reference node, and the other whose positive terminal is at the reference node and whose negative terminal is at $V_{SUPPLY-}$.

To showcase the practicality of vertical circuits, consider the circuit schematic shown in Fig. 1.15a. This circuit does not contain electronic components, nor does it contain a signal source, but it will serve the purpose of explaining the key points.

Looking at the vertical circuit in Fig. 1.15b, we can see that the visually highest voltage is $V_{SUPPLY+}$ and the visually lowest one is the reference. Therefore all the currents always go downwards.

By just looking at the visual placement of the voltages, we see that:

$$V_{SUPPLY+} > V_A > V_B > 0 \tag{1.1}$$

Therefore, all the voltages are positive, which is a direct consequence of taking the reference at the negative terminal of the power supply in a single supply circuit.

Additionally, we see that the lower part of the circuit has one level to the left (reference to V_A) and two levels to the right (reference to V_B to V_A). This creates an imbalance, since R_2 will have to handle the voltage drop handled by R_3 and R_4 combined in series.

All this combination of resistors constitutes the load of the power supply. Therefore, what the power supply sees is

$$R_{TOTAL} = R_1 + (R_2 // (R_3 + R_4)) \tag{1.2}$$

and I_{R1} coming out of the power supply passes through this combination.

Doing a KVL, we start from the reference, go up through the power supply, then go down through the circuit. As a result, we have

$$- V_{SUPPLY+} + I_{R1} R_{TOTAL} = 0 \tag{1.3}$$

In other words,

$$V_{SUPPLY+} = I_{R1} R_{TOTAL} \tag{1.4}$$

Out of this, given $V_{SUPPLY+}$ and R_{TOTAL}, we can get I_{R1}.

When dealing with impedances or resistors for that matter, we have three parameters: R, V, and I. In order to solve for these, two of them are needed.

Therefore, now that we have I_{R1}, we can see visually that

$$V_A = V_{SUPPLY+} - I_{R1} R_1 \tag{1.5}$$

In other words, we took a known voltage $V_{SUPPLY+}$ and dropped it by $I_{R1} R_1$.

We can do the same to find I_{R2} by starting from V_A as in

$$I_{R2} = \frac{V_A}{R_2} \tag{1.6}$$

As for I_{R34}, we can get it in one of two ways: using KCL or starting from the known voltage V_A

$$I_{R34} = I_{R1} - I_{R2} = \frac{V_A}{R_3 + R_4} \tag{1.7}$$

Therefore, V_B can now be deduced either starting from the known V_A or from the known reference:

$$V_B = V_A - I_{R34} R_3 = 0 + I_{R34} R_4 \tag{1.8}$$

The above observations, including the results, look simple, but for the trained eye, they are deduced instantly when looking at the vertical circuit. More training will lead to all these observations instantly, even when looking at the circular circuit while drawing the vertical circuit mentally. This is what we mean by actually reading the circuit.

1.5   Unit Conversion and Logarithmic Scales

The range of numbers encountered when dealing with electronic circuits is quite large. The engineering notation for numbers is used in order not to have to write very large or small numbers, and for these numbers to be quickly readable. It makes use of a subset of SI prefixes that relate only to the bases having a power of three. As a result, the most common prefixes used in the engineering notation are shown in Table 1.2.

The range of values depends on the quantity being addressed:

- Voltages range from millivolts to several volts.
- Resistors range from a few Ohms to kilo-Ohms. Therefore, milli-Ohms is in the realm of short circuits and mega-Ohms is in the realm of open circuits.
- As a result of the above, currents range from micro-Amperes to Amperes. Nano-Amperes is thus in the realm of very low current and several Amperes is in the realm of high current.
- Capacitors range from femto-Farads to micro-Farads depending on whether they are integrated capacitors (as in integrated circuits) or discrete components.
- Inductors range from nano-Henries to milli-Henries also depending on whether they are integrated inductors (as in integrated circuits) or discrete components.
- Frequencies range from Hz to Giga-Hz.

A visual depiction of the ranges above is shown in Fig. 1.16.

Table 1.2 Common prefixes used in the engineering notation

Prefix name	Prefix symbol	Base 10 exponent
Giga	G	+9
Mega	M	+6
Kilo	k	+3
–	–	0
Milli	m	−3
Micro	μ or u	−6
Nano	n	−9
Pico	p	−12
Femto	f	−15

Fig. 1.16 Usual ranges of voltages, resistors, currents, capacitors, inductors, and frequencies

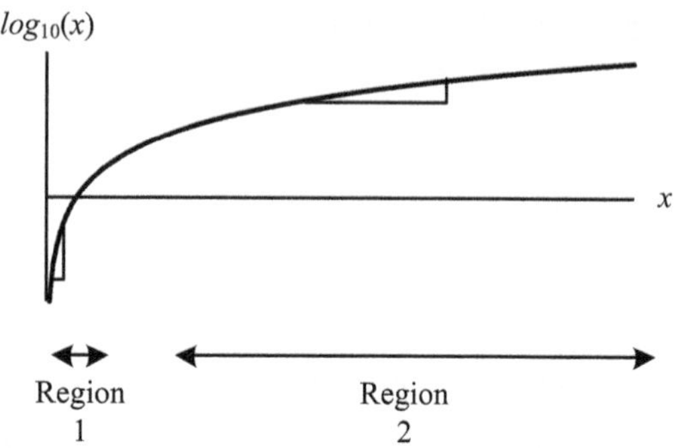

Fig. 1.17 Logarithmic scale

Another inconvenience that also exists when dealing with numbers shows up when we have a large range of variations for a certain quantity. In this case, the whole range needs to be preserved in order to see the big picture, while high precision is needed in only part of the range. If we plot, for example, this whole range and aim for high precision everywhere, the plot would be too large and very difficult to deal with. For this reason, it is usual in these cases to compress the range before reporting or plotting the values.

The most common function to compress a range is the logarithmic function, especially base ten. The plot is shown in Fig. 1.17.

As can be seen, in Region 1 where x is small, the slope is very high. This means that a small variation in x will show up as a large variation in its logarithmic value. Therefore, variations in x in Region 1 are well-observed.

On the other hand, in Region 2 where x is large, the slope is very small. This means that variations in x will not show up easily in its logarithmic value. Therefore, variations in x in Region 2 are subdued, or compressed.

The result is as if we are holding a magnifying glass over Region 1, while Region 2 looks blurred.

Classically, this is applied to relative power quantities and the unit is bel (B). Typically, this is multiplied by ten in order to avoid reporting many digits after the decimal point, resulting in the unit of decibel (dB). For example, if the power input to a system is P_{in} and the power output is P_{out}, then the power gain of the system in dB is

$$\text{Power Gain}\big|_{dB} = 10 \log\left(\frac{P_{out}}{P_{in}}\right) \tag{1.9}$$

Therefore, the unit-less ratio of powers in scalar translates to dB on a decibel scale.

Note that when the scalar power ratio is half, the power gain is -3 dB, which is useful in many applications.

$$\text{Power Gain}|_{\text{dB}} = 10 \log\left(\frac{1}{2}\right) = -3 \text{ dB} \tag{1.10}$$

This can be applied to ratio of voltages and currents also, knowing that power is proportional to the voltage squared or current squared, resulting in:

$$\text{Power Gain}|_{\text{dB}} = 10 \log\left(\frac{v_{out}^2}{v_{in}^2}\right) = 20 \log\left(\frac{v_{out}}{v_{in}}\right)$$
$$\text{Power Gain}|_{\text{dB}} = 10 \log\left(\frac{i_{out}^2}{i_{in}^2}\right) = 20 \log\left(\frac{i_{out}}{i_{in}}\right) \tag{1.11}$$

Therefore, when dealing with voltages and currents, the multiplication factor is 20 instead of 10.

Note that when the scalar voltage ratio or current ratio is half, the power gain is $-$ 6 dB.

In addition to unit-less quantities, the decibel scale can be used for quantities with an actual unit.

For example, if we want to express an amount of x mW using the decibel scale, we get the units of dBmW, which is usually shortened as dBm:

$$x|_{\text{dBm}} = 10 \log\left(\frac{x|_{\text{mW}}}{1 \text{ mW}}\right) = 10 \log(x) \tag{1.12}$$

Sometimes, when dealing with the unit of Watts, the 'W' is dropped and dBW becomes dB, but this is misleading since dB relates to unit-less scalar quantities. Therefore, it is best not to drop the 'W' when dealing with Watts and report the unit as dBW.

Another example, if we want to convert an amount of x Volts into the decibel scale, we get the units of dBV as in:

$$x|_{\text{dBV}} = 20 \log\left(\frac{x|_{\text{V}}}{1 \text{ V}}\right) = 20 \log(x) \tag{1.13}$$

1.6 Feedback Basics

Systems with feedback are frequently encountered in electronic circuits. Often we intentionally make use of feedback in our circuits in order to get a desirable effect. Systems with feedback are called closed-loop systems as opposed to open-loop systems that that do not contain feedback.

A simple system closed-loop system is shown in Fig. 1.18.

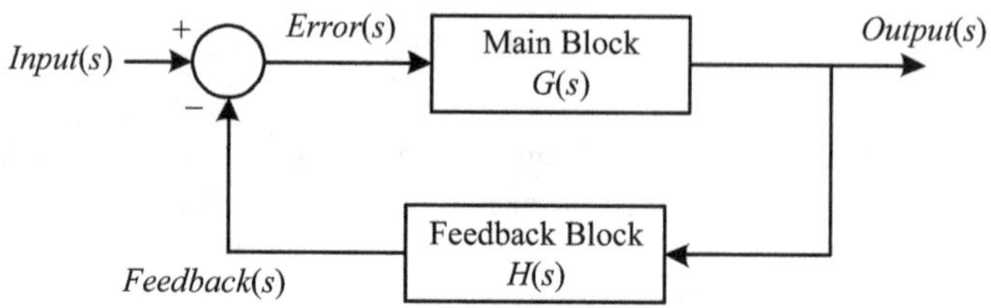

Fig. 1.18 Simple closed-loop system

In this analysis, for convenience, we will be using the *s*-domain. The blocks are shown with their transfer functions. They can be described as follows:

- The Main Block is the one that we are trying to control. Typically, it is an amplifier, but in principle it can be anything else. Its transfer function is called the open-loop transfer function. In this setup, it is

$$G(s) = \frac{Output(s)}{Error(s)} \tag{1.14}$$

- The Feedback Block is a circuit that takes the output, processes it, then feeds it back into the input. Usually, this block somehow attenuates the output signal before feeding it back into the input. If it is replaced by a short, then the amount that is fed back is a hundred percent. Its transfer function in this setup is

$$H(s) = \frac{Feedback(s)}{Output(s)} \tag{1.15}$$

- The subtraction block might be a stand-alone block, but more commonly, it is either part of the Main Block, or it simply is the way the Feedback signal is combined with the Input signal. Anyhow, its input–output relation is

$$Error(s) = Input(s) - Feedback(s) \tag{1.16}$$

An important assumption is made in this analysis and that is the blocks do not load each other. In other words, their input and output resistances are ideal irrespective of the nature of the signals being used. As a result, their behavior does not change in the presence of the surrounding blocks.

The way this system works is that for every change at the input, an error signal is generated, which is then processed by the Main Block, generating the Output signal. However, this Output signal is not stable yet because part of it is fed back through the Feedback Block, generating the Feedback signal, thus generating a new Error signal. This loop keeps on going until the Error signal becomes zero, at which point, the Output signal is stable.

It is the subtraction that ensures that the Error signal goes to zero eventually, stabilizing the Output signal. As a result, this system is called a negative feedback system. If for some reason the subtraction becomes an addition, the feedback becomes positive, and the loop becomes unstable resulting in either an oscillatory behavior at the output, or in the output latching up to a value and staying there without reacting to the input signal anymore. Therefore, instability should be avoided.

We will now derive the relations between the input and output of the system, also known as the closed-loop transfer function. Starting from the behavior of the individual blocks (1.14) and (1.15):

$$Output(s) = G(s)Error(s) = G(s)Input(s) - G(s)Feedback(s) \tag{1.17}$$

Now, substituting (1.16) in (1.17), we get

$$Output(s) = G(s)Input(s) - G(s)H(s)Output(s) \tag{1.18}$$

As a result, we have

$$G_f(s) = \frac{Output(s)}{Input(s)} = \frac{G(s)}{1 + G(s)H(s)} \tag{1.19}$$

The denominator $1 + G(s)H(s)$ is called the amount of feedback. As for $G(s)H(s)$ itself, it is called the open-loop transfer function.

An interesting situation arises when $G(s)$ is very large. In this case, $G_f(s)$ in (1.19) can be rewritten as:

$$G_f(s) = \frac{G(s)}{1 + G(s)H(s)} = \frac{1}{\frac{1}{G(s)} + H(s)} \approx \frac{1}{H(s)} \tag{1.20}$$

This shows that when $G(s)$ is very large, the behavior of the closed-loop system becomes independent of the Main Block and dependent only on the Feedback Block. This has a lot of practical implications. For example, when the Main Block is an amplifier with very high gain, then the behavior of the closed-loop system will be determined exclusively by the circuit in the feedback path.

To visualize whether the system is stable, we will plot the locus of the $G(s)H(s)$ in the GH plane, with $s = j2\pi f$. In practice, we are interested in situations where $G(s)$ $H(s)$ has no poles in the right-half side of the s-plane. In this case, and according to the Nyquist stability criterion, for the system to be stable, there should be no encirclement of the -1 point.

Figure 1.19 shows two samples of such plots, one for a stable loop in Fig. 1.19a and another for an unstable loop in Fig. 1.19b.

At zero frequency, $G(s)H(s)$ starts from the right-hand side of the plane on the real axis, and then works its way to the origin as the frequency goes to infinity. This is

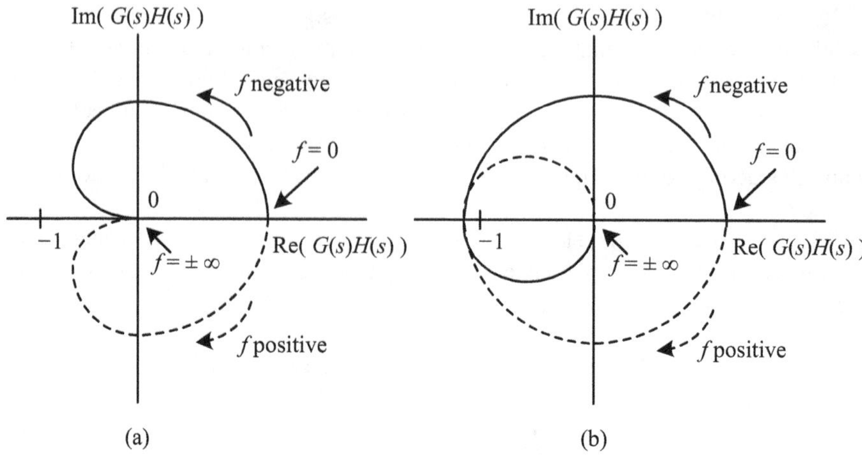

Fig. 1.19 Sample $G(s)H(s)$ locus plot of a stable loop (**a**) and unstable loop (**b**)

because at very high frequencies, all blocks will have no gain, and their transfer functions go to zero.

Note that we always aim for stability across all frequencies, even across the ones that are not within the range of our input. This is because noise and interference coupling to the system can have any frequency. As a result, if the loop is not completely stable, interference and noise alone might throw it out of stability.

Now we will look at the effects of negative feedback on the gain and the bandwidth by comparing those of the stand-alone Main Block to those of the whole closed-loop system.

For illustration purposes, and without any loss of generalization, let us assume that the Main Block transfer function has a low-frequency gain A and a single pole f_{high}, then

$$G(s) = \frac{A}{1 + s/2\pi f_{high}} \tag{1.21}$$

Also, let us assume that $H(s)$ is a constant with a value B.
Therefore, substituting (1.21) in (1.19) we get

$$G_f(s) = \frac{G(s)}{1 + G(s)H(s)} = \frac{\dfrac{A}{1 + s/2\pi f_{high}}}{1 + \dfrac{AB}{1 + s/2\pi f_{high}}} \tag{1.22}$$

$$= \frac{A}{1 + AB + s/2\pi f_{high}} = \frac{A/(1 + AB)}{1 + s/2\pi f_{high}(1 + AB)}$$

Comparing (1.22) to (1.21), we conclude the following:

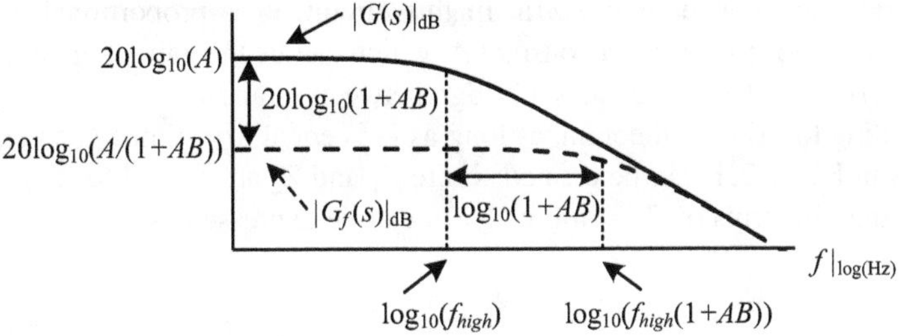

Fig. 1.20 Bode magnitude plots of $G(s)$ and $G_f(s)$

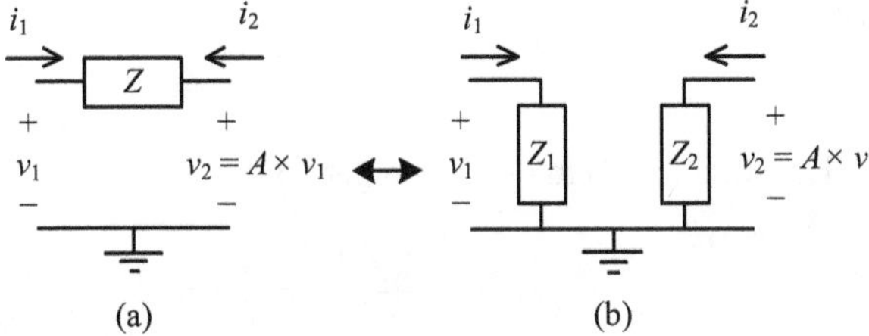

Fig. 1.21 Special closed-loop circuit (a) and its equivalent (b)

1. The low-frequency gain of the closed-loop system is lower than that of the stand-alone Main Block by a factor equal to the amount of feedback which is $1 + AB$.
2. The pole of the closed-loop system is higher than that of the stand-alone Main Block by a factor equal to the amount of feedback which is $1 + AB$.

As a result of the two observations above, the product between the low-frequency gain and the pole is the same in both situations, that is:

$$A \times f_{high} = (A/(1+AB)) \times \left(f_{high}(1+AB) \right) \qquad (1.23)$$

The pole is also the bandwidth of the system. For this reason, we say that the scalar gain-bandwidth product of a system is constant. Practically, this means that often we can trade gain for bandwidth and the other way around.

This can be seen visually when comparing the Bode magnitude plot of the stand-alone Main Block $G(s)$ with the Bode magnitude plot of the closed-loop system $G_f(s)$ as in Fig. 1.20.

1.7 Miller's Theorem

Closed-loop systems sometimes involve a certain configuration of components that can be easily analyzed using Miller's Theorem.

Consider the setup in Fig. 1.21a. In this circuit, v_2 is proportional to v_1. The constant of proportionality is a constant A, which can be the gain of an amplifier for example. An impedance Z creates a bridge between v_1 and v_2.

According to Miller's theorem, as long as v_2 is equal to v_1 times A, an equivalent circuit as in Fig. 1.21b can be created where Z_1 and Z_2 are related to Z.

To find Z_1 in terms of Z, i_1 in Fig. 1.21a can be expressed as

$$i_1 = \frac{v_1 - v_2}{Z} = \frac{v_1 - A \times v_1}{Z} = v_1 \left(\frac{1 - A}{Z}\right) \tag{1.24}$$

Also i_1 in Fig. 1.21b can be expressed as

$$i_1 = \frac{v_1}{Z_1} \tag{1.25}$$

Equating (1.24) to (1.25) we get

$$v_1 \left(\frac{1 - A}{Z}\right) = \frac{v_1}{Z_1} \Rightarrow Z_1 = \frac{Z}{1 - A} \tag{1.26}$$

The same procedure can be carried for Z_2. Current i_2 in Fig. 1.21a can be expressed as

$$i_2 = \frac{v_2 - v_1}{Z} = \frac{v_2 - v_2/A}{Z} = v_2 \left(\frac{1 - 1/A}{Z}\right) \tag{1.27}$$

Also i_2 in Fig. 1.21b can be expressed as

$$i_2 = \frac{v_2}{Z_2} \tag{1.28}$$

Equating (1.27) to (1.28) we get

$$v_2 \left(\frac{1 - 1/A}{Z}\right) = \frac{v_2}{Z_2} \Rightarrow Z_2 = \frac{Z}{1 - 1/A} \tag{1.29}$$

Note that, in order to get Z_1 and Z_2 in terms of Z, we left the original source at v_1 where it was originally, leaving the relation between v_1 and v_2 intact. As a result, Z_1 can be called the input impedance of the circuit. However, Z_2 is not the output impedance since that would require disabling the v_1 and inputting a test voltage at v_2, in which case the relation between v_1 and v_2 will collapse since electronic circuits are in general not bidirectional.

The results in (1.26) and (1.29) have a lot of practical implications. Here we will look into some of them.

If Z is a capacitor, we get

$$Z = \frac{1}{sC} \quad \text{and} \quad Z_1 = \frac{1}{sC_1} = \frac{Z}{1-A} = \frac{1}{sC(1-A)} \tag{1.30}$$

where C_1 which is $C(1-A)$ is the input capacitance of the circuit as seen by the source. The factor $(1-A)$ is called the Miller multiplier. If A is less than -1, then we see that the input capacitance is larger than C. Therefore, C will have a larger effect on the circuit than its physical size implies. This is very significant and can be used both ways. For example, if a circuit such as that in Fig. 1.21b requires a large capacitor C_1, we can substitute this circuit with an equivalent one as in Fig. 1.21a that requires a physically smaller capacitor, while maintaining the same result.

Intuitively, this effect can be foreseen by realizing that while C_1 is experiencing a voltage change of v_1 only, C is experiencing a voltage change of

$$v_1 - v_2 = v_1 - A \times v_1 = v_1(1-A) \tag{1.31}$$

which is a Miller multiplier times that experienced by C_1.

A more intriguing situation is if A is greater than 1. In this case, the resulting circuit will act as a negative capacitor C_1. This can be used to cancel an unwanted capacitance and is more interesting than using an inductor. This is because an inductor can resonate with a capacitor at only one frequency, thus canceling it at only that frequency. However, a negative capacitance can in theory cancel the unwanted capacitance at all frequencies. That is in theory, but in practice, we need to keep in mind the relation between v_1 and v_2, which might not hold for all frequencies. Additionally, there is a price to pay and that is in order to maintain the relation between v_1 and v_2, power needs to be consumed, unlike when using a simple inductor.

Similar analysis can be done by replacing Z with a resistor or an inductor, where in some cases, we might end up with a negative resistor and a negative inductor.

Chapter 2
Basic Applications

Three of the most basic and important applications of electronic circuits are signal amplification, signal filtering, and logic computation. This chapter starts by looking into the two main variations of amplifiers: regular and operational. We will model them behaviorally and look into their non-idealities. This is followed by continuous-time filtering where we will look into passive and active filters. Finally, logic gates for computation are briefly discussed along with their non-idealities.

2.1 Regarding Specifications

Applications dictate their own specifications to the circuit. These specifications consist of a list of various performance parameters and the ranges of values within which they should be. At the face of it, for a particular circuit, these parameters are independent. However designers understand that deep within the circuit, they are interrelated and usually there are compromises between them in that when one improves, another might deteriorate.

Circuits have inherent variability in them due to the manufacturing process in addition to the fact that their performance varies with temperature, age, etc. As a result, the specifications dictate a range of values, not only a single one, within which every parameter should be in order for the specifications to be met.

For a given circuit, the specifications include a variety of parameters such as gain, power consumption, and bandwidth. Three values of these parameters are usually specified: the minimum value, the typical value, and the maximum value, thus a range of values.

One way to report these parameters is to put them in a table format, which works well when looking into the parameters of one circuit. These are usually accompanied by the applicable conditions such as the ambient temperature. Table 2.1 shows a sample of such a table where T_A stands for ambient temperature.

J. G. Atallah, M. Ismail, *Integrated Electronic Circuits*, https://doi.org/10.1007/978-3-031-62707-1_2

Table 2.1 Sample table of specifications

Parameter	Min	Typ	Max	Units	Conditions
Gain	28	30	35	V/V	$T_A = -40\,°C$ to $+85\,°C$
Power consumption	0.9	1	1.1	mW	$T_A = +25\,°C$
Bandwidth	10	12	13	MHz	$T_A = +25\,°C$

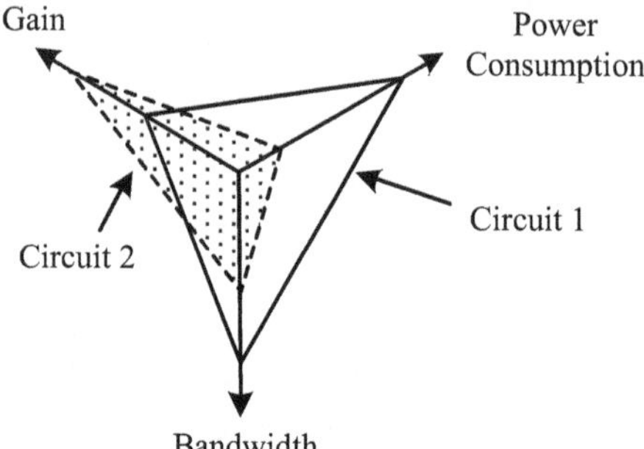

Fig. 2.1 Comparison of typical parameters for two circuits

When a comparison between the performance parameters of several similar circuits is needed, more columns can be added to Table 2.1. However, this quickly becomes impractical if the number of circuits is large. As a partial solution, in this case, only the typical values can be compared in order to narrow down the choice of circuits.

Another way to compare many circuits is to come up with a Figure-of-Merit (FoM), which is a single number that summarizes the performance of a whole circuit. FoMs are constructed in such a way that the circuit is perceived to be better if its FoM is higher than that of another circuit. In its simplest embodiment, the FoM can be a ratio, where the parameters that are deemed better if they are higher in value are in the numerator and the parameters that are deemed better when they are lower in value are in the denominator. For example, higher gain and bandwidths, but lower power consumption are desirable, so we have

$$\text{FoM} = \frac{\text{Gain} \times \text{Bandwidth}}{\text{Power Consumption}} \tag{2.1}$$

Another way to compare several circuits is graphically, by putting every performance parameter on its own axis in such a way that if the value on the axis increases, the performance related to this parameter is better, as in Fig. 2.1.

By having a quick look at Fig. 2.1, we can see that Circuit 1 has better power consumption and bandwidth compared to Circuit 2.

Also, by visually comparing the areas, covered by each circuit, we might be able to tell which circuit is better overall. For example, Circuit 1 in Fig. 2.1 is clearly better than Circuit 2. However, this last point might not be of practical use since, depending on the application, some parameters might be of higher importance than others. For example, if gain is of utmost importance to us, and the other parameters are not so, then we would choose Circuit 2 rather than Circuit 1.

2.2 Amplify

Information-bearing signals, especially of the continuous type, often require amplification so that they can be interpreted reliably by the receiver. The presence of a small signal in the first place might be due to several reasons. For example, if the signal comes from a sensor, such as a microphone, then this sensor might not be capable of generating a large signal, thus amplification is needed as in Fig. 2.2a. Another example is that all transmission media, such as copper wires and printed-circuit board traces, attenuate the signal, which then requires amplification as in Fig. 2.2b. This is because transmission media such as copper for example can be modeled as a string of resistors due to the nonzero resistivity of the copper. Therefore, the IR drop across these resistors attenuates the signal.

In this discussion, and without any loss of generality, we will assume that the signals are in the voltage domain.

Amplification can be done either using regular amplifiers or operational amplifiers as will be discussed.

2.2.1 Regular Amplifiers

Regular amplifiers are the simplest types of amplifiers. They take a signal from a signal source and use the power from a power source to output a signal with a higher signal amplitude and higher power than the one at its input. The general configuration is shown in Fig. 2.3.

The signal source stage preceding the amplifier is modeled using its Thevenin equivalent model. It consists of two sources in series and a finite output resistance

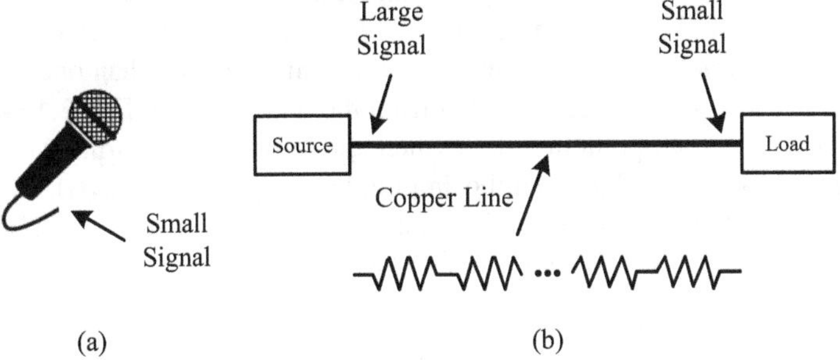

Fig. 2.2 Examples of scenarios where a small signal requiring amplification exists

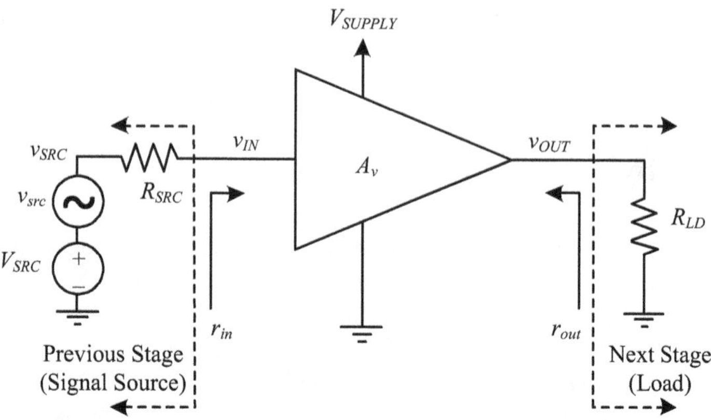

Fig. 2.3 General configuration of a regular amplifier

R_{SRC}. As a result, the v_{IN} that it outputs is composite. It consists of the information-bearing signal v_{in} on top of a DC V_{IN}. The DC component V_{IN}, although it contains no information, has the role of shifting the information-bearing signal v_{in} so that the resulting composite signal v_{IN} fits within the allowable range of the amplifier power supply, V_{SUPPLY} and ground.

The input resistance of the amplifier is denoted as r_{in} whereas the output resistance is r_{out}. Both of these resistances are seen only by the information-bearing part of the signal.

The output is also a composite signal v_{OUT} consisting of an information-bearing signal v_{out} on top of a DC V_{OUT}.

Sample signal waveforms of the regular amplifier are shown in Fig. 2.4.

The relation between the input and output is as follows, where the voltage gain A_v has the units of V/V:

$$v_{out} = A_v \times v_{in} \tag{2.2}$$

Notice that this relation applies to only the information-bearing parts of the signals.

As a result, the output versus input relation, also known as the transfer characteristic, is shown in Fig. 2.5. The gain A_v is the slope of the transfer characteristic. Since this is an amplifier, the magnitude of this gain is greater than one.

Practically, the gain can be extracted from the waveforms in Fig. 2.4 as follows. Take a certain arbitrary point in time t_1 where the information-bearing part v_{in} of the input signal is preferably large, let the amplitude of v_{in} be called $v_{in}(t_1)$. At the same point in time, let the amplitude of only the information-bearing part v_{out} of the output signal be called $v_{out}(t_1)$. The gain is defined as:

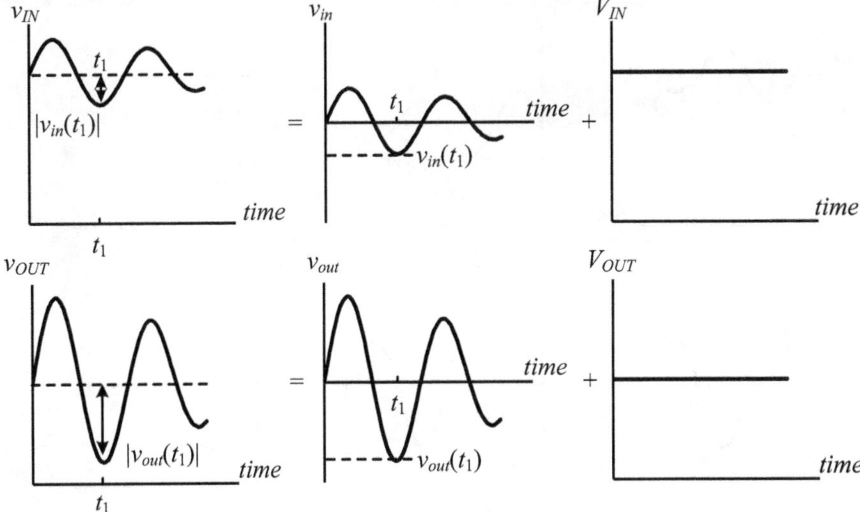

Fig. 2.4 Sample signal waveforms associated with a non-inverting amplifier

Fig. 2.5 Sample input–output transfer characteristic of a non-inverting amplifier

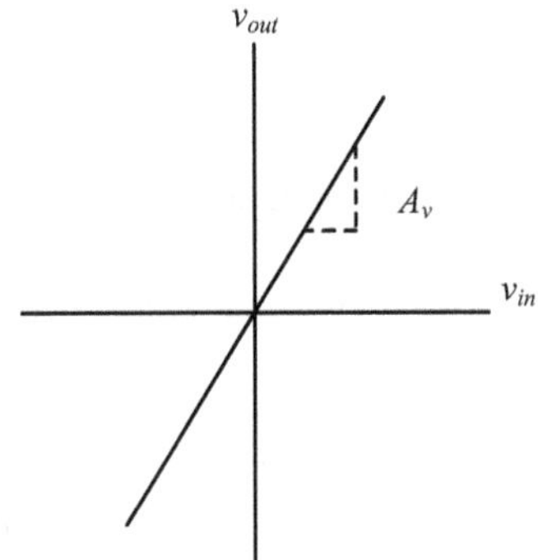

$$A_v = \frac{v_{out}(t_1)}{v_{in}(t_1)} \tag{2.3}$$

In this example, since $v_{in}(t_1)$ and $v_{out}(t_1)$ have the same sign (negative), they are in phase and the gain is positive. Consequently, the amplifier is non-inverting.

Alternatively, an inverting amplifier would have signal waveforms shown in Fig. 2.6.

In this case, $v_{in}(t_1)$ is negative and $v_{out}(t_1)$ is positive. Having opposite signs, they are out-of-phase and the gain is negative. Consequently, the amplifier is inverting.

In some cases, whether the gain is positive or negative is not important and its absolute value is used:

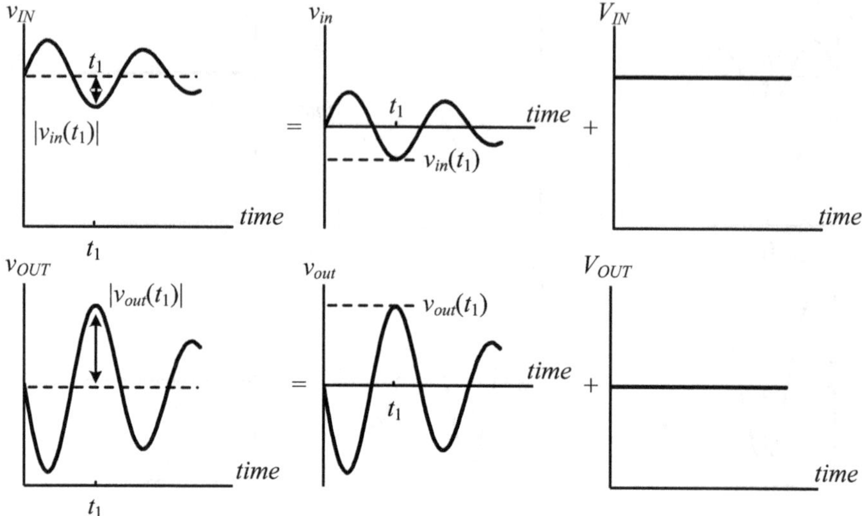

Fig. 2.6 Sample signal waveforms associated with an inverting amplifier

$$|A_v| = \frac{|v_{out}(t_1)|}{|v_{in}(t_1)|} \tag{2.4}$$

It is common to express this gain in decibels as

$$A_v|_{dB} = 20 \log \left(\frac{|v_{out}(t_1)|}{|v_{in}(t_1)|} \right) \tag{2.5}$$

This gain ideally is independent of frequency, which means that an ideal amplifier has an infinite frequency operating range, called bandwidth.

Note that the gain has nothing to do with the DC parts V_{IN} and V_{OUT} of the signals. The relation between these have to do with the way the amplifier is implemented internally.

Since the information is in the voltage domain, the ideal input resistance r_{in} in Fig. 2.3 is very high which leads to zero input current and to v_{IN} being equal to v_{SRC}. Also, ideally, the output resistance r_{out} is very small, which means that the amplifier will have the same gain, no matter what the load is.

In summary, an ideal voltage amplifier has the following properties:

1. The transfer characteristic is linear.
2. The gain is independent of frequency.
3. The input resistance is infinite.
4. The output resistance is zero.

Properties 1 and 2 culminate in the fact that the gain is a pure constant. Property 3 results in the fact that the input current into the amplifier is zero. Thus, the amplifier does not take any power from the previous stage, which means it does not load it. It only reads the voltage from the previous stage, which is where the information is. Property 4 results in the fact that the overall gain does not depend on the input resistance of the next stage. Therefore, any load can be used and the amplifier will maintain its gain.

2.2.1.1 Behavioral Modeling

An amplifier can be modeled as a dependent source. Since the input and the output in this case are a voltage signal, a voltage-controlled voltage source is used. The resulting functional model focuses only on the behavior of the information-bearing parts v_{in} and v_{out} of the signals as in Fig. 2.7.

Alternatively, a model based on a voltage-controlled current source can be used, which will prove very useful when the amplifier is implemented at the circuit level. The model is shown in Fig. 2.8.

The constant of proportionality between in the input voltage and the current generated by the dependent source is the transconductance G_m having the units of A/V. This current is then converted into an output voltage after passing through the output resistance.

$$v_{out} = (G_m \times v_{in}) \times r_{out} = (G_m \times r_{out}) \times v_{in} \tag{2.6}$$

Since the models in Figs. 2.7 and 2.8 are equivalent, then comparing (2.6) to (2.2) we get

$$A_v = G_m \times r_{out} \tag{2.7}$$

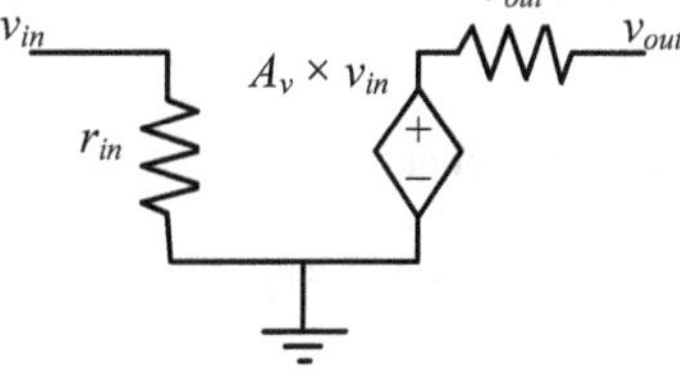

Fig. 2.7 Voltage source-based model of a regular voltage amplifier

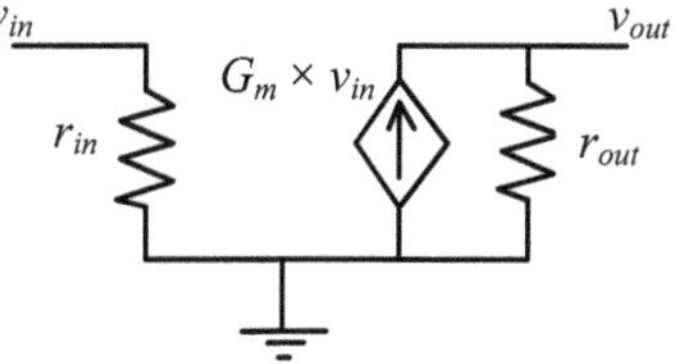

Fig. 2.8 Current source-based model of a regular voltage amplifier

2.2.1.2 Non-idealities

When amplifiers are implemented in practice, several non-idealities show up in their behavior. The extent to which we take these non-idealities into account depends on their effect on our application. If a certain non-ideality does not affect our application, then it can be ignored; otherwise, it should be taken into consideration.

We will look into a few of these potential non-idealities and see how they affect the ideal behavior of the amplifier.

The first non-ideality is related to the gain. This is seen when looking at the transfer characteristic of the amplifier as in Fig. 2.9 which shows the ideal case and the realistic case.

As can be seen, the gain, which is the slope of the transfer characteristic, is not constant across the whole range of the input. In particular, when the input reaches an extreme value, the gain tapers off and the amplifier will not be able to amplify the signal adequately. This reduction in the gain at the extreme values of the input is called gain compression. This nonlinearity between the input and output signals is due to the components that make up the amplifier, which are inherently nonlinear.

Note, however, that if the application allows us to confine the input to a small range around the origin within the dotted circle in Fig. 2.9, the two transfer characteristics would be very close and we can use the ideal behavior as an approximation for the realistic behavior.

Another non-ideality is related to the input and output resistances of the amplifier. Given the dependent voltage source model in Fig. 2.7, both the input and output signals are in the voltage domains. Consequently, ideally, r_{in} should be infinite, which means that the input current is zero, and r_{out} should be zero. In practice one or both might not be the case.

If both are not ideal, then replacing the amplifier in Fig. 2.3 with the dependent voltage source model in Fig. 2.7, we get the configuration in Fig. 2.10. Notice how this model focuses only on the information-bearing parts of the signal.

In this case, v_{in} and v_{src} are not the same owing to the finite value of r_{src}. Also, we have distinguished between v_{out} and v_{ld} because they are different owing to the nonzero value of r_{out}.

Fig. 2.9 Sample input–output transfer characteristic of an amplifier with gain non-idealities

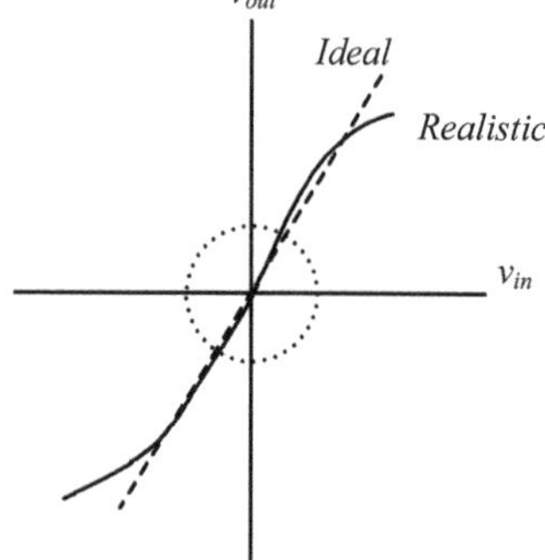

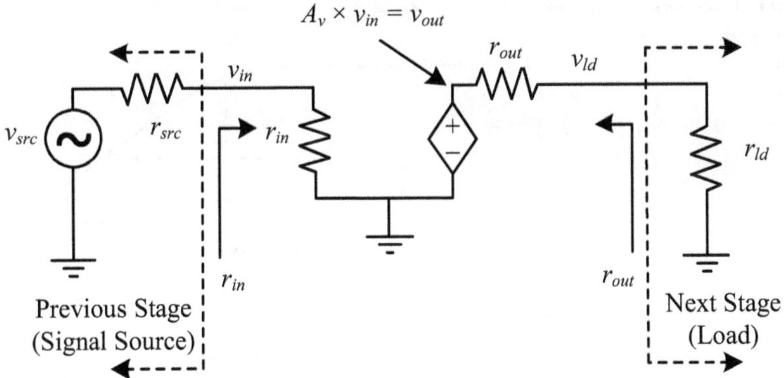

Fig. 2.10 Model of the amplifier with source and load

The relation now between the voltages is as follows, starting from the output:

$$\frac{v_{ld}}{v_{src}} = \left(\frac{v_{ld}}{v_{out}}\right)\left(\frac{v_{out}}{v_{in}}\right)\left(\frac{v_{in}}{v_{src}}\right)$$
$$= \left(\frac{r_{ld}}{r_{ld} + r_{out}}\right)(A_v)\left(\frac{r_{in}}{r_{in} + r_{src}}\right) \tag{2.8}$$

Ideally, this relation should be equal to A_v only. However, in this case, A_v is multiplied by two factors that are less than one, thus reducing the overall gain.

Looking at things more closely, r_{in} and r_{out} do not have to be ideal for the amplifier to behave ideally. In fact, if $r_{in} >> r_{out}$ and $r_{out} << r_{ld}$, then the overall gain can be approximated as in the ideal case.

As a result, the qualities of the preceding and the following stages affect the way an amplifier behaves, and this is why they should always be taken into consideration.

The last non-ideality has to do with the behavior of the amplifier as the input frequency changes. Until now, we might get the impression that the amplifier behaves the same no matter what the input frequency is. Unfortunately, in practice, if the input frequency exceeds a certain value, the gain of the amplifier starts to drop. As a result, a typical plot of the gain in decibels versus frequency on a logarithmic scale looks as in Fig. 2.11.

As can be seen, the gain starts by being almost constant with the value of $A_v Maximum$, but beyond f_{high}, it starts to drop significantly. A_v $Maximum$ is also called the "DC gain", but as we know, the information-bearing part of the signal does not contain a DC component. The x-axis in Fig. 2.11 is logarithmic, which means it never reaches zero, but it looks as if it does.

In general, f_{high} is determined by the application itself as being the frequency beyond which the gain is so low that the amplifier does not meet the required specifications. In the absence of such information, f_{high} can be calculated as the frequency at which the power of the output signal is half what it is supposed to be. As such, the power gain at f_{high} is half the maximum power gain. Consequently,

Fig. 2.11 Gain versus frequency of a typical amplifier

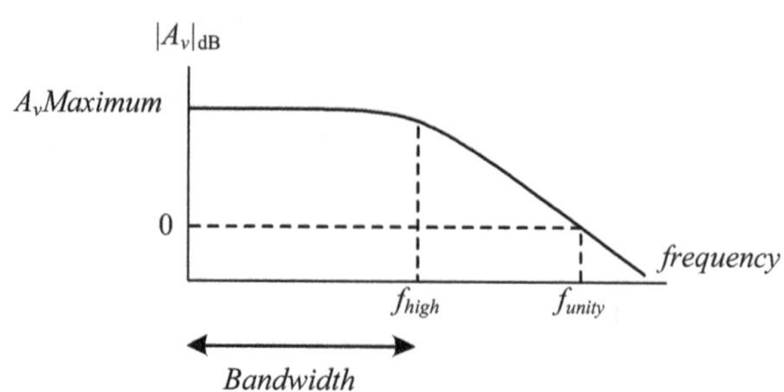

$$PowerGainRatio = \frac{|PowerGain\,(f_{high})|}{|PowerGain\,Maximum|} = \frac{\left(\frac{v_{out}\,(f_{high})}{vin}\right)^2}{\left(\frac{v_{out}\,Maximum}{vin}\right)^2} = \frac{1}{2}$$

$$\Rightarrow PowerGainRatio\big|_{dB} = 10\log\left(\frac{\left(\frac{v_{out}\,(f_{high})}{vin}\right)^2}{\left(\frac{v_{out}\,Maximum}{vin}\right)^2}\right) = 10\log\left(\frac{1}{2}\right) = -3dB \tag{2.9}$$

This is why f_{high} is also called the 3-dB frequency. This frequency practically defines the viable operating bandwidth of the amplifier.

Another frequency of interest is that when the output power is the same as the input power that is when the gain is unity (zero on a decibel scale). This frequency is denoted as f_{unity} in Fig. 2.11. Beyond this frequency, the amplifier attenuates the signal instead of amplifying it.

2.2.2 Operational Amplifiers and Operational Transconductance Amplifiers

The gain of the amplifiers discussed so far is built into them by design. Therefore, every time an application requires a certain gain, a new amplifier must be designed. A better approach for this is to design an amplifier whose gain can be configured externally. One way to do so is to have the overall gain depend only on components outside the amplifier. This way, the same amplifier can be used for several applications and the gain can be set by choosing the proper external components.

The Operational Amplifier (Op Amp) and the Operational Transconductance Amplifier (OTA) achieve exactly this and will be discussed in details. The symbol of an Op Amp is shown in Fig. 2.12a and that of an OTA is shown in Fig. 2.12b.

The input–output relation for an Op Amp is

Fig. 2.12 Op Amp symbol (**a**) and OTA symbol (**b**)

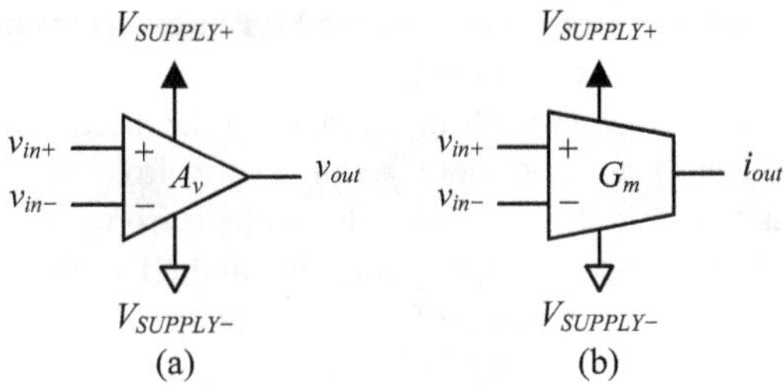

$$v_{out} = A_v \times (v_{in+} - v_{in-}) \tag{2.10}$$

where A_v is the voltage gain of the Op Amp. Of importance is the sequence of subtraction since v_{out} will be in phase with v_{in+} and out-of- phase with v_{in-}.

As for the OTA, the input–output relation is

$$i_{out} = G_m \times (v_{in+} - v_{in-}) \tag{2.11}$$

where G_m is the transconductance of the OTA.

The main difference between the Op Amp in Fig. 2.12a and the amplifier in Fig. 2.3 is that the Op Amp has two inputs and the output is a functions of those. Therefore, the source preceding the Op Amp does not need to have the same ground as the Op Amp itself. This is quite significant in practice since it makes the setup more robust in the presence of noise and interference.

Since both of these circuits, the Op Amp and the OTA deal with the difference between v_{in+} and v_{in-}, we will call this signal the differential input voltage v_{ind}.

An ideal Op Amp has the following properties:

1. The transfer characteristic is linear.
2. The gain is independent of frequency.
3. The gain is very large, a property that will be very useful later.
4. The input resistance is infinite.
5. The output resistance is zero.

Properties 1 and 2 culminate in the fact that the gain is a pure constant.

Property 3 will be used extensively later-on to analyze circuits built using Op Amps. This property results in the following observation

$$v_{out} = A_v \times (v_{in+} - v_{in-}) \Rightarrow A_v = \frac{v_{out}}{v_{in+} - v_{in-}} \tag{2.12}$$

Therefore

$$A_v \to \infty \Rightarrow v_{in+} - v_{in-} \to 0 \Rightarrow v_{in+} \to v_{in-} \tag{2.13}$$

To say that the two inputs become equal is counter-intuitive since it is the difference between them that the Op Amp amplifies. However, this is an approximation that works well for large values of A_v and simplifies the analysis and design, as we will see later.

Property 4 results in the fact that the input current into the amplifier is zero. Thus, the amplifier does not take any power from the previous stage, which means it does not load it. It only reads the voltage from the previous stage, which is where the information is. Property 5 results in the fact that the overall gain does not depend on the input resistance of the next stage. Therefore, any load can be used and the amplifier will maintain its gain.

A similar list for the ideal behavior of an OTA can be derived:

1. The transfer characteristic is linear.
2. The transconductance is independent of frequency.
3. The transconductance is very large.
4. The input resistance is infinite.
5. The output resistance is infinite.

2.2.2.1   Behavioral Modeling

The Op Amp can be modeled as a dependent voltage-controlled voltage source. The resulting functional model focuses only on the behavior of the information-bearing parts v_{ind} and v_{out} of the signals as in Fig. 2.13. Ideally, r_{in} is infinite and r_{out} is zero.

As for the OTA, it can be modeled as a dependent voltage-controlled current source. The resulting functional model focuses only on the behavior of the information-bearing parts v_{ind} and i_{out} of the signals as in Fig. 2.14. Ideally, r_{in} and r_{out} are infinite.

Since the primary output of an OTA is a current, then the load should have a finite input resistance in order to pass this current.

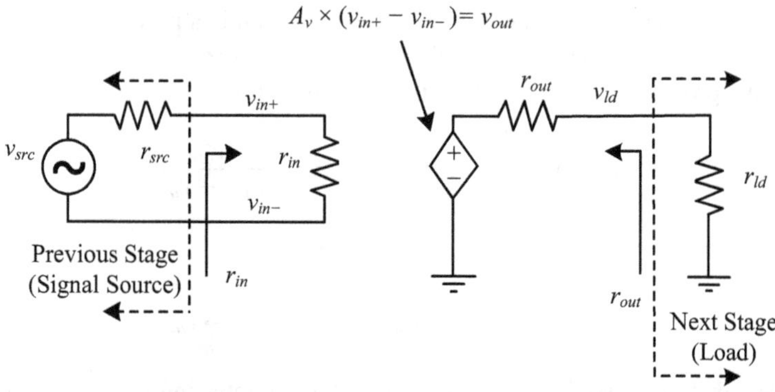

Fig. 2.13 Voltage source-based model of an Op Amp

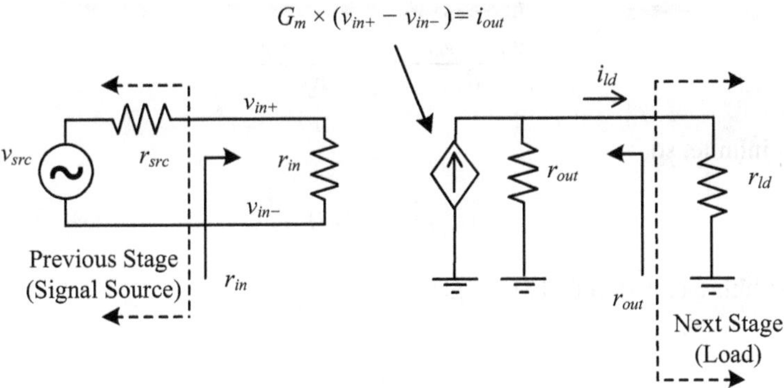

Fig. 2.14 Current source-based model of an OTA

Fig. 2.15 Op Amp
inverting configuration

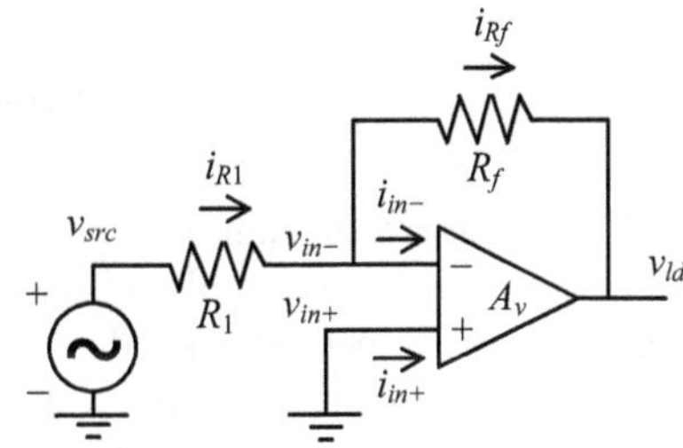

2.2.2.2 Basic Configurations

Op Amps and OTAs are not used primarily in a stand-alone setup. They are placed within configurations involving feedback loops that allow the designer to set the performance parameters based on the values of the external components.

The most important Op Amp configuration is the inverting configuration shown in Fig. 2.15. Notice how R_f, being in the feedback path, connects the output to the negative input of the Op Amp.

In this analysis, we will assume that the previous and next stages are ideal, and so is the Op Amp.

If the input resistance to the Op Amp is infinite then no current enters it and

$$i_{in+} = i_{in-} = 0 \tag{2.14}$$

Consequently, the current passing through R_1 is the same as that passing through R_f.

$$i_{R1} = i_{Rf}$$
$$\frac{v_{src} - v_{in-}}{R_1} = \frac{v_{in-} - v_{ld}}{R_f} \tag{2.15}$$

A_v is infinite, so

$$v_{in-} = v_{in+} = 0 \tag{2.16}$$

Substituting (2.16) in (2.15) we get

$$\frac{v_{src} - 0}{R_1} = \frac{0 - v_{ld}}{R_f} \tag{2.17}$$

Therefore, the closed-loop gain A_{vcl} will be

$$A_{vcl} = \frac{v_{ld}}{v_{src}} = -\frac{R_f}{R_1} \tag{2.18}$$

Notice how this gain is completely and exclusively defined by R_1 and R_f. Therefore, the designer only needs to choose the values of these external components in order to get the desired gain.

This gain is negative, so the output is out-of-phase compared to the input, thus the name inverting configuration. This stems from the fact that the positive terminal of the source is connected to the negative input of the Op Amp.

As a side note, if in an application, the source gives us a current, say i_{R1} in Fig. 2.15, then

$$\frac{v_{ld}}{i_{R1}} = -R_f \tag{2.19}$$

and the structure is called a Transimpedance Amplifier (TIA).

Considering the assumption that A_v is infinite as in (2.16), we will show how large it should be so that this assumption is accurate.

Therefore, revisiting the calculations for Fig. 2.15, the infinite input resistance assumption in (2.14) is kept the same but the infinite A_v in (2.16) is dropped, so (2.15) still applies.

Knowing that

$$A_v = \frac{v_{ld}}{v_{in+} - v_{in-}} \tag{2.20}$$

and that v_{in+} is zero, then

$$A_v = \frac{v_{ld}}{-v_{in-}} \Rightarrow v_{in-} = \frac{-v_{ld}}{A_v} \tag{2.21}$$

Replacing (2.21) in (2.15) we get

$$\frac{v_{src} + \frac{v_{ld}}{A_v}}{R_1} = \frac{-\frac{v_{ld}}{A_v} - v_{ld}}{R_f} \tag{2.22}$$

Rearranging the equation above, we get

$$\frac{v_{src}}{R_1} = -v_{ld}\left(\frac{1}{A_v R_1} + \frac{1}{A_v R_f} + \frac{1}{R_f}\right)$$
$$\Rightarrow \frac{v_{ld}}{v_{src}} = \frac{-1}{\frac{1}{A_v} + \frac{R_1}{A_v R_f} + \frac{R_1}{R_f}} = \frac{-A_v R_1 R_f}{R_1 R_f + R_1{}^2 + A_v R_1{}^2} = \frac{-A_v R_f}{R_f + R_1 + A_v R_1} \tag{2.23}$$

Dividing up and down by $A_v R_1$, we get

$$\frac{v_{ld}}{v_{src}} = \frac{-R_f/R_1}{1 + \frac{1}{A_v}\left(\frac{R_1 + R_f}{R_1}\right)} = \frac{-R_f/R_1}{1 + \frac{1}{A_v}\left(1 + \frac{R_f}{R_1}\right)} \tag{2.24}$$

Looking at (2.24), we can see that if

$$\frac{1}{A_v}\left(1 + \frac{R_f}{R_1}\right) << 1 \Rightarrow A_v >> 1 + \frac{R_f}{R_1} \tag{2.25}$$

then (2.24) will be the same as (2.18).

Therefore, the condition to be met is that in (2.25). A closer look shows that if

$$A_v >> 1 + \frac{R_f}{R_1} \Rightarrow A_v >> \frac{R_f}{R_1} \tag{2.26}$$

But, R_f/R_1 is just the magnitude of the closed-loop gain A_{vcl}. Therefore, a large A_v simply means that it should be much larger than the target A_{vcl} that the designer is aiming for.

The other important Op Amp configuration is the non-inverting configuration. It can be derived from the inverting configuration as shown in Fig. 2.16.

The resulting non-inverting configuration is shown in Fig. 2.17.

Note that the feedback resistor R_f is still connected to the negative Op Amp input terminal. As for the source, it is now connected to the positive Op Amp input terminal. Therefore, we expect the closed-loop gain to be positive, thus non-inverting.

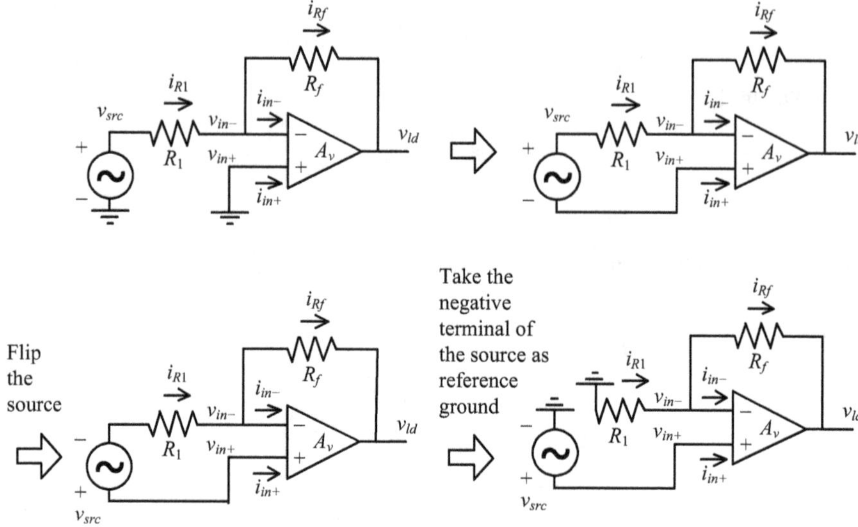

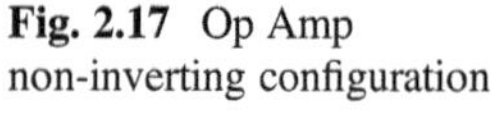

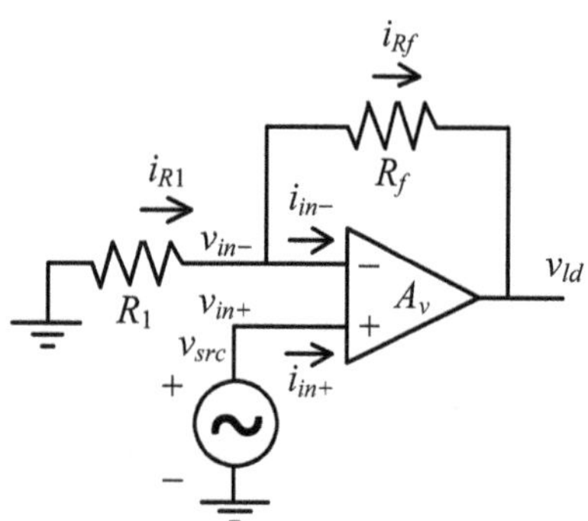

Fig. 2.16 Op Amp non-inverting configuration derivation

Fig. 2.17 Op Amp
non-inverting configuration

As with the analysis of the inverting configuration, we will assume that the previous and next stages are ideal, and so is the Op Amp.

The input resistance to the Op Amp is infinite. Therefore, no current enters it and

$$i_{in+} = i_{in-} = 0 \tag{2.27}$$

Consequently, the current passing through R_1 is the same as that passing through R_2:

$$i_{R1} = i_{Rf}$$
$$\frac{0 - v_{in-}}{R_1} = \frac{v_{in-} - v_{ld}}{R_f} \tag{2.28}$$

A_v is infinite, so

$$v_{in-} = v_{in+} = v_{src} \tag{2.29}$$

Substituting (2.29) in (2.28) we get

$$\frac{-v_{src}}{R_1} = \frac{v_{src} - v_{ld}}{R_f} \Rightarrow v_{src}\left(\frac{1}{R_1} + \frac{1}{R_f}\right) = \frac{v_{ld}}{R_f} \tag{2.30}$$

Therefore, the closed-loop gain A_{vcl} will be

$$A_{vcl} = \frac{v_{ld}}{v_{src}} = 1 + \frac{R_f}{R_1} \tag{2.31}$$

As expected, A_{vcl} is positive in this case, so the output is in phase with the input, thus the non-inversion. Also, A_{vcl} is a function of the resistors only.

As we did for the inverting configuration, and considering the assumption that A_v is infinite as in (2.29), we will show how large it should be so that this assumption is accurate.

Therefore, revisiting the calculations for Fig. 2.17, the infinite input resistance assumption in (2.27) is kept the same but the infinite A_v in (2.29) is dropped, so (2.28) still applies.

Knowing that

$$A_v = \frac{v_{ld}}{v_{in+} - v_{in-}} \tag{2.32}$$

and that v_{in+} is v_{src}, then

$$A_v = \frac{v_{ld}}{v_{src} - v_{in-}} \Rightarrow A_v v_{src} - A_v v_{in-} = v_{ld} \Rightarrow v_{in-} = v_{src} - \frac{v_{ld}}{A_v} \tag{2.33}$$

Replacing (2.21) in (2.28) we get

$$\frac{-v_{src} + \frac{v_{ld}}{A_v}}{R_1} = \frac{v_{src} - \frac{v_{ld}}{A_v} - v_{ld}}{R_f} \tag{2.34}$$

Rearranging the equation above, we get

Fig. 2.18 Op Amp
non-inverting configuration
with unity gain

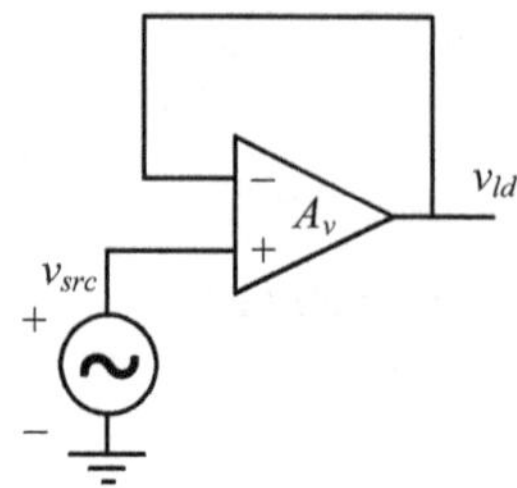

$$v_{ld}\left(\frac{1}{A_vR_1}+\frac{1}{A_vR_f}+\frac{1}{R_f}\right)=v_{src}\left(\frac{1}{R_1}+\frac{1}{R_f}\right)$$

$$\Rightarrow \frac{v_{ld}}{v_{src}}=\frac{A_v\left(R_1+R_f\right)}{R_1+R_f+A_vR_1}$$

$$(2.35)$$

Dividing up and down by A_vR_1, we get

$$\frac{v_{ld}}{v_{src}}=\frac{\frac{R_1+R_f}{R_1}}{1+\frac{1}{A_v}\left(\frac{R_1+R_f}{R_1}\right)}=\frac{1+\frac{R_f}{R_1}}{1+\frac{1}{A_v}\left(1+\frac{R_f}{R_1}\right)} \qquad (2.36)$$

Looking at (2.36), we can see that if

$$\frac{1}{A_v}\left(1+\frac{R_f}{R_1}\right)<<1 \Rightarrow A_v>>1+\frac{R_f}{R_1} \qquad (2.37)$$

then (2.36) will be the same as (2.31).

Therefore, the condition to be met is the one in (2.37), which is the same as that for the inverting case in (2.25).

As a result, similarly to the inverting configuration, in a non-inverting configuration, a large A_v simply means that it should be much larger than the target A_{vcl} that the designer is aiming for.

A special case of the non-inverting configuration arises if we take R_f to be zero and R_1 to be infinite as in Fig. 2.18. The resulting A_{vcl} will be unity and this circuit is used to isolate the load from the source.

The above two main configurations can be used with either one input, several inputs, or combined together.

As an example, consider the Op Amp inverting configuration with two inputs as in Fig. 2.19. This configuration is called the adder since the output consists of the sum of the inputs.

Assuming that the Op Amp is deal, its linear transfer characteristic (constant gain) allows us to use the concept of superposition. Therefore v_{ld} can be split into two parts:

Fig. 2.19 Op Amp
inverting configuration with
two inputs: adder

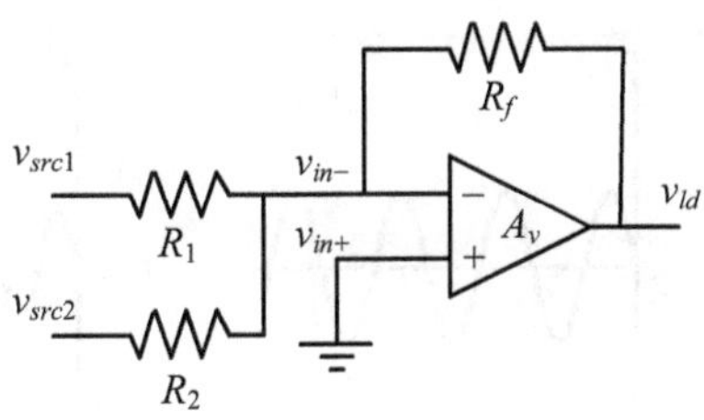

$$v_{ld} = v_{ld1} + v_{ld2} \tag{2.38}$$

where v_{ld1} is the reaction to v_{src1} while v_{src2} is disabled and v_{ld2} is the reaction to v_{src2} while v_{src1} is disabled.

Given that

$$v_{in+} = v_{in-} = 0 \tag{2.39}$$

when v_{src2} is disabled, no current passes through R_2 and we have

$$v_{ld1} = -\frac{R_f}{R_1} v_{src1} \tag{2.40}$$

and

$$v_{ld2} = -\frac{R_f}{R_2} v_{src2} \tag{2.41}$$

As a result

$$v_{ld} = -\frac{R_f}{R_1} v_{src1} - \frac{R_f}{R_2} v_{src2} = -\left(\frac{R_f}{R_1} v_{src1} + \frac{R_f}{R_2} v_{src2} \right) \tag{2.42}$$

As can be seen, the output consists of the inverted weighted sum of the inputs, with coefficients R_f / R_1 and R_f / R_2. This result can be extended to several inputs as needed.

Sample time and frequency domain adder signals are shown in Fig. 2.20a, b respectively.

If we do not want the inversion, just the addition, then we can follow the circuit in Fig. 2.19 with an inverting configuration as in Fig. 2.15.

Another example is if we combine an Inverting with a Non-inverting Configuration using one Op Amp. The resulting circuit will have two inputs, one at the positive terminal of the Op Amp and one at its negative terminal as in Fig. 2.21. This configuration is called the subtractor since the output consists of the difference between the inputs.

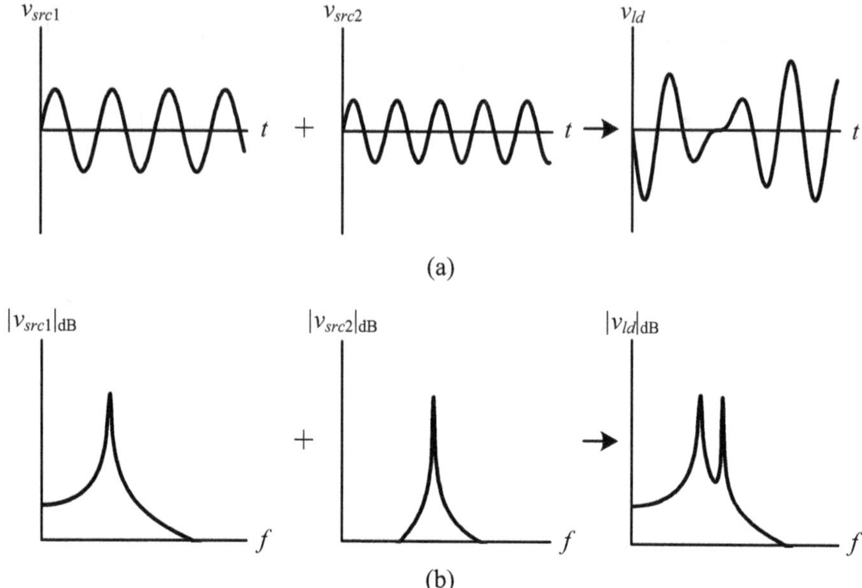

(a)

(b)

Fig. 2.20 Sample adder signals in the time domain (**a**) and frequency domain (**b**)

Fig. 2.21 Op Amp combined inverting and non-inverting configuration: subtractor

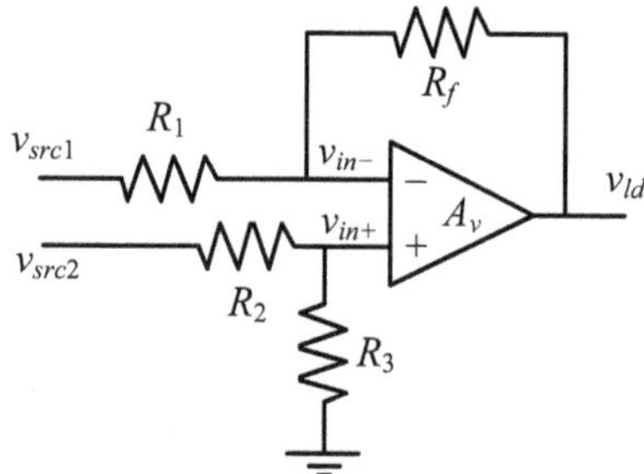

Notice how v_{src2} is not connected directly to the positive input terminal of the Op Amp. Instead, it is first attenuated using a voltage divider circuit consisting of R_2 and R_3. The reason for this is that the magnitude of the non-inverting configuration is slightly larger than that of the inverting configuration. By interposing this voltage divider, we can slightly attenuate v_{src2} so that we can balance the coefficients of v_{src1} and v_{src2}.

Sample time and frequency domain subtractor signals are shown in Fig. 2.22a, b respectively.

Notice how, for the same input signals, the magnitude spectra of the subtractor output signal in Fig. 2.22b is identical to that of the adder output signal in Fig. 2.20b.

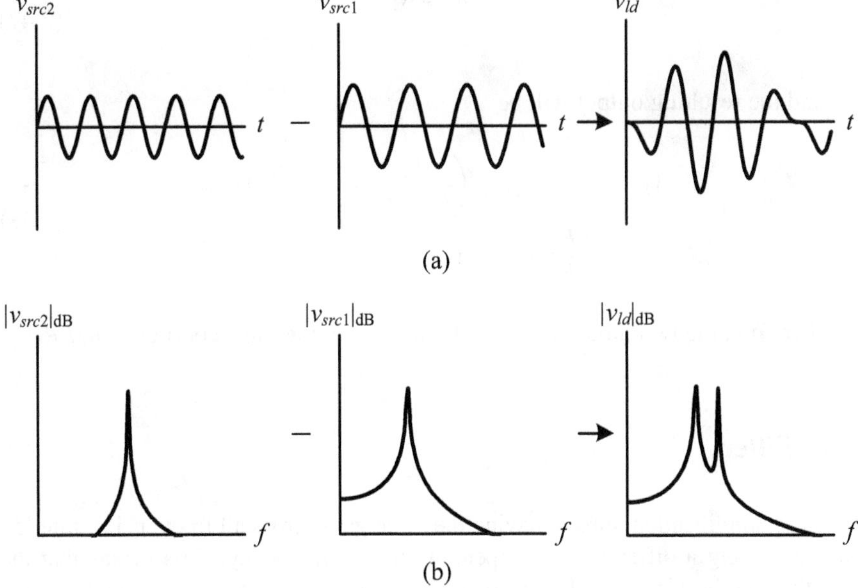

Fig. 2.22 Sample subtractor signals in the time domain (**a**) and frequency domain (**b**)

Using the superposition concept as before, the analysis goes as follows:

$$v_{ld} = -\frac{R_f}{R_1}v_{src1} + \left(\frac{R_3}{R_2 + R_3}\right)\left(1 + \frac{R_f}{R_1}\right)v_{src2} \tag{2.43}$$

Now, if we want to have the same coefficients, then

$$\frac{R_f}{R_1} = \left(\frac{R_3}{R_2 + R_3}\right)\left(1 + \frac{R_f}{R_1}\right) \tag{2.44}$$

Multiplying by R_1 and expanding, we get

$$R_f(R_2 + R_3) = R_3(R_1 + R_f)$$
$$R_2 R_f + R_3 R_f = R_1 R_3 + R_3 R_f \tag{2.45}$$

To satisfy the equality above, we need

$$R_2 R_f = R_1 R_3 \Rightarrow \frac{R_2}{R_1} = \frac{R_3}{R_f} \tag{2.46}$$

We can simply take

$$R_2 = R_1$$
$$R_3 = R_f \tag{2.47}$$

And the resulting output will be

$$v_{ld} = -\frac{R_f}{R_1} v_{src1} + \left(\frac{R_f}{R_1 + R_f}\right)\left(1 + \frac{R_f}{R_1}\right) v_{src2}$$
$$= \frac{R_f}{R_1}\left(v_{src2} - v_{src1}\right) \tag{2.48}$$

Therefore, the two inputs will be subtracted with the same coefficient R_f / R_1.

2.3   Filter

Whereas amplification aims at having the same gain across all frequencies, filtering aims at having a different gain depending on the frequency. This means that the variation of this gain versus frequency is what interests us. Therefore, as will be seen, the best way to analyze and design filters is to view things in the frequency domain.

The design of circuit-based filters follows the steps outlined in Fig. 2.23.

We will first give an overview of filters, transfer functions, then specifications, and finally look into some filter implementations.

2.3.1   General Description

There are several ways to implement a filter as seen in Fig. 2.24.

A filter can be implemented either algorithmically, using digital signal processing techniques or it can be implemented as a circuit.

Algorithmic implementations are suitable if the signal is already digital. Being implemented as an algorithm, these filters are versatile and easily programmable.

Circuit-based filters are implemented when algorithmic implementations are not suitable, for example, when the signal is not digital or when the signal contains very high frequencies. Circuit-based implementations can be either continuous or switched-capacitor-based using mixed-signal techniques. We will restrict this discussion to continuous filters.

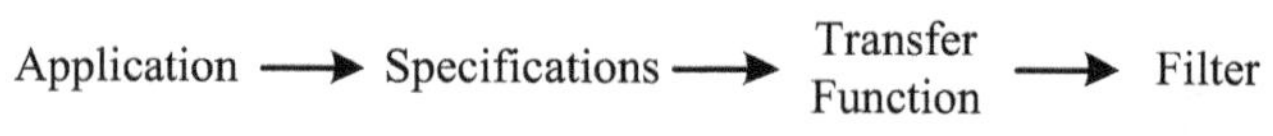

Application $\longrightarrow$ Specifications $\longrightarrow$ Transfer Function $\longrightarrow$ Filter

Fig. 2.23   Steps for circuit-based filter design

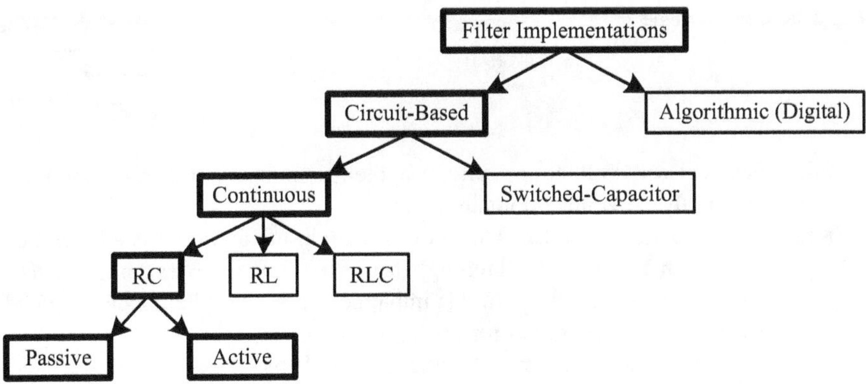

Fig. 2.24 Main filter implementations

Table 2.2 Brief Comparison between Passive and Active Filters

Passive filters	Active filters
+ Use only passive components	− Use passive components and one or more amplifiers
− Cannot provide signal gain	+ Can provide signal gain
+ Do not consume power from the power supply	− Consume power from the power supply
− Cannot be easily cascaded	+ Easier to cascade than passive filters
+ Do not inject too much noise into the circuit	− Might inject too much noise into the circuit
+ Can accept a large signal swing	− Signal swing is limited by the amplifier
+ There is no amplifier to limit the maximum bandwidth	− The amplifier might limit the maximum bandwidth

In order to have a filtering effect, the circuit should make use of components whose impedance is a function of frequency, in addition to the omnipresent resistor. Therefore, we can use either inductors or capacitors or both. In practice, capacitors are the first choice owing to their small physical size when implemented in an integrated circuit. Therefore, we will restrict this discussion to *RC* filters.

Finally, an *RC* filter can be implemented using either passive components only, or using passive components in addition to one or more amplifiers (active components), giving rise to passive and active implementations. Passive implementations do not consume power owing to the lack of amplifiers, but also cannot provide gain, only attenuation. Also, they are difficult to cascade owing to their typically far-from-ideal input and output impedances. Active implementations can provide gain and are easier to cascade, but they consume power from the power supply. Additionally, the presence of an amplifier in an active amplifier might inject too much electrical noise, which is detrimental if the signal is too small. Also, the amplifier might limit the maximum signal swing and frequency that the filter can handle. These do not apply to passive filters. A comparison of these implementations is shown in Table 2.2.

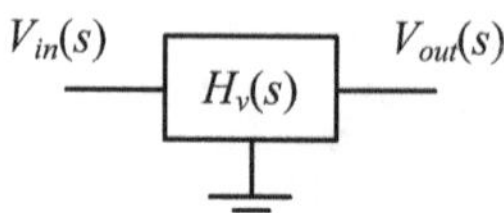

Fig. 2.25 General single-ended filter

Filters can deal with voltages or currents at their terminals. In this discussion, we will focus primarily on voltage-domain filters.

Keeping the practicality of the frequency domain in mind, a filter is a block that takes an input signal $V_{in}(s)$, passes it through its transfer function $H_v(s)$, and outputs a signal $V_{out}(s)$ as seen in Fig. 2.25. In this implementation, the filter is single-ended since there is one common ground for the input and output.

The relation between the input and output is as follows:

$$V_{out}(s) = H_v(s)V_{in}(s)$$
$$\Rightarrow H_v(s) = \frac{V_{out}(s)}{V_{in}(s)} \tag{2.49}$$

These quantities in general are complex numbers since they are a function of s. Therefore, they have a magnitude and a phase. The relation between them is as follows:

$$|V_{out}(s)| = |H_v(s)||V_{in}(s)|$$
$$\angle(V_{out}(s)) = \angle(H_v(s)) + \angle(V_{in}(s)) \tag{2.50}$$

In typical applications, the magnitude is more important than the phase, and filter types are categorized accordingly. Also, this magnitude is usually expressed in decibels. The magnitude of the filter transfer function $H_v(s)$ at a certain frequency is the gain of this filter at that frequency. Since we are looking at the steady-state behavior of filters, then

$$s = j\omega = j2\pi f \tag{2.51}$$

and all the quantities above are a function of the frequency f in Hz.

The application dictates the specifications, the most important of which is whether the filter is a Low-Pass Filter (LPF), Band-Pass Filter (BPF), High-Pass Filter (HPF), General Filter (GENF), All-pass Filter (APF), etc. In spite of these filters acting differently across frequencies, most of their specifications can be described in a unified manner.

2.3.2   Specifications

Taking an LPF for example, the specifications give rise to certain frequency-domain magnitude profile constraints that the filter should satisfy. A sample graph of these constraints is shown in Fig. 2.26. The specifications dictate a certain region, called the specification region, hashed region in Fig. 2.26, within which the magnitude of the filter transfer function should reside.

Being an LPF, f_{pass}, which is also the bandwidth of the LPF, is the limit of the passband and f_{stop} is the limit of the stopband. The region between them is the transition band. The selectivity factor of the filter is f_{stop} / f_{pass}. The closer the selectivity factor is to unity, the smaller is the transition band, the more selective is the filter, and the more difficult it is to design it.

Within the passband, H_{vmax} is the maximum required gain and H_{vmin} is the minimum required gain. Within the context of an LPF, they are called the DC gain. The difference between them is called H_{var}. In other words, H_{var} is the amount of variation allowed in the passband.

H_{diff} is the minimum difference allowed between what is passed and what is not passed. It is the difference between the minimum transfer function magnitude in the passband and the maximum transfer function magnitude in the stopband. Therefore, the constraint is

$$\left|H_v(f)\right|_{passband\_min} - \left|H_v(f)\right|_{stopband\_max} > H_{diff} \tag{2.52}$$

As a summary, the minimum set of constraints to be met for an LPF is shown in Table 2.3.

The situation is similar for other filters such as a BPF, and the sample graph of its specifications constraints is shown in Fig. 2.27. Minor differences exist such as the presence of two stopbands and two transition bands along with their edge frequencies.

Fig. 2.26 LPF specifications constraints

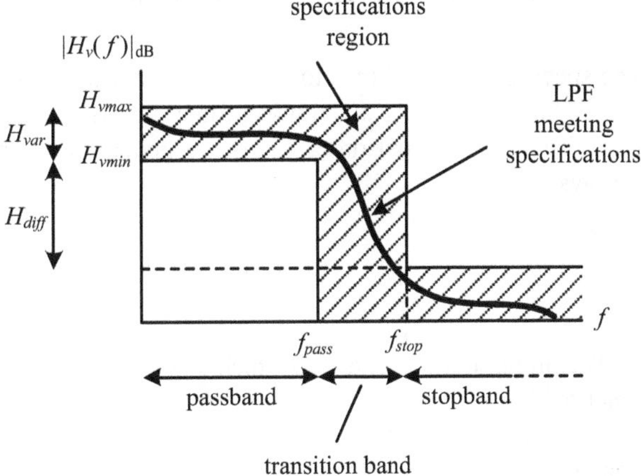

Table 2.3 Constraints to be met for an LPF

Constraint name	Description
Passband limit	The passband limit should be greater than f_{pass}.
Stopband limit	The stopband limit should be less than f_{stop}.
Maximum passband gain	The maximum passband gain should be less than H_{vmax}.
Minimum passband gain	The minimum passband gain should be greater than H_{vmin}.
The minimum difference between what is passed and what is not passed	$\|H_v(f)\|_{passband\_min} - \|H_v(f)\|_{stopband\_max} > H_{diff}$

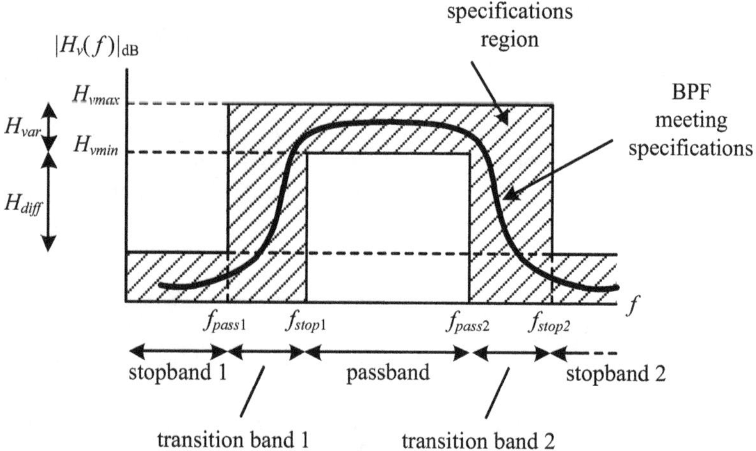

Fig. 2.27 BPF specifications constraints

2.3.3   Transfer Function and Implementation

The specifications will lead to a desirable filter transfer function. The highest power of s in the denominator of a transfer function is the order of the filter. We will restrict this discussion to first-order filters. A general first-order transfer function is as follows:

$$H_v(s) = \frac{V_{out}(s)}{V_{in}(s)} = \frac{as + b}{s + 2\pi f_p} \tag{2.53}$$

The numerator has one zero which is $-b/a$. This is called the zero of the transfer function.

The denominator also has one zero which is $-2\pi f_p$. This is called the pole of the transfer function.

First-order filters have only one pole and they can have a maximum of one zero in order to stay stable. If a is zero, then the filter will not have any finite zero, and it will be called an all-pole filter.

The presence and absence of a and b dictate the filter type (LPF, HPF...), and f_0 dictates the operating frequency limits of the filter as we will see.

We will look into all the combinations of a and b and see what type of first-order filters we can design.

2.3.3.1 Low-Pass Filter

If a is zero and b is not, we will get an all-pole filter, and the result is an LPF. In this case, the transfer function will be

$$H_v(s) = \frac{V_{out}(s)}{V_{in}(s)} = \frac{b}{s + 2\pi f_p} \tag{2.54}$$

To figure out the behavior, let

$$s = j\omega = j2\pi f \tag{2.55}$$

A plot of this transfer function is shown in Fig. 2.28. Note how, when the frequency goes to infinity, the magnitude goes to zero with a slope of -20 dB/

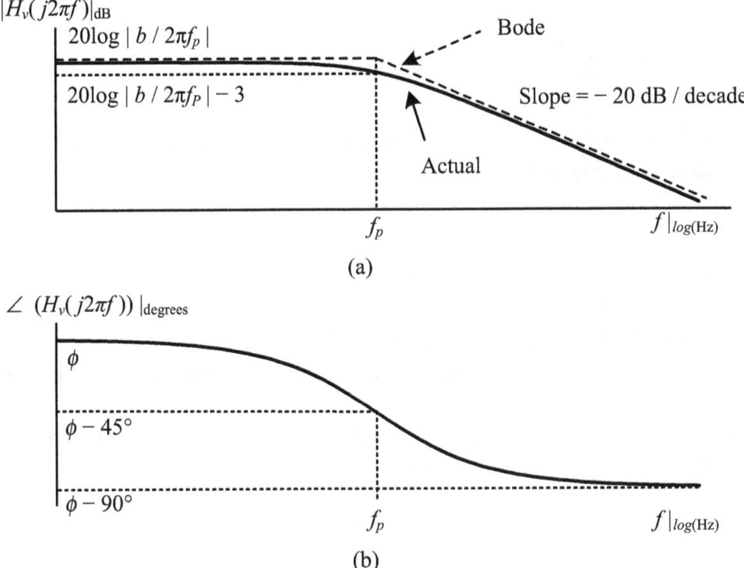

Fig. 2.28 LPF transfer function actual magnitude and Bode plots (**a**), and phase (**b**)

Fig. 2.29 LPF symbol

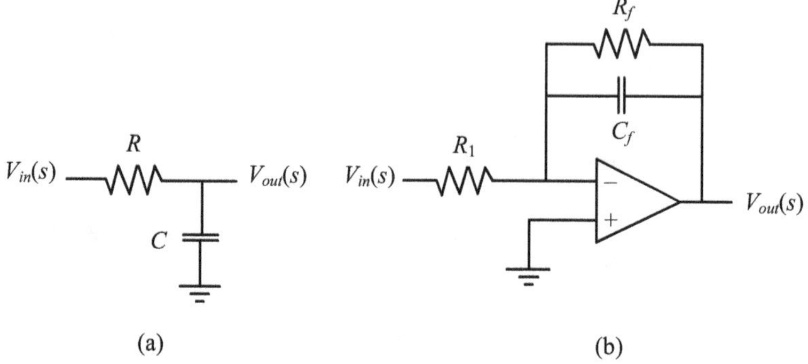

(a)   (b)

Fig. 2.30 Passive LPF (**a**) and active LPF (**b**)

decade. This is why we say that even though this transfer function does not have a finite zero, it has a zero at infinity. Also note that even within the filter passband, the phase of the signal is modified by as much as $-45°$. This might not be desirable for some applications.

The symbol for an LPF is shown in Fig. 2.29.

The implementation of this LPF is shown in Fig. 2.30.

Analyzing the passive LPF in Fig. 2.30a,

$$\frac{V_{out}(s)}{V_{in}(s)} = \frac{Z_C}{Z_R + Z_C} = \frac{1/sC}{R + 1/sC} = \frac{1}{1 + sRC} = \frac{1/RC}{s + 1/RC} \tag{2.56}$$

Z_R and Z_C are the impedances of the resistor and capacitor respectively. Comparing (2.56) to (2.54), we can see that

$$b = 1/RC$$
$$f_p = 1/2\pi RC \tag{2.57}$$

As a result, the DC gain of the passive LPF is one.

Regarding ϕ in Fig. 2.28b, in the passive case it is $0°$ since there is no inversion in the DC gain.

Analyzing the active LPF in Fig. 2.30b, and from the result of the inverting configuration in (2.18), we can see that

$$\frac{V_{out}(s)}{V_{in}(s)} = -\frac{Z_f}{Z_{R1}} \tag{2.58}$$

where Z_f is the equivalent impedance of the circuit in the feedback path.

$$Z_f = Z_{Rf}//Z_{Cf} = \frac{R_f \times 1/sC_f}{R_f + 1/sC_f} = \frac{R_f}{1 + sR_fC_f} \tag{2.59}$$

Substituting (2.59) in (2.58) we get

$$\frac{V_{out}(s)}{V_{in}(s)} = -\frac{\frac{R_f}{1+sR_fC_f}}{R_1} = \frac{-R_f/R_1}{1 + sR_fC_f} = \frac{\left(-R_f/R_1\right)\left(1/R_fC_f\right)}{s + 1/R_fC_f} = \frac{-1/R_1C_f}{s + 1/R_fC_f} \tag{2.60}$$

Comparing (2.60) to (2.54), we can see that

$$\begin{aligned} b &= -1/R_1C_f \\ f_p &= 1/2\pi R_fC_f \end{aligned} \tag{2.61}$$

As a result, the DC gain of the active LPF is $-R_f/R_1$, which is expected since at low frequencies, the capacitor will act as an open-circuit and the whole circuit will act as an inverting configuration amplifier as in Fig. 2.15.

As for ϕ in Fig. 2.28b, in the active case it is $180°$ since there is an inversion in the DC gain.

2.3.3.2   High-Pass Filter

If b is zero and a is not, we will get an HPF with a zero at zero. In this case, the transfer function will be

$$H_v(s) = \frac{V_{out}(s)}{V_{in}(s)} = \frac{as}{s + 2\pi f_p} \tag{2.62}$$

A plot of this transfer function is shown in Fig. 2.31. Note that even within the filter passband, the phase of the signal is modified by as much as $-45°$. This might not be desirable for some applications.

The symbol for an HPF is shown in Fig. 2.32.

The implementation of this HPF is shown in Fig. 2.33.

Analyzing the passive HPF in Fig. 2.33a,

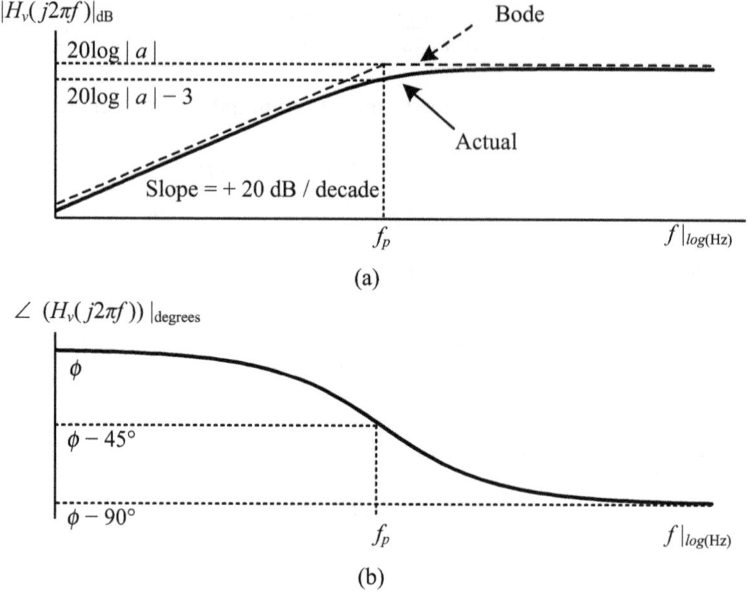

Fig. 2.31 HPF transfer function actual magnitude and Bode plots (**a**), and phase (**b**)

Fig. 2.32 HPF symbol

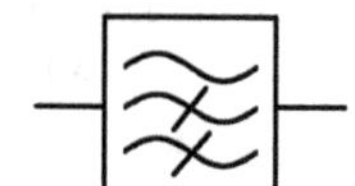

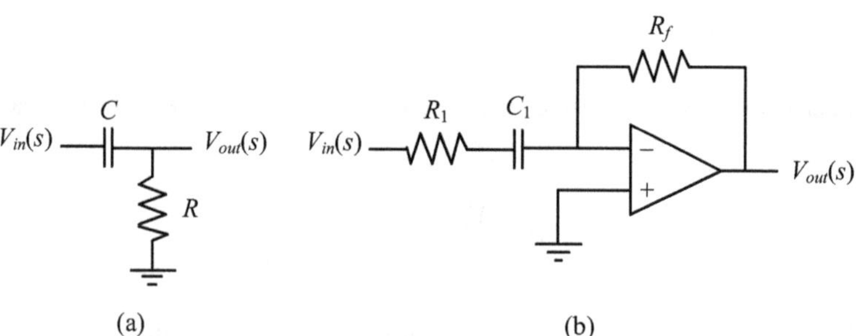

Fig. 2.33 Passive HPF (**a**) and active HPF (**b**)

$$\frac{V_{out}(s)}{V_{in}(s)} = \frac{Z_R}{Z_R + Z_C} = \frac{R}{R + 1/sC} = \frac{sRC}{1 + sRC} = \frac{s}{s + 1/RC} \tag{2.63}$$

Z_R and Z_C are the impedances of the resistor and capacitor respectively. Comparing (2.64) to (2.62), we can see that

$$a = 1$$
$$f_p = 1/2\pi RC \tag{2.64}$$

As a result, the high-frequency gain is one.

As for ϕ in Fig. 2.31b, in the passive case it is $90°$ since there is no inversion in the high-frequency gain.

Analyzing the active HPF in Fig. 2.33b, and from the result of the inverting configuration in (2.18), we can see that

$$\frac{V_{out}(s)}{V_{in}(s)} = -\frac{Z_f}{Z_1} \tag{2.65}$$

where Z_1 is the equivalent impedance of the circuit at the input.

$$Z_1 = Z_{R1} + Z_{C1} = R_1 + 1/sC_1 \tag{2.66}$$

Substituting (2.59) in (2.65) we get

$$\frac{V_{out}(s)}{V_{in}(s)} = -\frac{R_f}{R_1 + 1/sC_1} = \frac{-sR_fC_1}{1 + sR_1C_1} = \frac{-sR_f/R_1}{s + 1/R_1C_1} \tag{2.67}$$

Comparing (2.67) to (2.62), we can see that

$$a = -R_f/R_1$$
$$f_p = 1/2\pi R_1 C_1 \tag{2.68}$$

As a result, the high-frequency gain is $-R_f/R_1$, which is expected since at high frequencies, the capacitor will act as a short-circuit and the whole circuit will act as an inverting configuration amplifier as in Fig. 2.15.

As for ϕ in Fig. 2.31b, in the active case it is $-90°$ since there is an inversion in the high-frequency gain.

2.3.3.3   General Filter

If a and b are both not zero, we will get a GENF with a non-zero zero. In this case, the transfer function will be

$$H_v(s) = \frac{V_{out}(s)}{V_{in}(s)} = \frac{as + b}{s + 2\pi f_p} \tag{2.69}$$

A plot of this transfer function is shown in Fig. 2.34. It shows the case when the pole frequency is lower than the zero frequency.

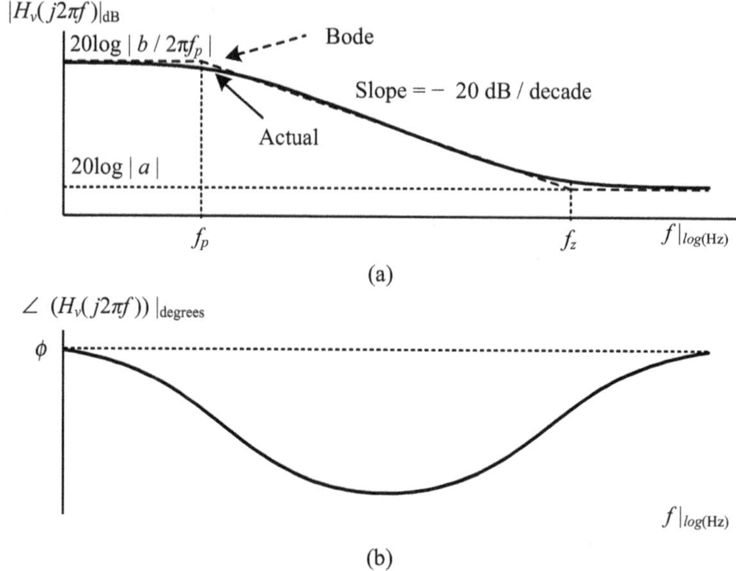

Fig. 2.34 GENF transfer function actual magnitude and Bode plots (**a**), and phase (**b**)

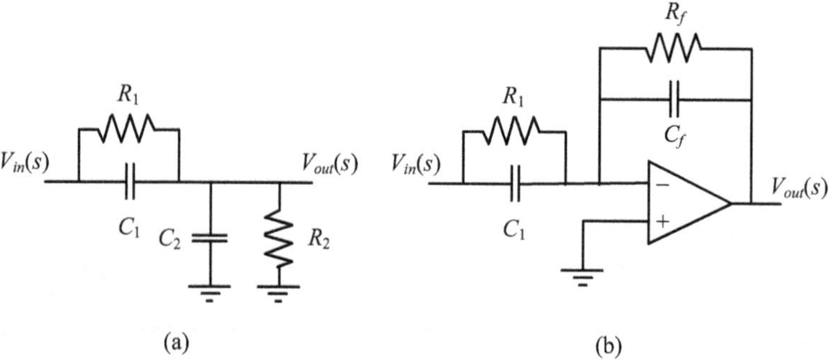

Fig. 2.35 Passive GENF (**a**) and active GENF (**b**)

The implementation of this GENF is shown in Fig. 2.35.
Analyzing the passive GENF in Fig. 2.35a,

$$\frac{V_{out}(s)}{V_{in}(s)} = \frac{Z_2}{Z_1 + Z_2} \tag{2.70}$$

where Z_1 is the parallel combination of R_1 and C_1, and Z_2 is the parallel combination of R_2 and C_2 resulting in

$$Z_1 = R_1 // 1/sC_1 = \frac{R_1/sC_1}{R_1 + 1/sC_1} = \frac{R_1}{1 + sR_1C_1}$$
$$Z_2 = R_2 // 1/sC_2 = \frac{R_2/sC_2}{R_2 + 1/sC_2} = \frac{R_2}{1 + sR_2C_2} \tag{2.71}$$

Combining (2.71) with (2.70) we get

$$
\begin{aligned}
\frac{V_{out}(s)}{V_{in}(s)} &= \frac{\dfrac{R_2/sC_2}{R_2 + 1/sC_2}}{\dfrac{R_1}{1 + sR_1C_1} + \dfrac{R_2/sC_2}{R_2 + 1/sC_2}} = \frac{(R_2/sC_2)(1 + sR_1C_1)}{R_1(R_2 + 1/sC_2) + (R_2/sC_2)(1 + sR_1C_1)} \\[2ex]
&= \frac{R_2(1 + sR_1C_1)}{R_1(1 + sR_2C_2) + R_2(1 + sR_1C_1)} = \frac{R_2 + sR_1R_2C_1}{R_1 + R_2 + sR_1R_2(C_1 + C_2)} \\[2ex]
&= \frac{\left(\dfrac{C_1}{C_1 + C_2}\right)s + \left(\dfrac{1}{R_1(C_1 + C_2)}\right)}{s + \left(\dfrac{R_1 + R_2}{R_1R_2(C_1 + C_2)}\right)}
\end{aligned}
\tag{2.72}
$$

Comparing (2.72) to (2.69), we can see that

$$
\begin{aligned}
a &= \frac{C_1}{C_1 + C_2} \\[1.5ex]
b &= \frac{1}{R_1(C_1 + C_2)} \\[1.5ex]
f_p &= \frac{1}{2\pi}\frac{R_1 + R_2}{R_1R_2(C_1 + C_2)} = \frac{1}{2\pi(R_1//R_2)(C_1 + C_2)}
\end{aligned}
\tag{2.73}
$$

As a result, for the passive GENF, we have

$$
\begin{aligned}
\text{DC Gain} &= \frac{R_2}{R_1 + R_2} \\[1.5ex]
\text{High Frequency Gain} &= \frac{C_1}{C_1 + C_2} \\[1.5ex]
f_z &= \frac{1}{2\pi}\frac{\dfrac{1}{R_1(C_1 + C_2)}}{\dfrac{C_1}{C_1 + C_2}} = 1/2\pi R_1 C_1
\end{aligned}
\tag{2.74}
$$

This is not surprising since at low frequencies, the capacitors act as open-circuits and at high frequencies, the impedance of the capacitors is so small that they dominate the behavior in a parallel combination with resistors.

As for ϕ in Fig. 2.34b, in the passive case it is $0°$ as can be seen from the non-inversion in (2.74).

Analyzing the active GENF in Fig. 2.35b, and from the result of the inverting configuration in (2.18), we can see that

$$\frac{V_{out}(s)}{V_{in}(s)} = -\frac{Z_f}{Z_1} \tag{2.75}$$

where Z_1 is the equivalent impedance of the circuit at the input and Z_f is the equivalent impedance of the circuit in the feedback path.

$$Z_1 = Z_{R1}//Z_{C1} = \frac{R_1/sC_1}{R_1 + 1/sC_1} = \frac{R_1}{1 + sR_1C_1}$$

$$Z_f = Z_{Rf}//Z_{Cf} = \frac{R_f/sC_f}{R_1 + 1/sC_f} = \frac{R_f}{1 + sR_fC_f} \tag{2.76}$$

Substituting (2.76) in (2.75) we get

$$\frac{V_{out}(s)}{V_{in}(s)} = -\frac{\frac{R_f}{1+sR_fC_f}}{\frac{R_1}{1+sR_1C_1}} = -\frac{R_f(1 + sR_1C_1)}{R_1(1 + sR_fC_f)} = \frac{-\left(\frac{C_1}{C_f}\right)s - \frac{1}{R_1C_f}}{s + \frac{1}{R_fC_f}} \tag{2.77}$$

Comparing (2.77) to (2.69), we can see that

$$a = -\frac{C_1}{C_f}$$

$$b = -\frac{1}{R_1C_f} \tag{2.78}$$

$$f_p = 1/2\pi R_f C_f$$

As a result, for the active GENF, we have

$$\text{DC Gain} = -\frac{R_f}{R_1}$$

$$\text{High Frequency Gain} = -\frac{C_1}{C_f} \tag{2.79}$$

$$f_z = 1/2\pi R_1 C_1$$

This is not surprising, since at low frequencies, the capacitors act as open-circuits and at high frequencies, the impedance of the capacitors is so small that they dominate the behavior in a parallel combination with resistors.

As for ϕ in Fig. 2.34b, in the active case it is $180°$ as can be seen from the inversion in (2.79).

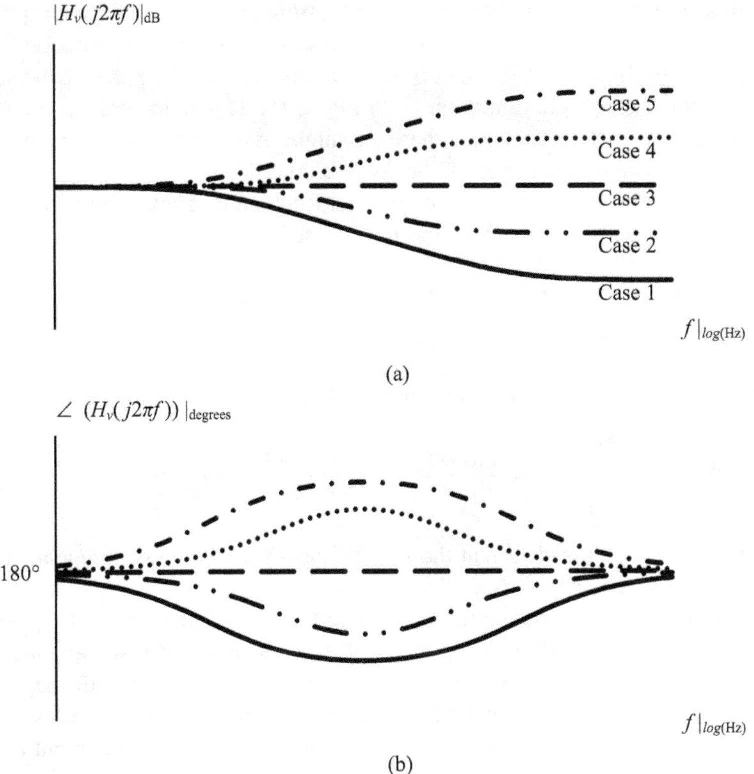

Fig. 2.36 Active GENF transfer function magnitude (**a**) and phase (**b**) for different positions of zeros and poles

What would happen to the plots in Fig. 2.34 if f_p increases and f_z decreases? Also, what if f_p is equal to f_z? For an active GENF for example, the results for such a sweep in frequencies are shown in Fig. 2.36. Each case corresponds to a certain position of zero and pole.

For example, Case 1 corresponds to the one in Fig. 2.34, which is where the pole frequency is much lower than the zero frequency. On the other extreme, there is Case 5 where the zero frequency is much lower than the pole frequency. Notice how the phase changes between these extreme cases.

In Case 3, an interesting phenomenon happens. The pole frequency and the zero frequency overlap, that is they cancel each other as follows:

$$f_z = 1/2\pi R_1 C_1 = f_p = 1/2\pi R_f C_f$$

$$\Rightarrow R_1 C_1 = R_f C_f \Rightarrow \frac{R_1}{R_f} = \frac{C_f}{C_1} \tag{2.80}$$

In this case, the DC gain is equal to the high-frequency gain as given in (2.79). The magnitude is constant and so is the phase at 180°. In this situation, all the frequencies are shifted by 180°, that is, they are inverted. In other words, the setup is acting as an inverting configuration as in Fig. 2.15. This pole-zero cancellation is quite important in other contexts where we might have a pole for example and we cancel it with a zero, leading to the behavior above.

What if we want to have a constant transfer function magnitude, but with a phase different than 180°? This is where the APF comes in.

2.3.3.4 All-Pass Filter

An APF has the following transfer function shape

$$H_v(s) = \frac{V_{out}(s)}{V_{in}(s)} = -a\frac{s - 2\pi f_0}{s + 2\pi f_0} \qquad a > 0 \tag{2.81}$$

In this situation, it is clear that the zero frequency and the pole frequency are the same, thereby referred to as f_0.

A plot of this transfer function is shown in Fig. 2.37. Note how at frequency f_0, the filter gives a phase shift of $-90°$. At other frequencies, it gives another phase shift, while always maintaining the same magnitude. This phase shift has a maximum of $-180°$. Therefore, an APF is used for its phase properties, thus called a phase shifter. The value of $-90°$ is particularly important, since if the input is called

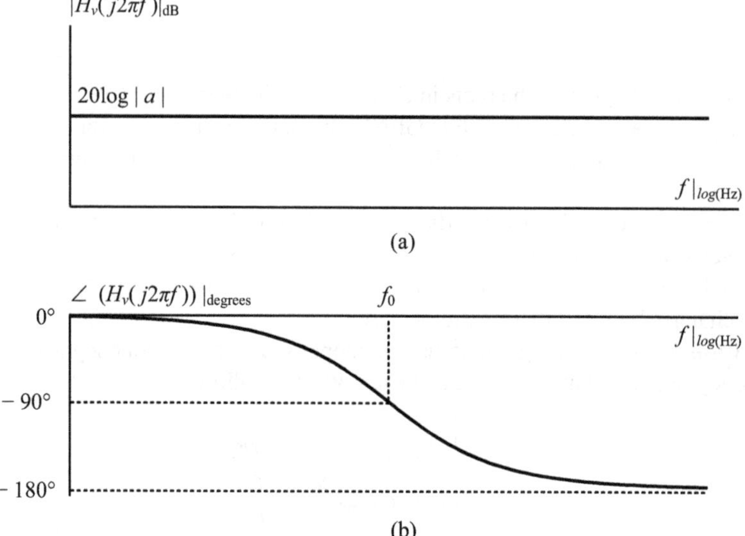

Fig. 2.37 APF transfer function magnitude (**a**) and phase (**b**)

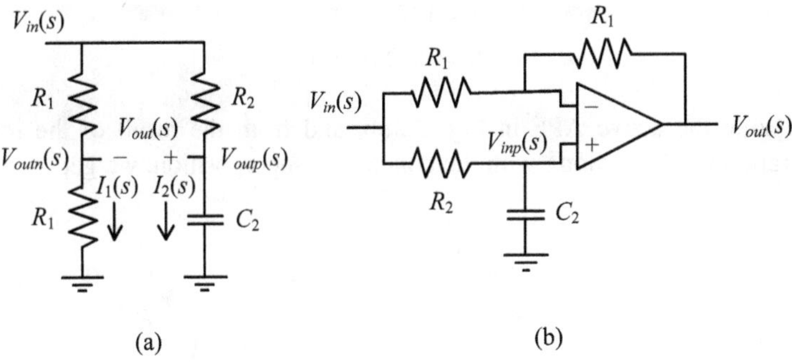

Fig. 2.38 Passive APF (a) and active APF (b)

a cosine, then the output is called a sine, resulting in a very useful pair of signals for signal processing applications.

The implementation of this APF is shown in Fig. 2.38.

Analyzing the passive APF in Fig. 2.38a and starting with the currents, we have

$$I_1(s) = \frac{V_{in}(s)}{2R_1}$$
$$I_2(s) = \frac{V_{in}(s)}{R_2 + 1/sC_2} \tag{2.82}$$

As for the voltages, we have

$$V_{outn}(s) = I_1(s) \times R_1$$
$$V_{outp}(s) = I_2(s) \times 1/sC_2 \tag{2.83}$$

Substituting (2.83) in (2.82) we get

$$V_{outn}(s) = V_{in}(s)\frac{R_1}{2R_1} = \frac{V_{in}(s)}{2}$$
$$V_{outp}(s) = V_{in}(s)\frac{1/sC_2}{R_2 + 1/sC_2} = \frac{V_{in}(s)}{1 + sR_2C_2} \tag{2.84}$$

The complete output is

$$V_{out}(s) = V_{outp}(s) - V_{outn}(s)$$
$$\frac{V_{out}(s)}{V_{in}(s)} = \frac{1}{1 + sR_2C_2} - \frac{1}{2} = \frac{1 - sR_2C_2}{2(1 + sR_2C_2)} = -\frac{1}{2}\left(\frac{s - 1/R_2C_2}{s + 1/R_2C_2}\right) \tag{2.85}$$

Comparing (2.85) to (2.81), we can see that

$$a = 1/2$$
$$f_0 = 1/2\pi R_2 C_2 \tag{2.86}$$

Analyzing the active APF in Fig. 2.38b, and from the result of the inverting configuration in (2.18) along with the concept of superposition, we get

$$\begin{aligned}
V_{out}(s) &= V_{inp}(s)\left(1 + \frac{R_1}{R_1}\right) + V_{in}(s)\left(-\frac{R_1}{R_1}\right) \\
&= V_{in}(s)\left(\frac{1/sC_2}{R_2 + 1/sC_2}\right)\left(1 + \frac{R_1}{R_1}\right) + V_{in}(s)\left(-\frac{R_1}{R_1}\right) \\
&= V_{in}(s)\left(\frac{2}{1 + sR_2C_2} + 1\right) = V_{in}(s)\left(\frac{1 - sR_2C_2}{1 + sR_2C_2}\right)
\end{aligned} \tag{2.87}$$

Dividing by R_2C_2, we get

$$\frac{V_{out}(s)}{V_{in}(s)} = -\frac{s - 1/sR_2C_2}{s + 1/sR_2C_2} \tag{2.88}$$

Comparing (2.88) to (2.81), we can see that

$$a = 1$$
$$f_p = 1/2\pi R_2 C_2 \tag{2.89}$$

2.4 Switch and Compute

One of the most important applications of electronic circuits is computation. This application makes use of logic gates, which are mainly composed of switches. As a result, the ability to create a good switch is at the foundation of computing.

Based on an input voltage, which can be either high or low, a switch either closes to pass the current, or opens to prevent the current from passing. As a result, two main types of switches exist. The first type will turn ON and pass a current when the input is high, and will turn OFF and prevent the current from passing when the input is low. We will call this an n-type switch. The other type operates in the opposite manner, and will be called a p-type switch. Table 2.4 shows a summary of these switches.

Table 2.4 Type-N and Type-P switches

Input	N-Type switch	P-Type switch
High	ON	OFF
Low	OFF	ON

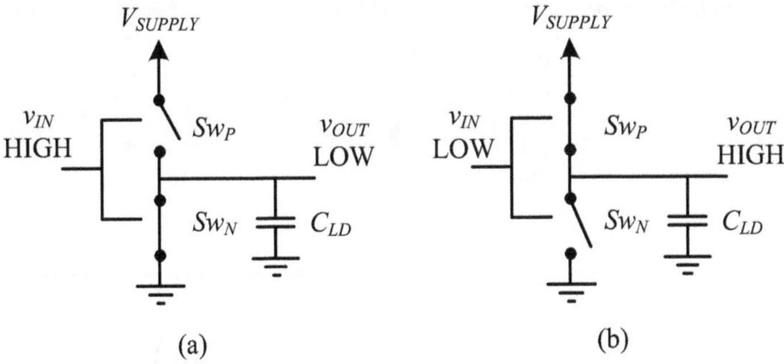

Fig. 2.39 Switches implementing an inverter logic gate, when the input is high (**a**) and low (**b**)

Table 2.5 Truth table for the inverter

V_{IN}	V_{OUT}
0	1
1	0

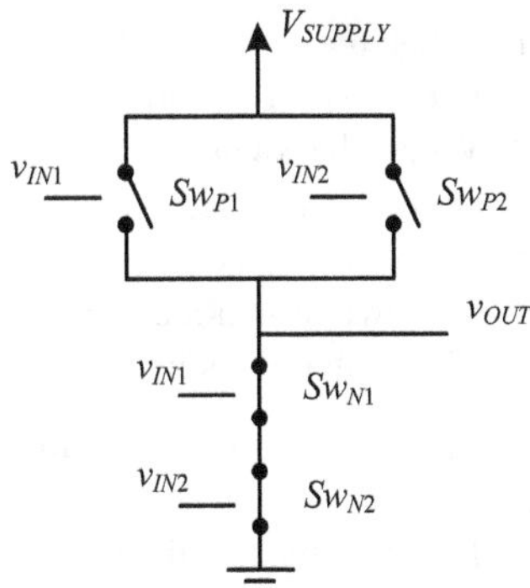

Fig. 2.40 Switches implementing a NAND logic gate

Placing a p-type switch above an n-type switch will give the configuration shown in Fig. 2.39. In this circuit, the load is modeled as a capacitor, which is the case in logic circuits.

When v_{IN} is high, the p-type switch will open and the n-type switch will close, resulting in a low v_{OUT} equal to ground across the load as in Fig. 2.39a. When v_{IN} is low, the p-type switch will close and the n-type switch will open, resulting in a high v_{OUT} equal to V_{SUPPLY} across the load as in Fig. 2.39b. With high interpreted as a 1 and low interpreted as a 0, the truth table is shown in Table 2.5. As a result, this circuit is an inverter logic gate.

Another example is shown in Fig. 2.40.

With high interpreted as a 1 and low interpreted as a 0, the truth table is shown in Table 2.6. As can be seen, this circuit implements a NAND logic gate.

In the above explanation, it is assumed that the switches behave in an ideal manner, which means at least the following:

Table 2.6 Truth table for the NAND gate

v_{IN1}	v_{IN2}	v_{OUT}
0	0	1
0	1	1
1	0	1
1	1	0

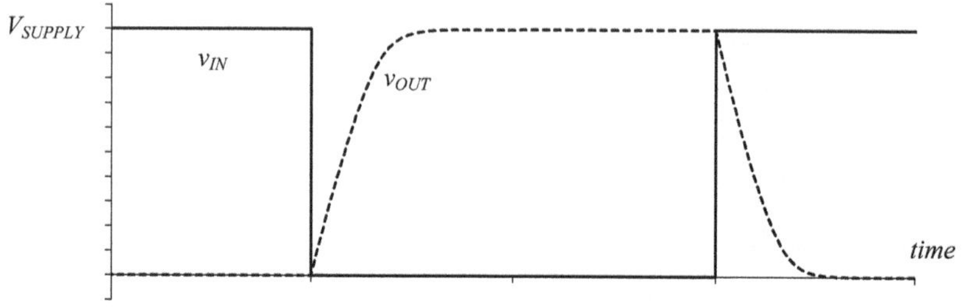

Fig. 2.41 Waveforms for an inverter driving a large load

1. When the switch is ON, it presents a zero resistance to the current, thus acting as a perfect short circuit.
2. When the switch is OFF, it presents an infinite resistance to the current, thus acting as a perfect open circuit.
3. The switch reacts instantaneously to the input.
4. The switch is not affected by the size of the load.
5. The switch does not take any current from the input.

The absence of any of the criteria above renders the gate less ideal, thus more realistic.

For example, for the inverter in Fig. 2.39, if the load capacitance is too large, thus affecting the behavior of the gate, the output will require a lot of time to reach its final result as shown in Fig. 2.41.

As can be seen, enough time should be given to the inverter output to stabilize. This results in a minimum switching period, thus a maximum switching frequency, which limits the speed of the gate, a very important performance criterion in today's implementations.

As another example, if the switches do not act as perfect open circuits when OFF, which is the case in practice, then a small current, called a leakage current, will still pass through them. Combining this with the huge number of switches, which is typical in today's circuits, results in very high stand-by power consumption, a definite undesirable situation. For this reason, additional measures are typically taken to reduce this power consumption when the circuit is in stand-by mode, such as disabling it by disconnecting it from the power supply.

Chapter 3
Diodes

Resistors, capacitors, and inductors are linear components since their current versus voltage plot consists of a straight line. However, there are many applications, such as switching, that require a nonlinear behavior. The simplest nonlinear component is a diode. This chapter presents several ways to model a diode. Some of these models are simpler than the others, albeit less accurate. The method to devise models and how to choose between them given a certain context applies not only to diodes but also to all electric and electronic circuits. As a result, these are important skills to acquire and hone in the long run. This chapter also presents some important sample applications of diodes such as AC-to-DC conversion, lighting, and photodetection. In fact, we are constantly seeking new applications for diodes and all other electronic components.

3.1 Diode Models

Resistors, inductors, and capacitors are passive components. The impedances of these components are as follows:

$$\text{Impedance of a resistor } Z_R = R \tag{3.1}$$

$$\text{Impedance of an inductor } Z_L = j\omega L = j2\pi f L \tag{3.2}$$

$$\text{Impedance of a capacitor } Z_C = \frac{-j}{\omega C} = \frac{-j}{2\pi f C} \tag{3.3}$$

In the above equations, ω is the angular frequency in rad/sec and f is the frequency in Hz.

© The Editor(s) (if applicable) and The Author(s), under exclusive license to Springer Nature Switzerland AG 2024

J. G. Atallah, M. Ismail, *Integrated Electronic Circuits*,

https://doi.org/10.1007/978-3-031-62707-1_3

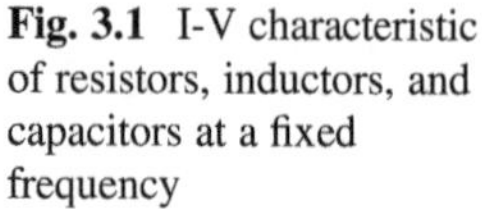

Fig. 3.1 I-V characteristic of resistors, inductors, and capacitors at a fixed frequency

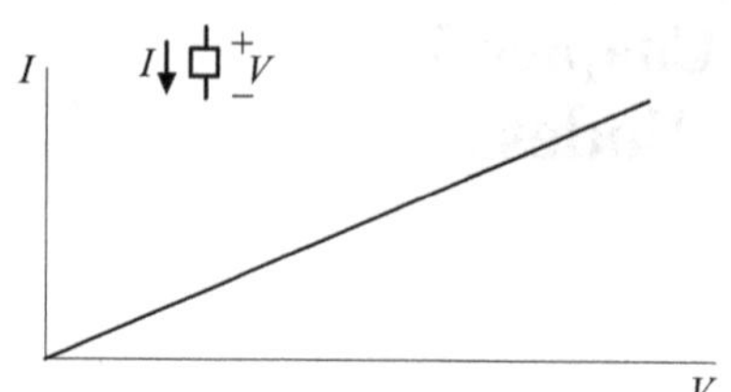

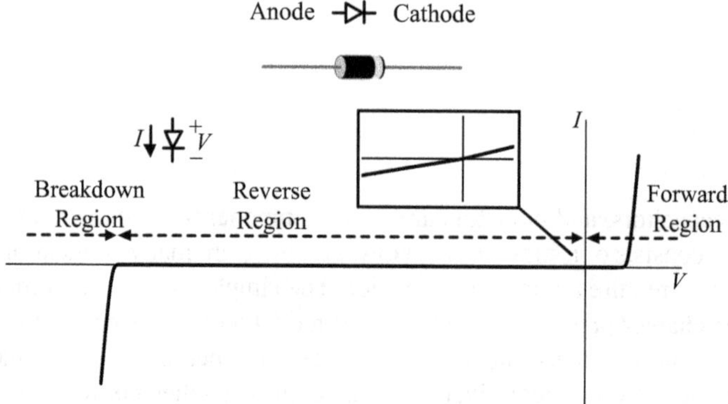

Fig. 3.2 Diode symbol, component, and I-V characteristic

Note that at a specific frequency, all these impedances are constant. Therefore, at a specific frequency, the I-V characteristic of all these components is as shown in Fig. 3.1.

The I-V characteristic consists of a line. This is why these components are called linear.

These components can be used to build useful circuits, but they are not enough. For some applications, such as switching, we need a component which is nonlinear. The simplest nonlinear component is a diode, whose I-V characteristic is shown in Fig. 3.2 along with the shape of a discrete diode and the diode symbol. As can be seen, the diode I-V characteristic is nonlinear. The positive terminal of the diode is called the anode and the negative one is called the cathode. The diode I-V characteristic can be divided into three main regions: the forward region, which is the most important, the reverse region, and the breakdown region.

We are going to approximate the above behavior using mathematical curves and lines. Each approximation will be called a model. Most of these models will deal with the part of the I-V characteristic which is around the origin and positive. Each model will differ from the others in terms of simplicity and accuracy.

Therefore, when analyzing or designing a circuit, the first decision to make is which model to use. This choice affects the speed with which the analysis or design is done as well as the accuracy of the results. In general the simpler the model is, the less is the effort required to use it, and the less accurate are the results that it yields.

Some of these models are suitable for hand calculations and others are more suitable for simulations.

We are going to describe several models starting with the simplest, least accurate one, and reaching the most complicated and most accurate one. First, we are going to look into the forward region, then the reverse and breakdown regions.

3.1.1 Forward Region Diode Models

In this subsection, we are going to discuss three diode models in the forward region.

3.1.1.1 Diode Model 1: Constant-Voltage Drop (C-VD) Model

The Constant-Voltage Drop (C-VD) model consists of the approximation shown in Fig. 3.3.

When the voltage across the diode is below V_{diode}, no current passes and the diode is OFF, thus acting as an open circuit. Also, V_{diode} itself can be zero.

When the voltage across the diode is V_{diode}, the current can take any value and the diode is ON, thus acting as a voltage source whose value is V_{diode}. Note that when the diode is ON, the current is limited by the other components in the circuit. The diode behavior is summarized in Table 3.1.

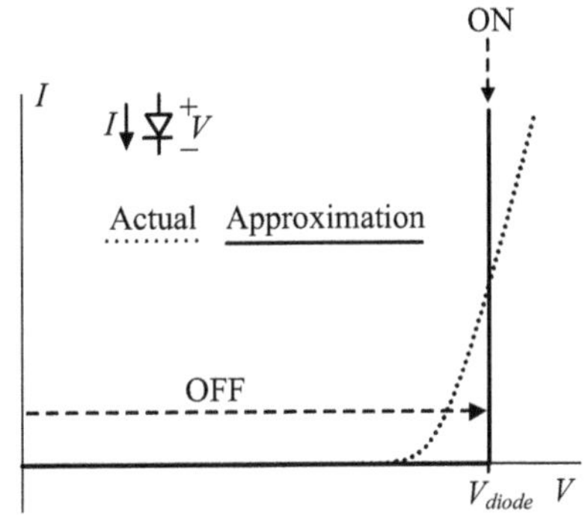

Fig. 3.3 C-VD diode model

Table 3.1 Diode behavior using C-VD model

Condition	Diode operating mode	Diode circuit behavior
$V < V_{diode}$	OFF	
$V = V_{diode}$	ON	

3.1.1.2 Diode Model 2: Variable-Voltage Drop (V-VD) Model

The Variable-Voltage Drop (V-VD) model consists of the approximation shown in Fig. 3.4.

When the voltage across the diode is below V_{on}, no current passes and the diode is OFF, thus acting as an open circuit.

When the voltage across the diode is above V_{on}, the diode behavior looks like that of a passive component, which, in its simplest form, is a resistor r_d. Note that this resistor will be in series with the V_{on} voltage source. This diode behavior is summarized in Table 3.2.

Note that using the V-VD model, the voltage across a diode does not have a maximum value.

This model is only used in cases where the C-VD model does not yield accurate enough results.

3.1.1.3 Diode Model 3: Exponential Model

The exponential model consists of the approximation shown in Fig. 3.5.

This model consists of approximating the diode behavior using an exponential curve. It is valid for diode voltages up to about 0.9 V. To motivate this curve, we will plot the logarithm of the current versus the voltage as in Fig. 3.6. As it can be seen, the logarithmic value of the current is almost linearly related to the voltage.

The current-voltage relation is called the Shockley diode equation and is as follows.

Fig. 3.4 V-VD diode model

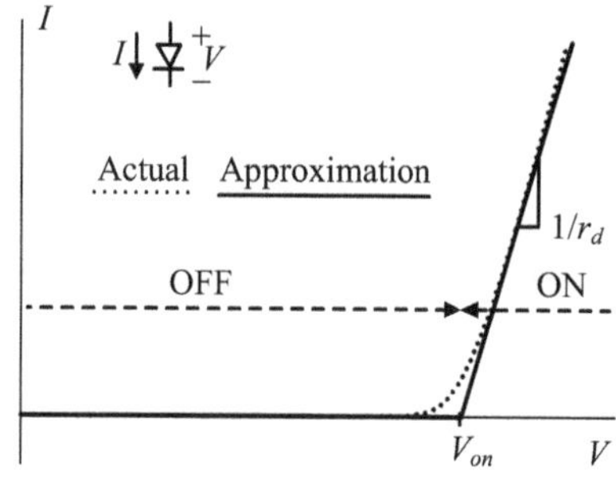

Table 3.2 Diode behavior using V-VD model

Condition	Diode operating mode	Diode circuit behavior
$V < V_{ON}$	OFF	
$V \geq V_{ON}$	ON	

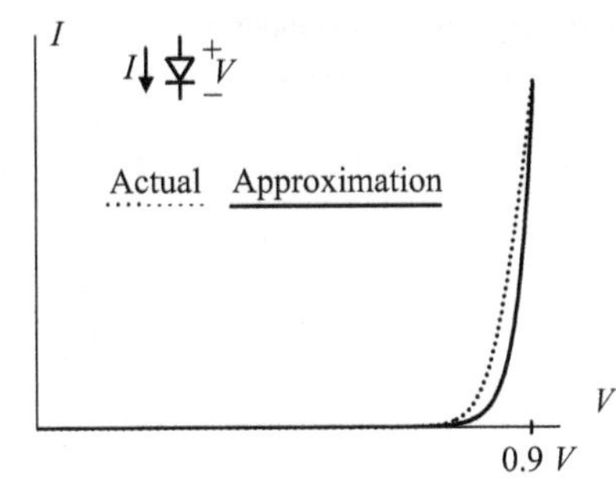

Fig. 3.5 Exponential diode model

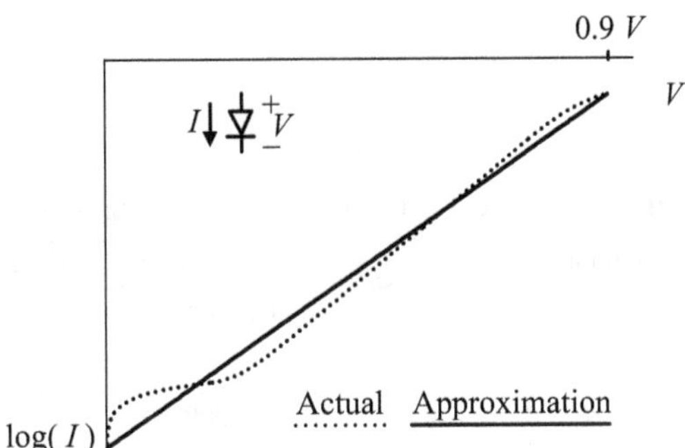

Fig. 3.6 Exponential diode model, logarithmic scale

$$I = I_s \left(e^{\frac{V}{nV_T}} - 1 \right) \approx I_s e^{\frac{V}{nV_T}} \tag{3.4}$$

I_s is a parameter of the diode that depends on its cross-section (junction) area. The larger the area, the higher is the I_s. I_s is called the saturation or scale current. Its value is much smaller than I, which is why the term (-1) in the above equation is usually ignored.

V_T is another parameter. It is called the thermal voltage. Its value is around 25 mV at room temperature which is around 20 °C (293 K).

n is called the ideality factor, which varies from 1 to 2, with 1 being the ideal and default value. For this reason, n is usually omitted.

Note that with this model, the diode is always ON within our region of interest. Also, this model is not typically used for hand-calculations, and is best used by simulators. It gives the most accurate results compared to the previous models.

3.1.2 Reverse and Breakdown Regions Diode Model

In the reverse and breakdown regions, we will describe only one model. This model relies on the observation that as the voltage decreases below zero the diode enters the reverse region and the diode reverse current increases very gradually in the opposite

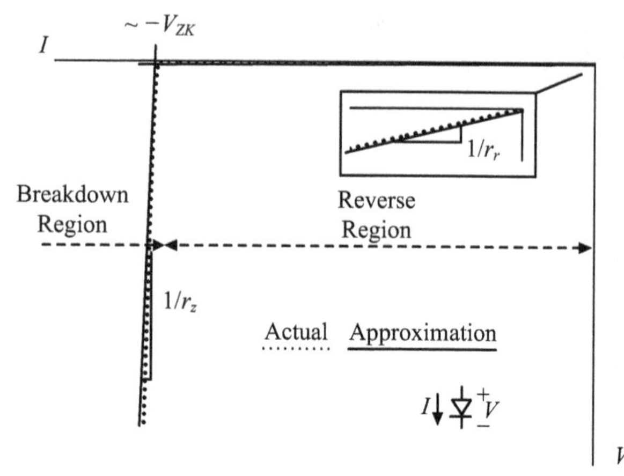

Fig. 3.7 Exponential diode model, logarithmic scale

Table 3.3 Diode behavior in the reverse and breakdown regions

Condition	Diode operating region	Diode circuit behavior
$-V_{ZK} \leq V < 0$	Reverse	r_r
$-\infty < V < -V_{ZK}$	Breakdown	r_z / $\sim V_{ZK}$

direction. After this reverse voltage exceeds a value called $-V_{ZK}$, the diode enters the breakdown region and the diode current increases very rapidly as in Fig. 3.7.

In these two regions, the behavior is approximated by two straight lines.

In the reverse region, the diode acts as a resistor whose value is the inverse of the slope of this line.

In the breakdown region, the diode acts as a resistor in series with a voltage source.

This diode behavior is summarized in Table 3.3.

Note that the voltage source is flipped since it represents $-V_{ZK}$ and that its value is approximately V_{ZK}.

3.1.3 Additional Notes Regarding the Diode Modeling

Here we will describe a couple of effects that might be useful when modeling diodes.

3.1.3.1 Temperature Effects

In the forward region, we described I_s and V_T for the exponential model. Both of these parameters increase with temperature. The combined result is that, when the ambient temperature increases, given a certain diode voltage, the current increases. This is useful for building thermometers. This effect is shown in Fig. 3.8. Note that the change in current or voltage is not linear with temperature. In fact, the change from $-10\ °C$ to $+20\ °C$ is larger than from $+20\ °C$ to $+40\ °C$.

3.1.3.2 Capacitive Effects

In the reverse region, the diode exhibits some capacitance between its terminals. This capacitance is a function of the reverse voltage. As the reverse voltage magnitude increases, the capacitance decreases.

3.2 Diode Applications

In this section, some important diode applications are described.

3.2.1 Clipping

Clipping or limiting a signal's range is needed in many applications, such as when driving a load whose input should not exceed a certain value.

A positive clipping circuit and its waveforms are shown in Fig. 3.9. We are going to use the C-VD diode model to describe the behavior of this circuit using the time intervals indicated in the figure.

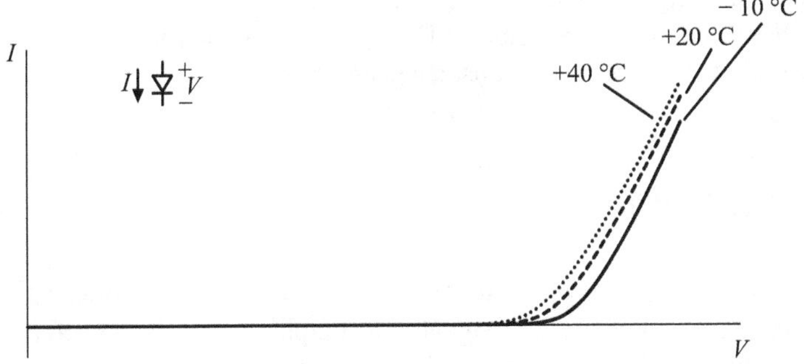

Fig. 3.8 Temperature effect I-V characteristic of diodes

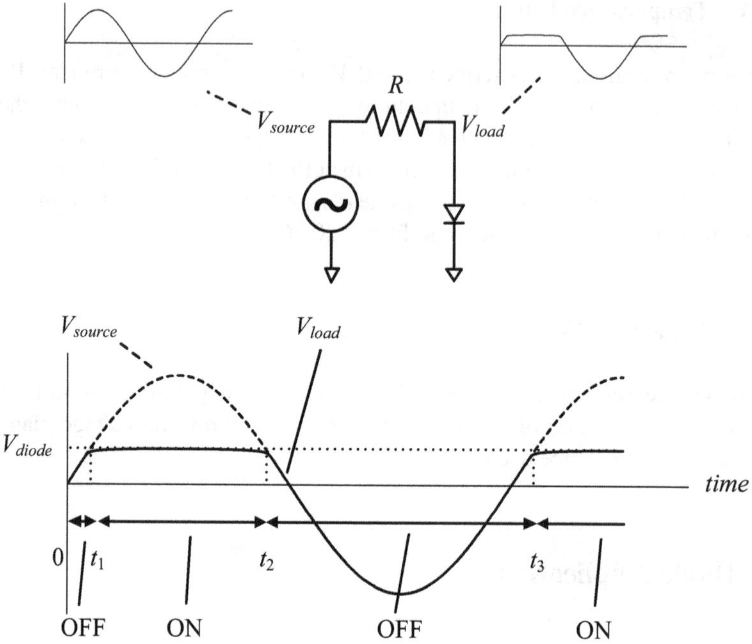

Fig. 3.9 Positive clipping circuit

As V_{source} increases from zero, the diode stays OFF until V_{source} reaches V_{diode} at t_1. At this instance, the diode turns ON and stays ON until V_{source} starts to go below V_{diode} at t_2. After that, the diode stays OFF until V_{source} reaches V_{diode} again at t_3. As a result, this circuit clips the signal and limits its maximum value to V_{diode}.

If we want to limit the signal to a maximum value higher than V_{diode}, we can put a DC voltage source in series with the diode as in Fig. 3.10.

If we want to clip the negative excursion of the signal, we can flip the diode (and the DC voltage source if it is used), resulting in a negative clipping circuit.

Additionally, we can limit both the positive and the negative excursions of the signal by combining both circuits as in Fig. 3.11.

Note that if the maximum value of V_{source} is much higher than $V_{diode} + V_{DC}$, the resulting output signal will be almost a square wave.

3.2.2 Rectification

A rectifier is a building block of an AC-DC converter. The converter takes an AC signal at its input and gives a DC signal at its output. These signals are typically in the voltage domain. The converter itself has many uses such as in a power supply unit (PSU) or an AC adaptor to power electronics devices from an AC power outlet

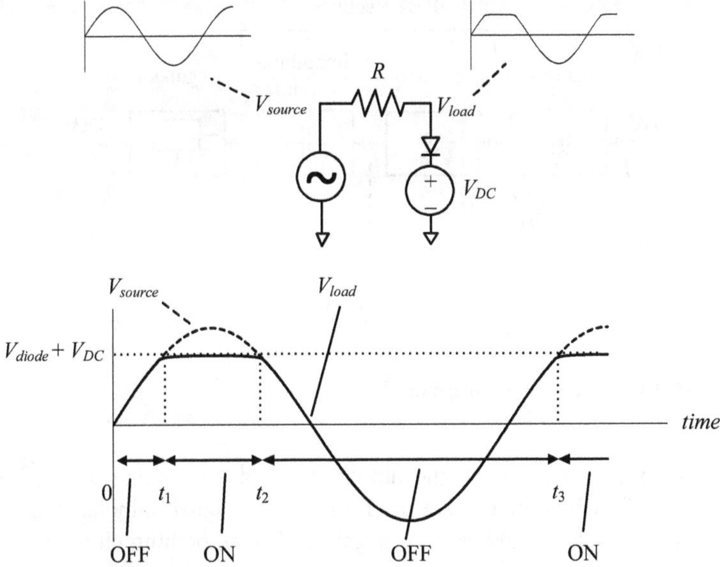

Fig. 3.10 Positive clipping circuit with an additional DC source

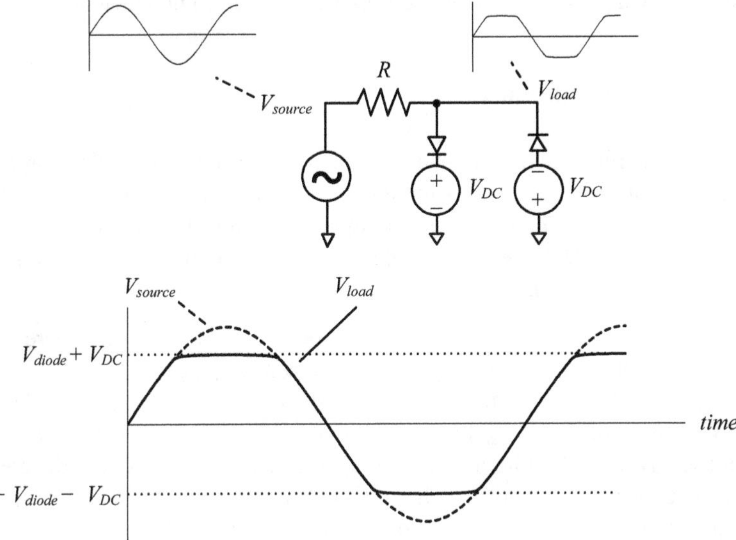

Fig. 3.11 Positive and negative clipping circuit with DC sources

or in a battery charger to charge a battery from an AC power source. As a result, the diodes in these circuits are exposed to signals that are not small.

The AC-DC converter consists of many stages as in Fig. 3.12.

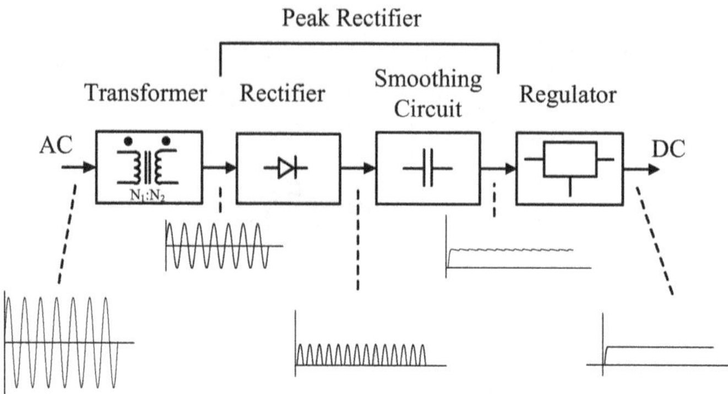

Fig. 3.12 AC-DC converter block diagram

The transformer steps down the amplitude of the incoming AC signal. The rectifier, which is where diodes are used, makes the signal unipolar, that is either positive, which is usually the case, or negative. The smoothing circuit reduces the variations in the output of the rectifier. The combination of a rectifier and a smoothing circuit is called a peak rectifier. The regulator maintains the output signal constant irrespective of input, load, and temperature variations.

In the following, we will showcase important rectifier implementations.

3.2.2.1 Half-Wave Rectifier

A half-wave rectifier circuit is shown in Fig. 3.13. The source is modeled as an ideal AC voltage source and the load is modeled as a resistor R_{load}. This circuit is called half-wave because the output consists of about only half of the input signal. We are going to use the C-VD diode model to describe the behavior of this circuit using the time intervals indicated in the figure.

- $0 \rightarrow t_1$: When V_{source} starts at zero, the diode is OFF. As V_{source} increases, the diode stays OFF until the voltage across it reaches V_{diode}.
- $t_1 \rightarrow t_2$: At t_1, the diode turns ON. At this instance of time, V_{source} is equal to V_{diode} and V_{load} is zero. As V_{source} increases above V_{diode}, the diode stays ON and V_{load} increases. Note that while the diode is ON, there will always be a difference of V_{diode} between V_{source} and V_{load} if the C-VD model is used. However, in reality, and as can be seen in Fig. 3.13, the difference increases as V_{source} increases. When V_{source} starts to decrease from its maximum value, V_{load} starts to decrease until it is equal to zero and V_{source} is equal to V_{diode}.
- $t_2 \rightarrow t_3$: At t_2, and as V_{source} keeps on decreasing, the diode turns OFF and stays OFF until V_{source} increases above V_{diode} again.

An important issue is related to the diode behavior when the voltage across it is negative as in Fig. 3.2. The negative voltage magnitude should not exceed that of

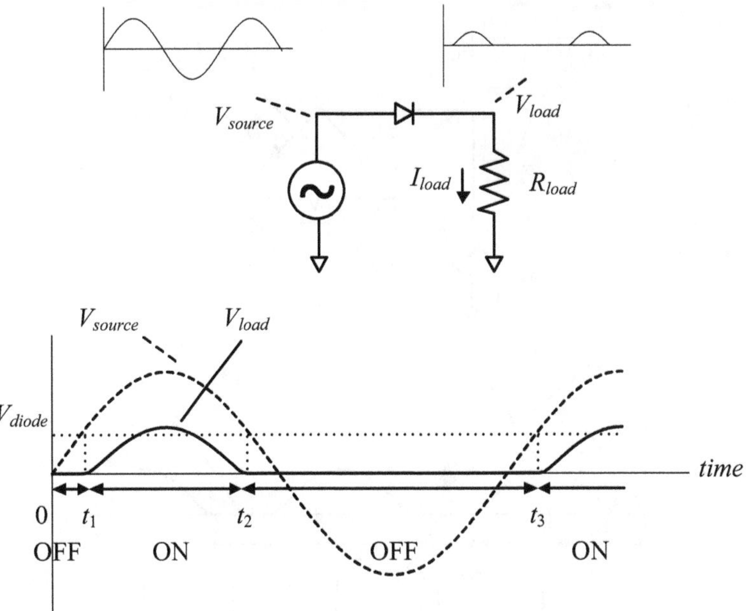

Fig. 3.13 Half-wave rectifier

$-V_{ZK}$, otherwise the diode will enter the breakdown region and will conduct significant current in the opposite direction. For the half-wave rectifier, the maximum negative voltage, also called the Peak Inverse Voltage (PIV), that the diode has to be able to sustain while staying in the reverse region is $V_{source}(max)$.

3.2.2.2   Bridge Rectifier

A bridge rectifier circuit is shown in Fig. 3.14. The source is modeled as an ideal AC voltage source and the load is modeled as a resistor R_{load}. This circuit belongs to the family of full-wave rectifiers because the output consists of about the complete input signal. We are going to use the C-VD diode model to describe the behavior of this circuit using the time intervals indicated in the figure.

- $0 \to t_1$: When V_{source} starts at zero, the diodes are OFF. As V_{source} increases, the diodes stay OFF until the voltage across each of D_1 and D_2 reaches V_{diode} at t_1.
- $t_1 \to t_2$: At t_1, diodes D_1 and D_2 turn ON. At this instance of time, V_{source} is equal to $2V_{diode}$ and V_{load} is zero. As V_{source} increases above $2V_{diode}$, D_1 and D_2 stay ON and V_{load} increases. Note that while D_1 and D_2 are ON, there will always be a difference of $2V_{diode}$ between V_{source} and V_{load}. When V_{source} starts to decrease from its maximum value, V_{load} starts to decrease until it is equal to zero and V_{source} is equal to $2V_{diode}$ at t_2.
- $t_2 \to t_3$: At t_2, and as V_{source} keeps on decreasing, D_1 and D_2 turn OFF.

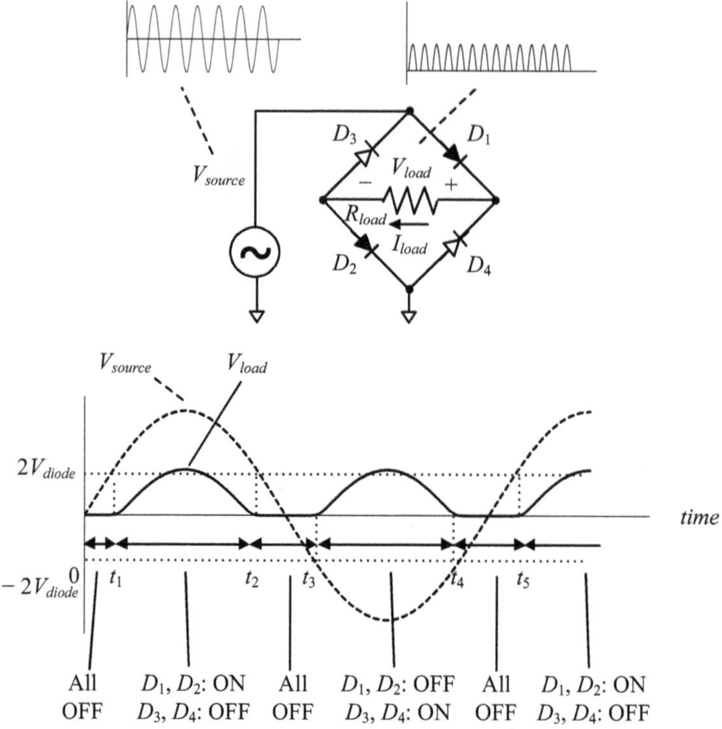

Fig. 3.14 Bridge rectifier

- $t_3 \rightarrow t_4$: At t_3, when V_{source} goes below $-2V_{diode}$, D_3 and D_4 turn ON. It is important to note that when this happens, and as V_{source} goes more negative, I_{load} passes through the load in the same direction as when D_1 and D_2 were ON during the $t_1 \rightarrow t_2$ interval, thus the rectification.
- $t_4 \rightarrow t_5$: At t_4, when V_{source} goes above $-2V_{diode}$, all diodes turn OFF.

As for the PIV, let us take the situation when V_{source} is maximum and positive. In this case, D_1 and D_2 are very much ON and D_3 and D_4 are very much OFF while experiencing their PIV. Doing a mesh starting from the source, then D_4, R_{load}, and D_3, we get:

$$-V_{source} + V_{D4\_inv} - V_{load} + V_{D3\_inv} = 0$$
$$V_{D4\_inv} + V_{D3\_inv} = V_{source} + V_{load}$$
$$V_{D4\_PIV} + V_{D3\_PIV} = V_{source\_MAX} + V_{load\_MAX}$$
$$= V_{source\_MAX} + \left(V_{source\_MAX} - 2V_{diode}\right)$$
$$= 2\left(V_{source\_MAX} - V_{diode}\right)$$
$$\therefore \mathrm{PIV} = V_{source\_MAX} - V_{diode}$$

$$(3.5)$$

The PIV of every diode in a bridge rectifier is less than that in a half-wave rectifier.

3.2.2.3 Peak Rectifier

A peak rectifier is a rectifier, half-wave or full-wave, followed by a smoothing circuit. The smoothing circuit is typically implemented as a capacitor in parallel with the load, effectively creating a low-pass filter. The net effect is that the ripples from the rectifier, irrespective of its type, will be dramatically reduced, although not eliminated. To showcase this fact, a peak bridge rectifier circuit is shown in Fig. 3.15.

Figure 3.16 shows the signals in the peak rectifier of Fig. 3.15 without C_{load} and with C_{load} although the actual output is the signal with C_{load}. This is done in order to make the explanation easier. We are going to use the C-VD diode model to describe the behavior of this circuit using the time intervals indicated in the figure.

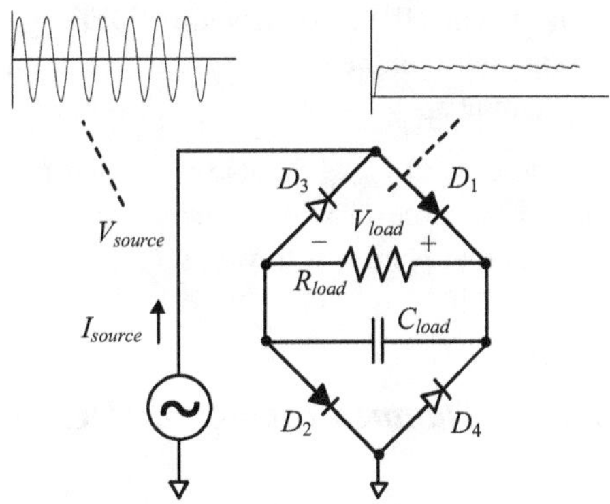

Fig. 3.15 Peak rectifier consisting of a bridge rectifier followed by a capacitor

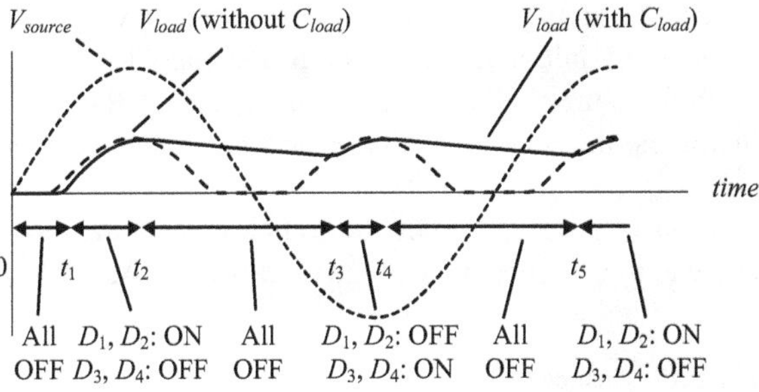

Fig. 3.16 Signals in the peak rectifier of Fig. 3.15

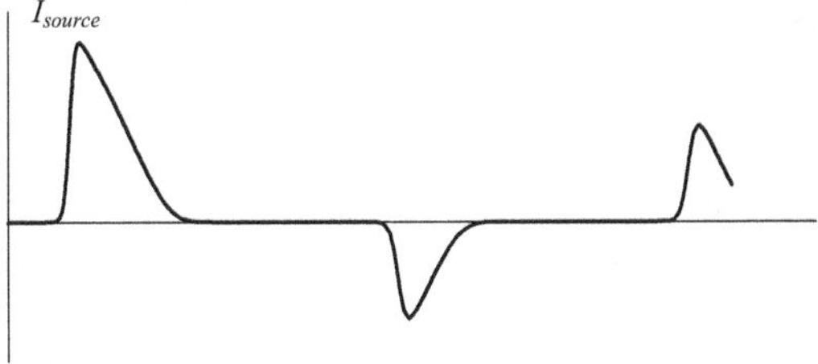

- $0 \rightarrow t_1$: When V_{source} starts at zero, the diodes are OFF. As V_{source} increases, the diodes stay OFF until the voltage across each of D_1 and D_2 reaches V_{diode} at t_1.

- $t_1 \rightarrow t_2$: At t_1, diodes D_1 and D_2 turn ON. At this instance of time, V_{source} is equal to $2V_{diode}$ and V_{load} is zero. As V_{source} increases above $2V_{diode}$, D_1 and D_2 stay ON and V_{load} increases, charging C_{load}. Note that while D_1 and D_2 are ON, there will always be a difference of $2V_{diode}$ between V_{source} and V_{load}, so C_{load} will be charged to a maximum of $V_{source} - 2V_{diode}$ at t_2.

- $t_2 \rightarrow t_3$: At t_2, V_{source} starts to decrease, but V_{load} is held at its maximum value due to C_{load}. Therefore, all diodes turn OFF, and C_{load} is left in parallel with R_{load}, disconnected from the source. As a result, from t_2 to t_3, C_{load} starts to discharge at a rate governed by the time constant $R_{load} \times C_{load}$. This continues until the rectified V_{source} (i.e. V_{load} without C_{load} in Fig. 3.16) exceeds V_{load} at t_3.

- $t_3 \rightarrow t_4$: At t_3, D_3 and D_4 turn ON again, and the circuit starts to charge C_{load} in order to recover the charges that were lost between t_2 and t_3. This continues until t_4, when all the diodes turn OFF again. As a result, at t_3, the current from the source increases very quickly from zero up to a peak value, down to zero again at t_4.

The current I_{source} is also shown in Fig. 3.16. As can be seen, when diodes are ON, the source provides current, thus charges, into C_{load} to compensate for the charges that were lost through R_{load}.

3.2.3 Clamped Capacitor (DC Restorer)

A DC restorer circuit DC-shifts the input AC signal while keeping its amplitude the same. A DC restorer is shown in Fig. 3.17.

The different signals in this circuit are shown in Fig. 3.18.

$0 \rightarrow t_1$: Initially, V_{source} is at a certain level and all other voltages are zero. Therefore, the diode is OFF and V_{cap} is at the same level as V_{source}. As V_{source} increases, the diode stays OFF and V_{cap} stays constant leading to V_{load} increasing with V_{source}.

$t_1 \rightarrow t_2$: As V_{source} starts to go below V_{cap}, the diode turns ON. This makes V_{load} equal to the inverse ON voltage of the diode. As for V_{cap}, it follows V_{source} since after

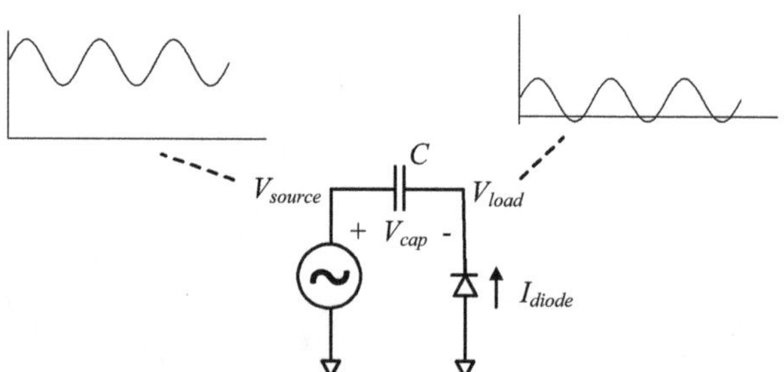

Fig. 3.17 Clamped capacitor circuit (DC Restorer)

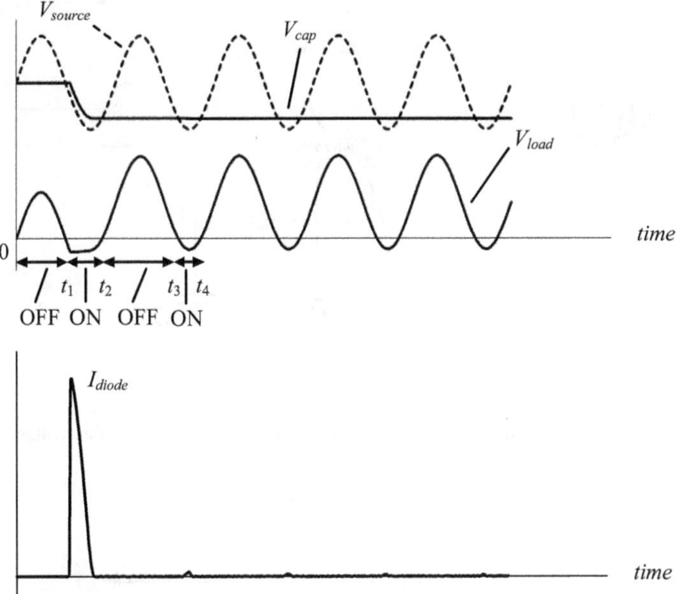

Fig. 3.18 Signals in the clamped capacitor circuit of Fig. 3.17

all it is equal to the difference between V_{source} and V_{load}. Note that during this interval, when the diode turns ON, the current peaks in order to charge the capacitor for the first time.

$t_2 \rightarrow t_3$: As V_{source} increases above V_{cap}, the diode turns OFF again keeping V_{cap} constant and allowing V_{load} to vary with V_{source} since V_{load} is equal to the difference between V_{source} and V_{cap}.

$t_3 \rightarrow t_4$: As V_{source} starts to go below V_{cap}, the diode turns ON. This makes V_{load} equal to the inverse ON voltage of the diode. As for V_{cap}, it stays as it is since it is already equal to the difference between V_{source} and V_{load}. Note that during this interval, when the diode turns ON, the current does not peak as much as during the t_1 to t_2 interval since the capacitor is already charged.

If V_{source} does not carry a DC component, V_{load} will still be the same. However, V_{cap} will be different since it is equal to the difference between V_{source} and V_{load}. As a result, no matter what the input DC is, the output DC will be the same, related only to the input peak-to-peak voltage, thus the name DC Restorer.

3.2.4   AC-to-DC Voltage Multiplication

An AC-to-DC voltage multiplier circuit converts the AC signal into a DC signal whose level is higher than the amplitude of the AC signal.

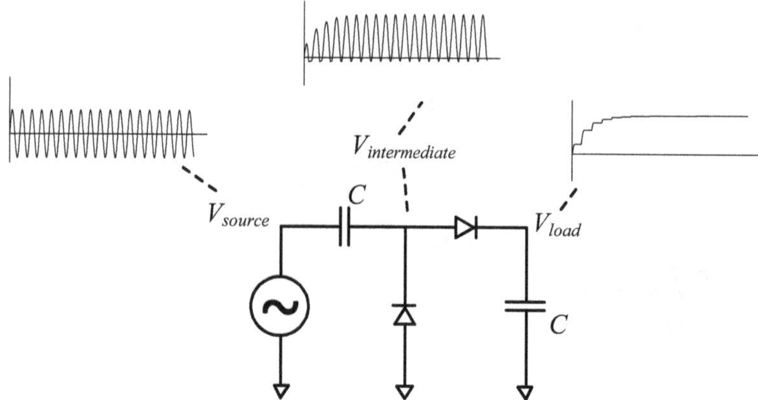

Fig. 3.19 Single-stage Greinacher/Cockcroft Walton multiplier also known as voltage doubler

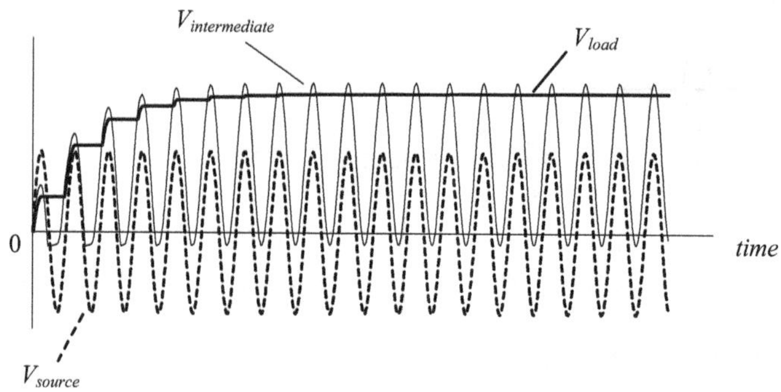

Fig. 3.20 Signals in the voltage doubler of Fig. 3.19

An example of such a multiplier is the Greinacher/Cockcroft-Walton multiplier sometimes called the Villard cascade voltage multiplier. This multiplier is made of several identical stages. Its single-stage version is shown in Fig. 3.19 and is commonly known as a voltage doubler. This circuit can be interpreted as a clamped capacitor circuit followed by a peak rectifier.

The signals in the circuit are shown in Fig. 3.20.

This circuit is called a voltage doubler since the DC output voltage has a value which is almost equal to double the amplitude of the input voltage.

Figure 3.21 shows the progression from a single-stage Greinacher/Cockcroft Walton multiplier to a double-stage stage Greinacher/Cockcroft Walton multiplier.

The signals in a double-stage Greinacher/Cockcroft Walton multiplier are shown in Fig. 3.22.

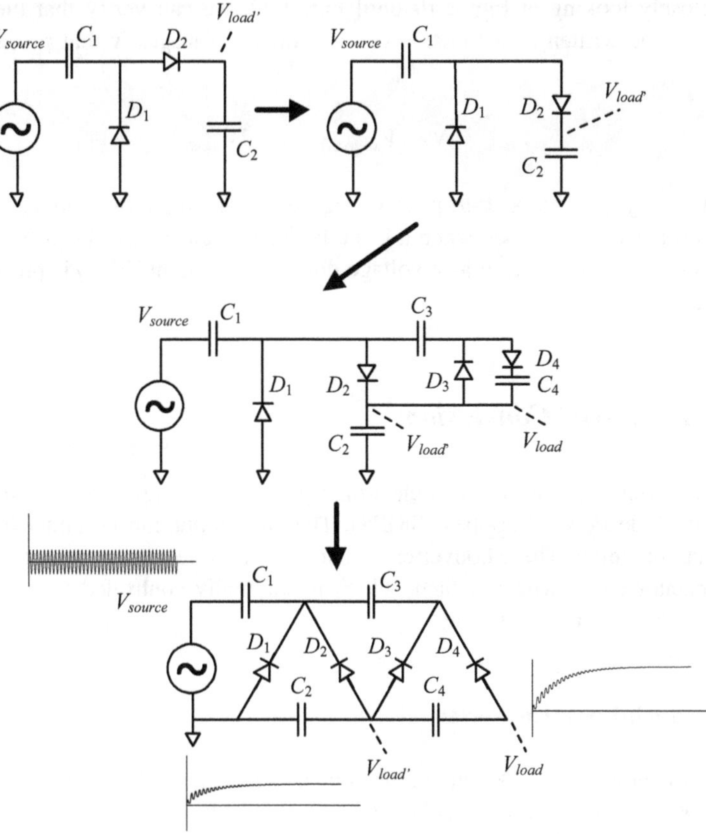

Fig. 3.21 progression from a single-stage Greinacher/Cockcroft Walton multiplier to a double-stage stage Greinacher/Cockcroft Walton multiplier

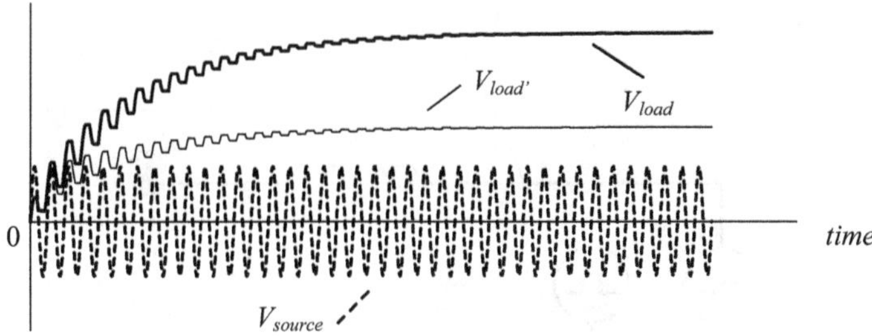

Fig. 3.22 Signals in the double-stage Greinacher/Cockcroft Walton multiplier of Fig. 3.21

By closely looking at Fig. 3.19 until Fig. 3.22, we can verify that the output voltage can be written as a function of the number of stages N that precede it as follows:

$$V_{load} = N \times \left(V_{source(p-p)} - 2V_{diode} \right) \tag{3.6}$$

where $V_{source(p-p)}$ is the peak-to-peak voltage of the source input and V_{diode} is the voltage drop across the diode when it is on. In this particular case, $V_{load'}$ is preceded by a single stage, thus acting as a voltage doubler output, and V_{load} is preceded by two stages.

3.2.5   DC-to-DC Conversion

DC-to-DC converters are used in electronic power supply applications such as in Switched-Mode Power Supplies (SMPS). The DC output can be either higher or lower than the input. These converters employ at least one storage component such as a capacitor along with a switch, which is externally controlled using a Pulse-Width-Modulated (PWM) signal.

3.2.5.1   Buck-Boost Converter

A Buck converter gives a DC output lower than its DC input while a Boost converter gives a DC output higher than its DC input.

A Buck converter is shown in Fig. 3.23.

The signals in the Buck converter are shown in Fig. 3.24.

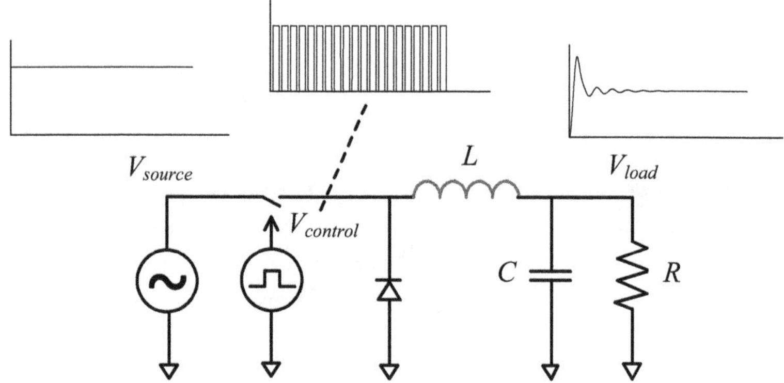

Fig. 3.23   A Buck converter

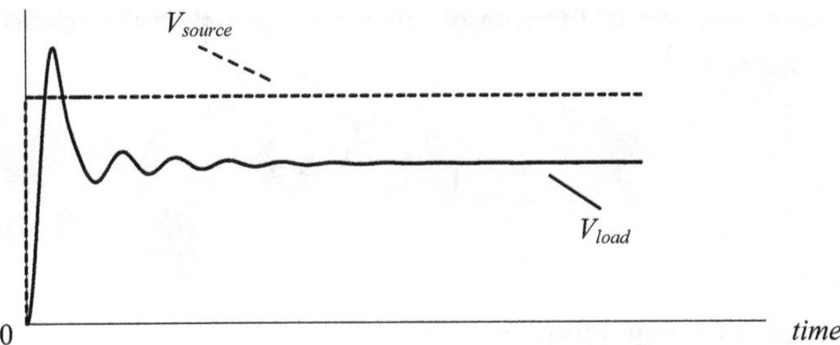

Fig. 3.24   Signals in a Buck converter

Fig. 3.25   A Boost converter

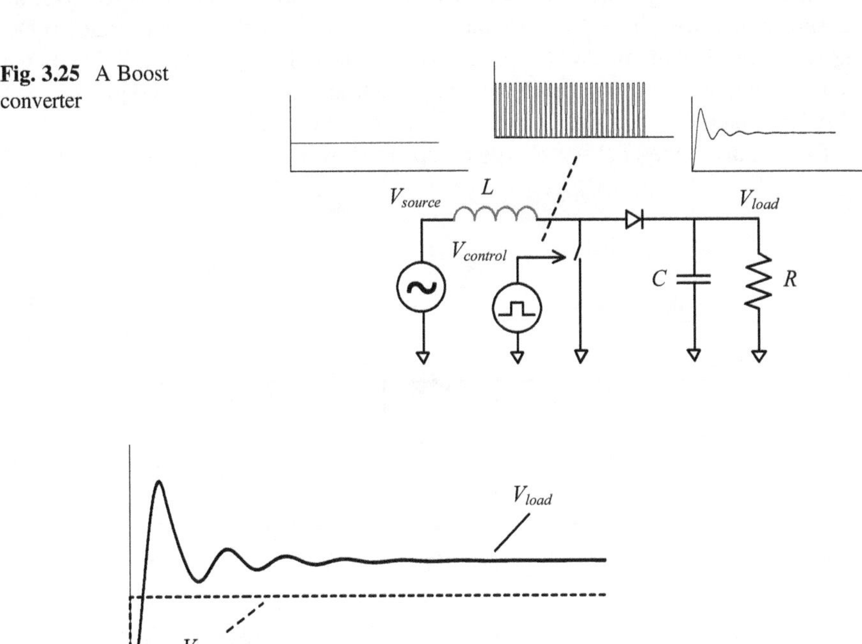

Fig. 3.26   Signals in a Buck converter

If D is the duty cycle of the control voltage $V_{control}$, then the relation between V_{source} and V_{load} is as follows:

$$V_{load} = D \times V_{source} \tag{3.7}$$

A Boost converter is shown in Fig. 3.25.
The signals in the Boost converter are shown in Fig. 3.26.

If D is the duty cycle of the control voltage $V_{control}$, then the relation between V_{source} and V_{load} is as follows:

$$V_{load} = \frac{1}{1-D} \times V_{source} \tag{3.8}$$

3.2.5.2   Dickson Charge Pump

A Dickson charge pump is a modification to the Greinacher/Cockcroft-Walton multiplier described in 3.2.4. This modification allows it to become a DC-to-DC converter instead of an AC-to-DC converter. In addition to the DC input, the Dickson charge pump requires two clock pulse trains that do not overlap.

A four-stage Dickson charge pump is shown in Fig. 3.27.

The signals in the Dickson charge pump are shown in Fig. 3.28.

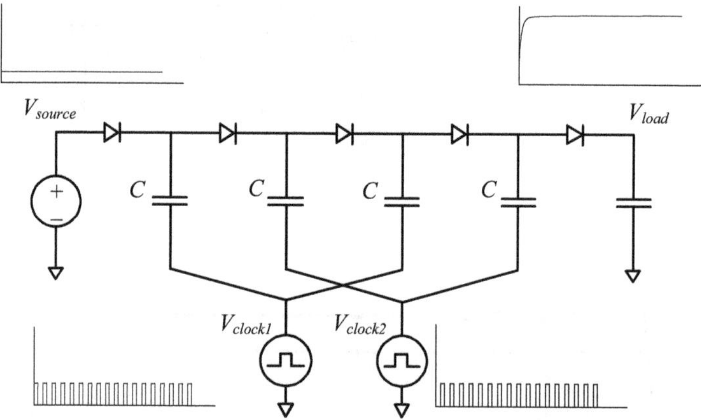

Fig. 3.27   A Dickson charge pump

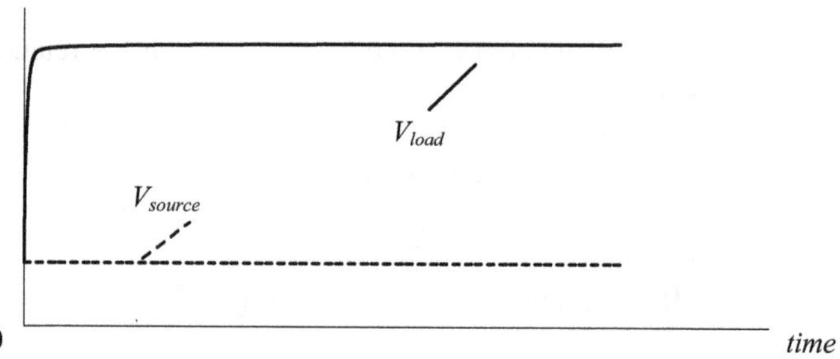

Fig. 3.28   Signals in a Dickson charge pump

If N is the number of stages in the Dickson charge pump, V_{clock} is the voltage of the clocks, which is assumed to be the same, and V_D is the voltage drop across the diode when it is ON, then the relation between V_{source} and V_{load} is approximately as follows:

$$V_{load} \approx V_{source} + N(V_{clock} - V_D) \tag{3.9}$$

3.2.6   DC Voltage Stabilization

An important application of diodes is their ability to keep the voltage almost constant, even if the current through them changes. This is called voltage regulation, and in particular DC voltage stabilization. This is a direct result of the steep diode I-V characteristic in the forward as well as the breakdown region.

Typically, and since the breakdown region curve is steeper than that in the forward region, the breakdown region is better-suited for stabilization than the forward region. Special diodes that operate in the breakdown region are used, such as Zener diodes and Avalanche Breakdown diodes.

We will concentrate for this application on Zener diodes. Zener diodes have the symbol shown in Fig. 3.29.

Unlike the diodes described above, the Zener diode operates normally in the breakdown region, which means that it experiences a negative voltage across it. Therefore, a Zener diode is connected in a circuit in the opposite manner compared to a regular diode with the cathode being at a higher voltage than an anode.

The breakdown region model is used when using a Zener diode. In fact, the letter 'z' in $-V_{zk}$ stands for Zener. A typical circuit using a Zener diode is shown in Fig. 3.30.

Anode   —▷⊢   Cathode

Fig. 3.29 Zener diode symbol

Fig. 3.30 Voltage stabilization using Zener diode

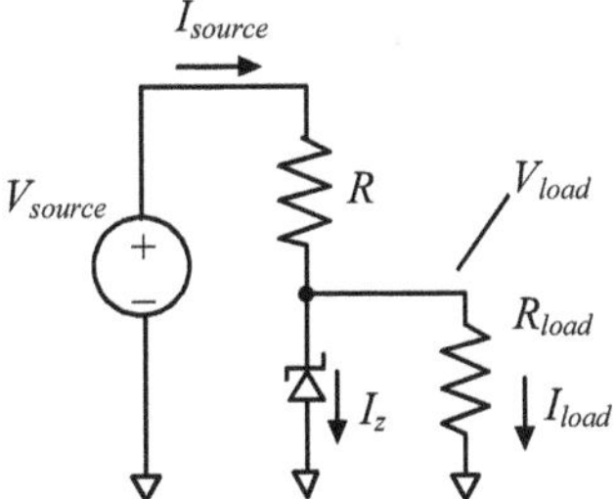

Variations in V_{source} result in variations in I_{source}, which result in variations in I_z. However, owing to the Zener diode steep I-V transfer characteristic, these variations in I_z will result in very small, if not negligible, variations in V_{load}, thus keeping V_{load} almost constant.

A drawback of this circuit is that V_{load} is restricted to a value around V_z. However, if a higher V_{load} is needed, then more than one Zener diode can be connected in series.

Another drawback is that the Zener diode itself will consume power due to V_{load} combined with I_z. This has a tendency to drain the power from the source even if no load is connected (i.e. R_{load} is infinite). Additionally, the power consumed by the Zener diode will be transformed into heat, which, if not dealt with appropriately, might damage the components.

3.2.7 Lighting

A light-emitting diode (LED) acts as a light source when a current flows through it. The LED symbol and a discrete LED component are shown in Fig. 3.31.

LEDs require a DC source to drive them. The simplest LED-driving circuit using a DC source is shown in Fig. 3.32.

The LED has its own V_{diode}. Of course, if a higher luminous intensity is needed, then either a larger LED can be used or several LEDs can be placed in series.

As with a regular diode, if the C-VD model is used, the LED would be either ON or OFF depending on the value of V_{DC}. However, in many applications such as in

Fig. 3.31 LED symbol and component

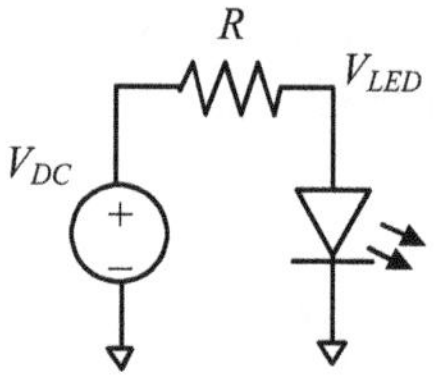

Fig. 3.32 LED-driving circuit

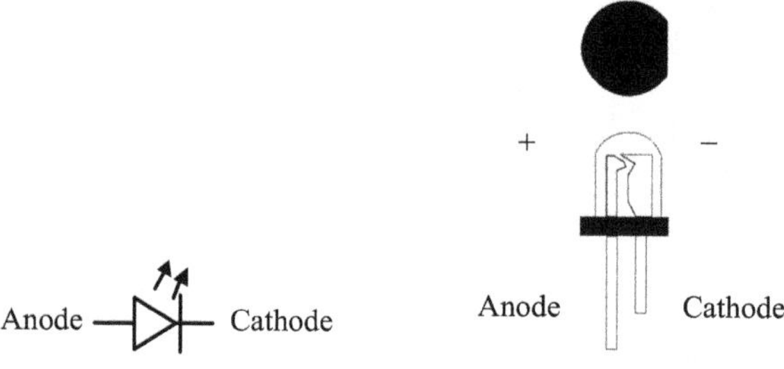

display screens, we might need to reduce the intensity of the light, commonly called the brightness. This can be done by driving the circuit with a pulse-width-modulated source. In this manner, the average value of the source can be changed by changing the width of the pulse wave and keeping the amplitude the same. This method has its drawbacks, especially at low intensity, where the LED's flickering starts to be visible.

Advanced circuits have been conceived in order to drive the LED directly with a user-variable DC voltage, thus taking advantage of the actual exponential behavior of the LED. This requires sources whose DC value is very accurately controlled. The final result is a flicker-free display which is easier on the eyes.

3.2.8 *Photodetection*

Photodiodes convert light into an electrical current, thus acting in an opposite manner compared to LEDs. Their symbol is shown in Fig. 3.33.

Photodiodes are used to detect the presence or absence of light. That light might either be an ambient light, in applications such as ambient light sensors, or it might be light emitted by an LED, in applications such as automatic door openers.

Photodiodes produce a reverse (leakage) current which is directly proportional to the intensity of the light. Therefore they are operated in the reverse region. A typical circuit used to convert that current into a voltage involves a transimpedance amplifier as shown in Fig. 3.34.

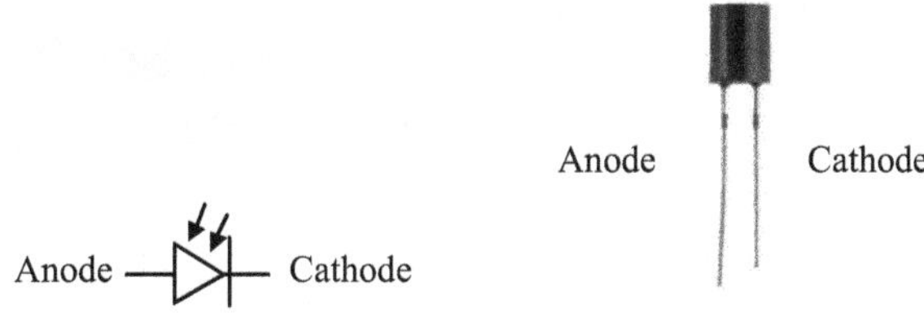

Fig. 3.33 Photodiode symbol and component

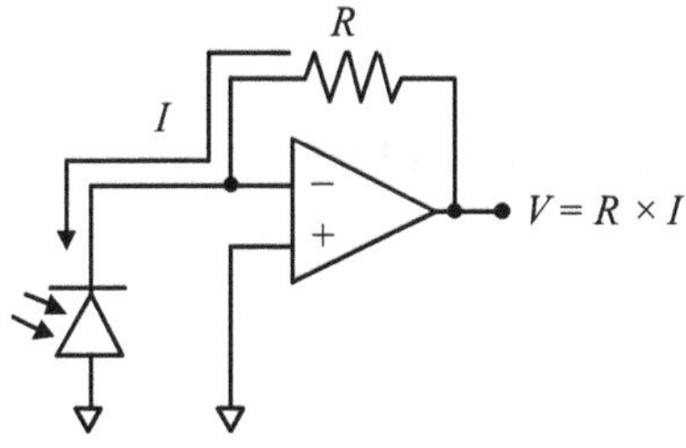

Fig. 3.34 Photodiode circuit using a transimpedance amplifier

3.2.9 *Electrostatic Discharge (ESD) Protection*

Diodes can be used to protect a circuit from the abrupt flow of charges due to the unintended contact between a charged object and the circuit itself. This flow of charges is called electrostatic discharge (ESD) and is very detrimental to electronic circuits.

This scenario typically happens when the bare connections of a circuit, usually its terminals, are accidentally touched by the hand. Since the human body accumulates charges, this results in an abrupt flow of these charges into the circuit causing its destruction. For this reason, ESD protection is used at the terminals of the circuit along with the good practice of not touching the terminals of a circuit with the hands. In laboratory setups, when dealing with sensitive circuits, ESD wrist straps such as the one shown in Fig. 3.35 are additionally used in case the terminals of that circuit are accidentally touched by the hands. On one side, these straps connect to the hand, and on the other side they connect to an earthing terminal, such as the one in an electrical outlet.

One way to implement ESD protection on the inputs and outputs of a circuit is shown in Fig. 3.36.

The diodes used in this circuit can be either regular diodes operating like rectifiers or Transient-Voltage-Suppression (TVS) diodes. They all operate based on the principle of protecting circuits from voltage spikes by passing the excess current when the voltage across them exceeds a certain value.

TVS diodes can be either unidirectional or bidirectional. The symbol of a bidirectional TVS diode is shown in Fig. 3.37.

Fig. 3.35 ESD wrist strap

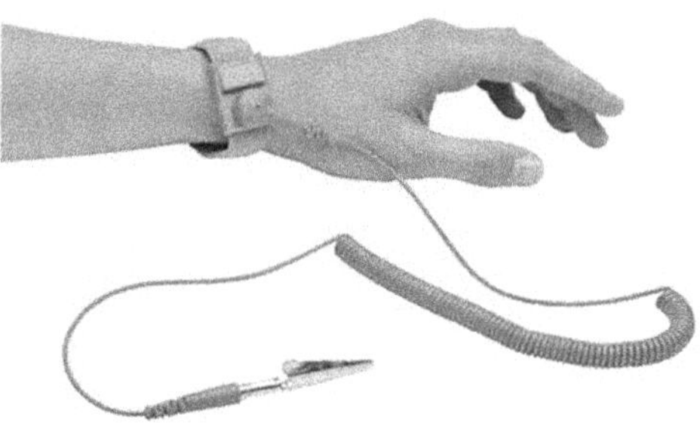

Fig. 3.36 Simple ESD
protection circuit

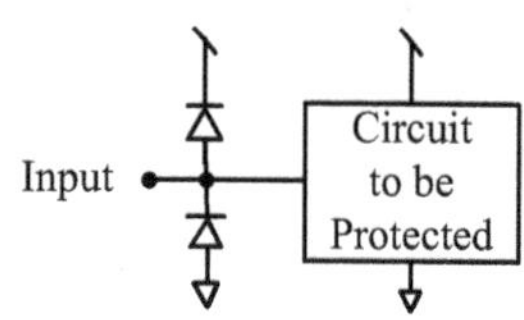

Fig. 3.37 Bidirectional TVS diode symbol

Fig. 3.38 Temperature sensor circuit

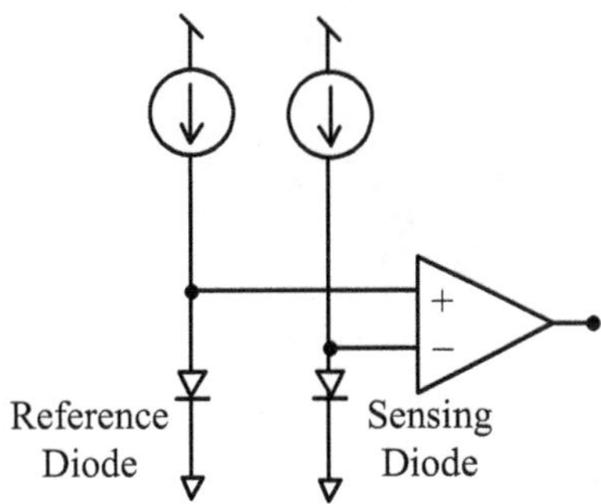

Anode ⟶⊳⊢ Cathode

Fig. 3.39 Varactor diode symbol

3.2.10 Temperature Measurements

The temperature dependence of the diode behavior as described in 3.1.3.1 gives rise to an interesting application, namely the temperature sensor which is used in electronic thermometers and in various electronic devices to sense extreme temperatures and take appropriate actions.

In its simplest implementation, a temperature sensor circuit looks as in Fig. 3.38.

The two diodes have the same current passing through them. However, since they are placed in different parts of the circuit, they experience different temperatures resulting in different voltages across them. This voltage difference is measured by the amplifier.

3.2.11 Variable Capacitance

Owing to the capacitive behavior of the diode in the reverse region as a function of the voltage which was described in 3.1.3.2, a diode can be used as a variable capacitor. A special diode type used for this application is called a varactor diode whose symbol is shown in Fig. 3.39.

A simple sample circuit that makes use of such an effect is shown in Fig. 3.40.

In this circuit, the resonance frequency between the inductor and the equivalent capacitance created by the diodes can be changed, i.e., tuned, by varying the DC voltage connected to the diodes.

Fig. 3.40 Varactor diode
tuning circuit

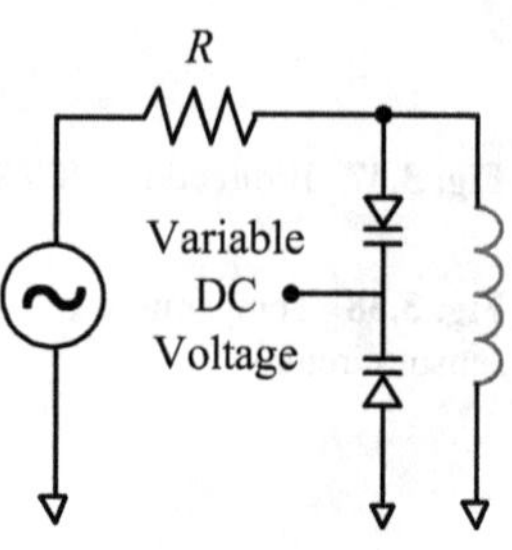

Chapter 4
Transistors

Diodes are 2-terminal nonlinear devices that can be controlled using either their voltage or their current. In either case, both of these quantities show up at their input and at their output. On the other hand, a transistor is fundamentally a 3-terminal nonlinear device giving rise to the ability to decouple the input from the output which makes it, for example, a better implementation of a single pole, single throw switch than a diode. This chapter discusses the general behavior of a transistor. This is followed by the behavior of technology-specific transistors, namely FETs and BJTs including their characterization. Afterwards, basic circuit applications are presented.

4.1 Transistor Models

Transistors can be modeled as 3-terminal devices. They typically exist in two types, an n-type and a p-type. Generic symbols of these transistor types are shown in Fig. 4.1.

In order to characterize these transistors, we are going to designate the Z terminal as the reference and take the other voltages with respect to it. As a result, an n-type transistor has the reference voltage at the bottom, with the current i going into the X terminal and the p-type transistor has the reference voltage at the top with i going out of the X terminal. In both cases, this current is the same or almost the same as the one passing through the Z terminal, depending on the specific transistor technology used.

We will then end up with two voltages: v_{XZ} and v_{YZ}. Since these voltages are positive in the case of an n-type transistor and negative in the case of a p-type transistor, we are going to take their magnitude. As a result, the terminal characteristics of a transistor will be three-dimensional with $|v_{XZ}|$ and $|v_{YZ}|$ as inputs and i as output as seen in Fig. 4.2.

© The Editor(s) (if applicable) and The Author(s), under exclusive license to
Springer Nature Switzerland AG 2024
J. G. Atallah, M. Ismail, *Integrated Electronic Circuits*,
https://doi.org/10.1007/978-3-031-62707-1_4

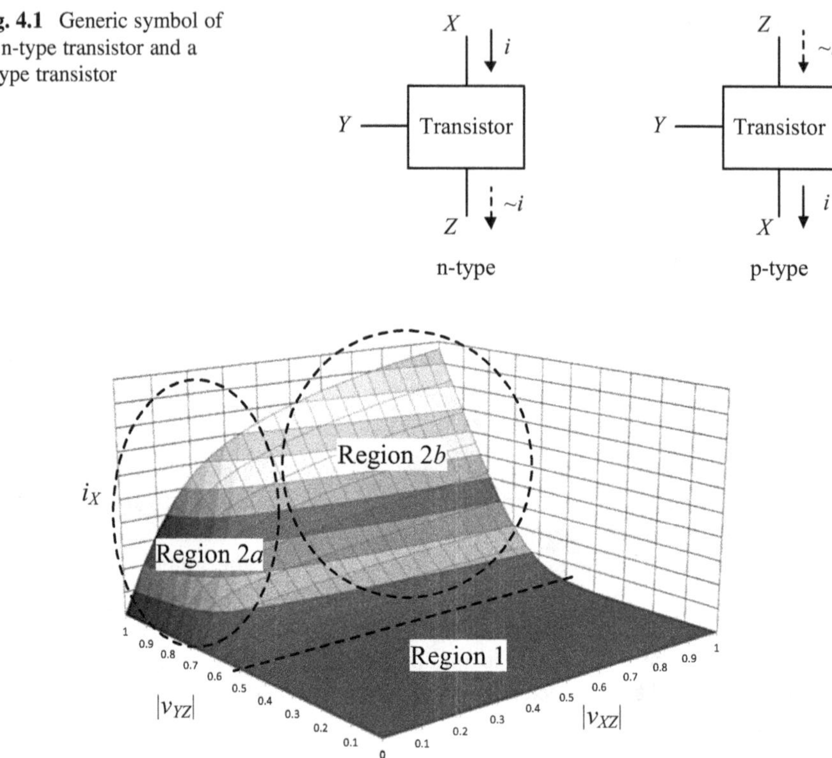

Fig. 4.1 Generic symbol of an n-type transistor and a p-type transistor

Fig. 4.2 General I-V characteristic of transistors

In order to be able to make use of this characteristic, we are going to partition the transistor behavior into regions.

- Region 1: The current in this region is almost zero, so the transistor is OFF.
- Region 2: In this region, the current is nonzero, so the transistor is ON. Furthermore, this region can be divided into two subregions:

  - Region 2a: In this region, the current is a strong linear function of both $|v_{XZ}|$ and $|v_{YZ}|$.
  - Region 2b: In this region, the current is a weak linear function of $|v_{XZ}|$ but a strong nonlinear function (power or exponential) of $|v_{YZ}|$.

Note that the above regions are approximate and so is their description. However, as long as this approximation serves our modeling purpose to come up with applications, it is good enough.

In order to better-perceive the behavior, we will cut the above three-dimensional graph using two planes. One keeps the value of $|v_{XZ}|$ fixed and shows how i varies with $|v_{YZ}|$, and the other keeps the value of $|v_{YZ}|$ fixed and shows how i varies with $|v_{XZ}|$.

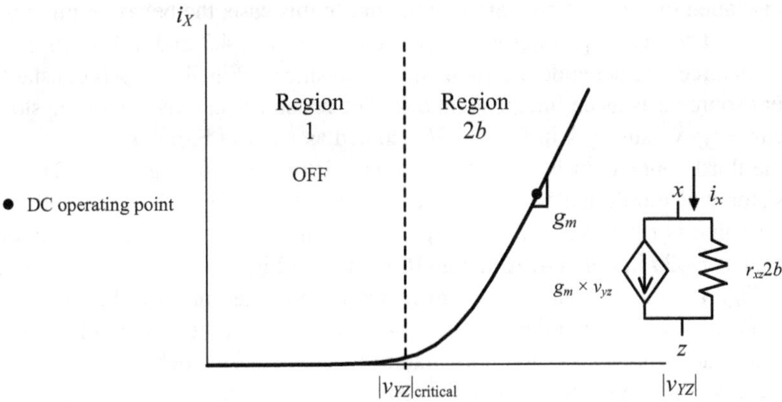

Fig. 4.3 Two-dimensional i_X versus $|v_{YZ}|$ cut of the plot in Fig. 4.2

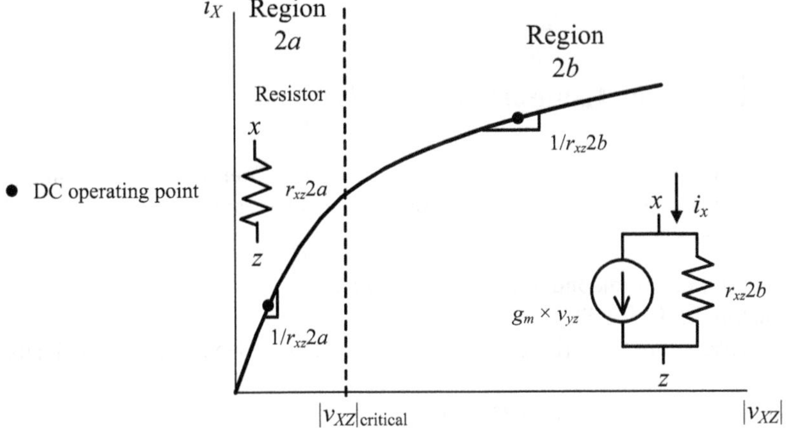

Fig. 4.4 Two-dimensional i_X versus $|v_{XZ}|$ cut of the plot in Fig. 4.2

The result is two graphs: one is i_X versus $|v_{YZ}|$ and the other is i_X versus $|v_{XZ}|$. The cuts in Figs. 4.3 and 4.4 are made at high values of $|v_{XZ}|$ and $|v_{YZ}|$.

With these plots at hand, it is convenient to come up with applications based on the transistor behavior.

The first application is one that moves the transistor between Region 1 and Region 2. In this case, the transistor is controlled by the voltage $|v_{YZ}|$ and the output is the current i_{XZ}. When $|v_{YZ}|$ is low, the transistor is OFF and when $|v_{YZ}|$ is high, the transistor is ON. This effectively mimics the behavior of a switch.

The second application is one where the transistor stays in Region 2b. There, the current is a linear function of $|v_{XZ}|$ but a strong nonlinear function (power or exponential) of $|v_{YZ}|$. However, if the voltages and currents are kept within the vicinity of a DC operating point, the curve can be approximated as a straight line.

This is called the small-signal approximation. In this case, the behavior mimics that of a current source in parallel with a resistor as in Figs. 4.3 and 4.4. Both of these current sources are dependent on $|v_{YZ}|$; however, since in Fig. 4.4 $|v_{YZ}|$ is constant, the current source acts as an independent one. The resistor is the inverse of the slope of the curve i_{XZ} versus $|v_{XZ}|$ in Region $2b$ denoted as $r_{xz}2b$ in Fig. 4.4.

The third application is one where the transistor stays in Region $2a$. There, the transistor will mimic the behavior of a resistor whose terminals are X and Z, and whose value is the inverse of the slope of the curve i_{XZ} versus $|v_{XZ}|$ in Region $2a$ denoted as $r_{XZ}2a$ in Fig. 4.4. Although Region $2a$ in Fig. 4.4 looks similar to Region $2b$ in Fig. 4.3, the behavior is very different since the terminals in Fig. 4.4 are the same ones through which the current passes, which is not the case in Fig. 4.3.

Note that owing to the nonlinear nature of the transistor behavior, $r_{xz}2a$, $r_{xz}2b$, and g_m are valid around a DC operating point. If this DC value of the signal changes, these values will change.

Following the generic presentation of transistor behavior, transistor examples from specific technologies will be presented in the following sections.

4.2 Transistor Example 1: MOSFET

In this section, we will describe the Field-Effect Transistor (FET) family and particularly the MOSFET transistor. In general, there are several types of FETs. Some of these are:

- Metal-Oxide-Semiconductor FET (MOSFET).
- Junction FET (JFET).
- Heterostructure FET (HFET) also known as High-Electron-Mobility Transistor (HEMT).
- Metal-Semiconductor FET (MESFET).
- Carbon Nanotube FET (CNTFET).
- Fin FET (FinFET).
- Gate-all-around FET (GAAFET) (Nanowire).
- Multi-bridge Channel FET (MBCFET) (Nanosheet).

These types differ in their structure and many times in the materials used, but have behaviorally two basic features in common:

1. At moderate to low frequencies, almost no current flows through their Y terminal. This means that the current through their X terminal is the same as that through their Y terminal.
2. Their i_X versus $|v_{YZ}|$ behavior in region $2b$ can be approximated as quadratic.

Now we will look specifically into MOSFETs and present their model.

The MOSFET is the most common type of transistors, particularly for digital and mixed-signal applications. It constitutes the bulk of the fabricated ICs in terms of

Table 4.1 FET terminal names

Generic terminal name	FET terminal name
X	Drain (*D*)
Y	Gate (*G*)
Z	Source (*S*)

Table 4.2 FET region names

Generic region name	FET region name
2*a*	Triode (linear or ohmic)
2*b*	Saturation (active)

Table 4.3 MOSFET Symbols (without the fourth terminal)

NMOS	PMOS
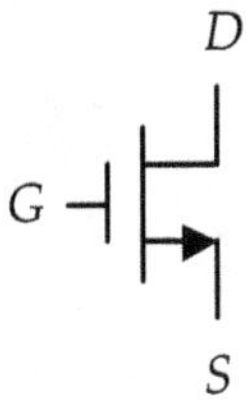 *D* — *G* — *S*	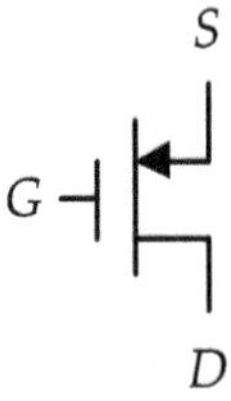 *S* — *G* — *D*
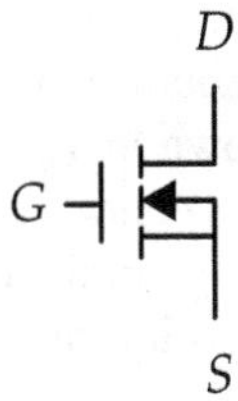 *D* — *G* — *S*	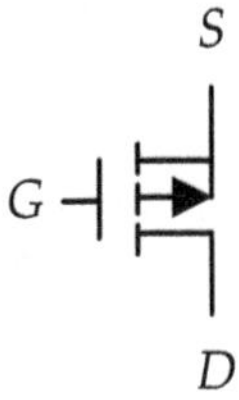 *S* — *G* — *D*
 *D* — *G* — *S*	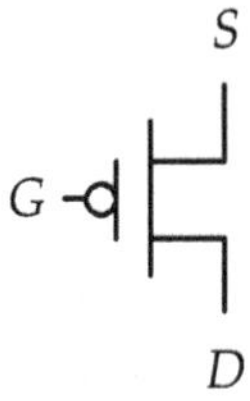 *S* — *G* — *D*

volume. The n-type MOSFET is called the NMOS and the p-type is called the PMOS.

The terminal names of MOSFETs are as in Table 4.1:

The region names of the MOSFETs are as in Table 4.2.

The symbols used for MOSFETs are shown in Table 4.3.

The arrows inside the symbols in the first row show the direction of the current. This current comes out of the source in the NMOS and into the source for the PMOS. In both cases, the current through the drain is the same as that through the source since there is no current through the gate. Note that the symbols in the last row do not

Table 4.4 MOSFET Symbols (with the fourth terminal)

NMOS	PMOS
D, G, B, S	S, G, B, D

Table 4.5 Critical voltages

Generic voltage	FET voltage
$|v_{XZ}|_{critical}$	Threshold voltage $|V_{th}|$
$|v_{YZ}|_{critical}$	Overdrive voltage $|v_{OV}| = |v_{GS}| - |V_{th}|$ where V_{th} is the threshold voltage.

differentiate between the drain and the source terminals. These are used in such applications as digital designs.

Additionally, sometimes a fourth terminal (B) is added to the middle of the transistor to indicate the connection to the bulk, many times called the substrate. In such a case, the symbols would be as shown in Table 4.4.

In most applications, the fourth terminal is connected to the lowest voltage in the circuit for an NMOS and to the highest voltage in the circuit for a PMOS.

For a MOSFET, the critical voltages are as in Table 4.5.

4.2.1 MOSFET Large-Signal Model

Regarding the voltage-current relations of a MOSFET, these are going to be presented against the backdrop of the general transistor discussions presented previously.

4.2.1.1 MOSFET Large-Signal Model: Saturation Region

In the saturation region, the current is related to the voltages as follows. Note that the relation between i_D and v_{GS} is parabolic.

$$i_D = \frac{1}{2}\mu C_{ox}\left(\frac{W}{L}\right)(|v_{GS}| - |V_{th}|)^2(1 + \lambda|v_{DS}|) \tag{4.1}$$

The term $(1 + \lambda|v_{DS}|)$ is included to account for the slope that i_D has when plotted versus $|v_{DS}|$. This effect is called channel-length modulation.

4.2.1.2   MOSFET Large-Signal Model: Triode Region

In the triode region, the current is related to the voltages as follows. Note that the relation between i_D and $|v_{GS}|$ is linear.

$$i_D = \mu C_{ox} \left(\frac{W}{L}\right) \left[(|v_{GS}| - |V_{th}|)\, |v_{DS}| - \frac{1}{2} |v_{DS}|^2 \right] \tag{4.2}$$

i_D is the drain current (A), which is the same as that at the source.

μ is the carrier mobility (m^2/V.s). It is called the electron mobility μ_n for NMOS and hole mobility μ_p for PMOS. Usually $\mu_n > \mu_p$.

C_{ox} is the oxide capacitance (F/m^2).

W is the width of the transistor (m).

L is the length of the transistor (m). The name of the technology is taken from this parameter. Therefore, a 2 nm technology has an L of 2 nm.

λ is the channel-length modulation parameter (1/V) and its inverse is the voltage V_A.

$$\lambda = \frac{1}{|V_A|} \tag{4.3}$$

V_A is related to L as follows:

$$V_A = V'_A L \tag{4.4}$$

V'_A is a process parameter (V/m).

Usually the carrier mobility μ is combined with the oxide capacitance C_{ox} to form k' the process transconductance parameter (F/V.s = A/V^2).

$$k' = \mu C_{ox} \tag{4.5}$$

Additionally, we can define another process transconductance parameter called k (A/V^2) that combines several previous parameters as follows:

$$k = \frac{1}{2} \mu C_{ox} \left(\frac{W}{L}\right) \tag{4.6}$$

As a result, Figs. 4.3 and 4.4 for MOSFETs become as in Figs. 4.5 and 4.6.

4.2.2   MOSFET Small-Signal Model

In Figs. 4.5 and 4.6, if v_{GS} and v_{DS} swing around a DC operating point and are kept small enough in amplitude, we can model the behavior as a straight line tangent to

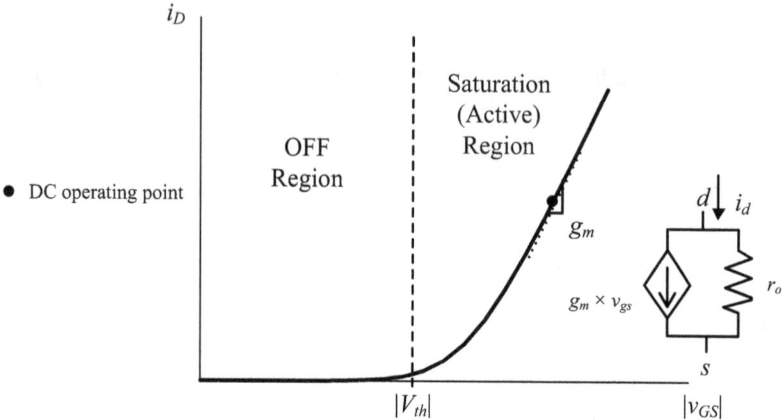

Fig. 4.5 MOSFET i_D versus $|v_{GS}|$

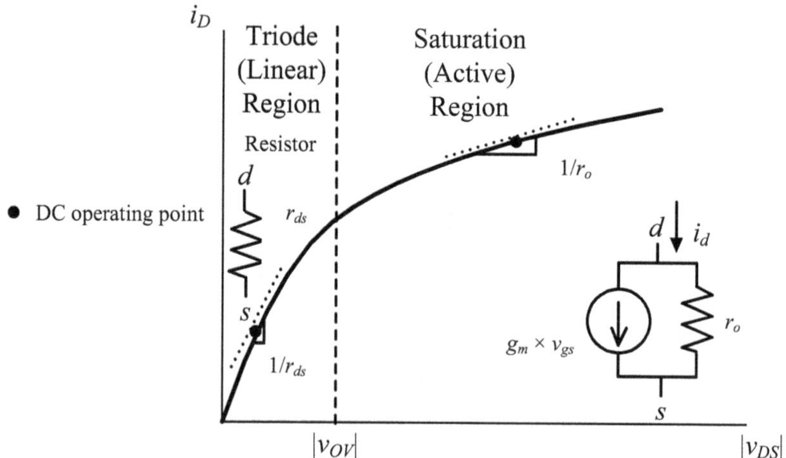

Fig. 4.6 MOSFET i_D versus $|v_{DS}|$

the original curve at that DC point. This is shown as a dotted line in Figs. 4.5 and 4.6. This is called the small-signal behavior of the transistor.

Now, we will derive the expressions for g_m, r_o, and r_{ds}.

4.2.2.1   MOSFET Small-Signal Model: Saturation Region

In the saturation region, and as can be seen in Fig. 4.5, g_m is the derivative of i_D with respect to $|v_{GS}|$ at the DC operating point. Therefore:

$$g_m = \left.\frac{\partial i_D}{\partial |v_{GS}|}\right|_{\text{at DC point}} = \left.\frac{\partial\left\{\frac{1}{2}k'\left(\frac{W}{L}\right)(|v_{GS}| - |V_{th}|)^2(1 + \lambda|v_{DS}|)\right\}}{\partial |v_{GS}|}\right|_{\text{at DC point}}$$

$$= \left.k'\left(\frac{W}{L}\right)(|v_{GS}| - |V_{th}|)(1 + \lambda|v_{DS}|)\right|_{\text{at DC point}} \tag{4.7}$$

$$= k'\left(\frac{W}{L}\right)(|V_{GS}| - |V_{th}|)(1 + \lambda|V_{DS}|)$$

$$= k'\left(\frac{W}{L}\right)|V_{OV}|(1 + \lambda|V_{DS}|) = \frac{2I_D}{|V_{OV}|}$$

Also in the saturation region, and as can be seen in Fig. 4.6, r_o is the inverse of the derivative of i_D with respect to $|v_{DS}|$ at the DC operating point. Therefore:

$$\frac{1}{r_o} = \left.\frac{\partial i_D}{\partial |v_{DS}|}\right|_{\text{at DC point}} = \left.\frac{\partial\left\{\frac{1}{2}k'\left(\frac{W}{L}\right)(|v_{GS}| - |V_{th}|)^2(1 + \lambda|v_{DS}|)\right\}}{\partial |v_{DS}|}\right|_{\text{at DC point}}$$

$$= \left.\lambda \times \frac{1}{2}k'\left(\frac{W}{L}\right)(|v_{GS}| - |V_{th}|)^2\right|_{\text{at DC point}} = \lambda \times \frac{1}{2}k'\left(\frac{W}{L}\right)(|V_{GS}| - |V_{th}|)^2 \tag{4.8}$$

$$= \lambda \times \frac{1}{2}k'\left(\frac{W}{L}\right)|V_{OV}|^2 = \lambda \times I_D'$$

where I_D' is the DC current without the channel-length modulation effect. Many times, this is approximated as I_D itself and as a result. This approximation becomes more accurate if λ is small (or conversely if V_A is large):

$$r_o = \frac{1}{\lambda \times |I_D'|} = \frac{|V_A|}{|I_D'|} \approx \frac{|V_A|}{|I_D|} \tag{4.9}$$

As a result of this, when operating around a DC point and keeping the amplitude small, the transistor in the saturation region can be modeled as a voltage-controlled current source in parallel with a resistor as in the PI-Model in Fig. 4.7a or equivalently the T-Model in Fig. 4.7b. Note that both small-signal models are valid without modifications for NMOS and PMOS transistors. Also, in the T-Model, it is important to indicate visually that there is no current through the gate terminal.

Additionally, in these models, the internal parasitic capacitances of the transistor are added. These capacitances are due to the internal structure of the transistor. They play an important role only when dealing with high-frequency signals since this is when their impedance $Z_C = 1 / (j\, 2\pi f\, C)$ is small. At low to mid frequencies, $Z_C \approx \infty$ so these capacitances will act as open circuits.

These capacitances play an important role in determining the maximum theoretical frequency at which the transistor can operate. There are several of these frequencies with different definitions. One of them is f_T, which is defined as the

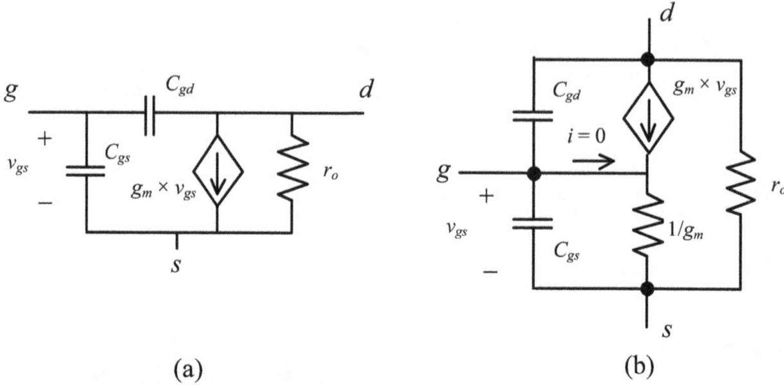

Fig. 4.7 MOSFET small-signal saturation region PI-model (**a**) and T-model (**b**)

Fig. 4.8 Conceptual circuit to test f_T.

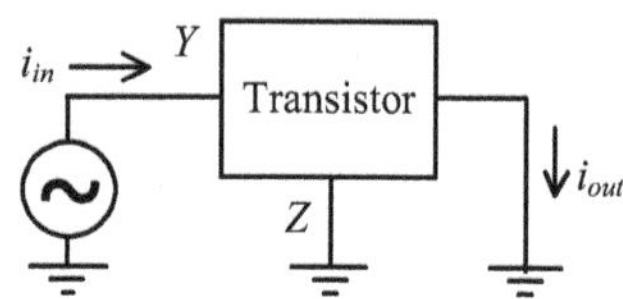

Fig. 4.9 Small-signal equivalent circuit model to test f_T.

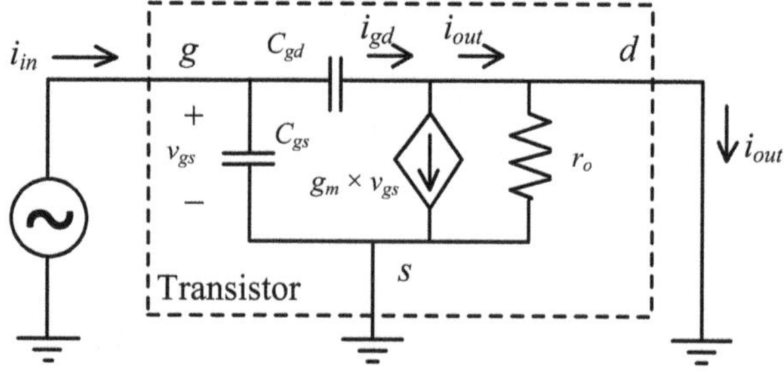

frequency at which the current gain of the transistor is one. Conceptually, the setup looks as in Fig. 4.8.

Note that at high frequencies, the 'Y' terminal will start to accept currents in MOSFETs due to the internal capacitances. Also, the load is a short circuit, which is ideal in the current mode.

The main idea is that the transistor is supposed to amplify the input current if it is operating within its proper frequency range. If the output current drops and starts to get close to the input current, then the transistor is leaving its proper operating frequency range. The frequency at which this happens is the f_T.

Therefore, we will observe the ratio i_{out} / i_{in} while increasing the frequency of i_{in}. The frequency at which i_{out} / i_{in} is equal to one is called f_T. To analyze this circuit, we will draw the small-signal model of the setup as in Fig. 4.9 and express i_{out} / i_{in} in terms of the internal components of the transistor.

The analysis goes as follows:

$$v_{gs} = i_{in}\left(Z_{Cgs}//Z_{Cgd}\right) = i_{gd}Z_{Cgd}$$
$$\Rightarrow i_{gd} = i_{in}\left(\frac{Z_{Cgs}//Z_{Cgd}}{Z_{Cgd}}\right) \tag{4.10}$$

$$i_{out} = i_{gd} - g_m \times v_{gs}$$
$$i_{out} = i_{in}\left(\frac{Z_{Cgs}//Z_{Cgd}}{Z_{Cgd}}\right) - g_m\left[i_{in}\left(Z_{Cgs}//Z_{Cgd}\right)\right]$$
$$\frac{i_{out}}{i_{in}} = \left[\frac{Z_{Cgs}//Z_{Cgd}}{Z_{Cgd}} - g_m\left(Z_{Cgs}//Z_{Cgd}\right)\right] \tag{4.11}$$
$$\frac{i_{out}}{i_{in}} = \frac{sC_{gd}}{s\left(C_{gs} + C_{gd}\right)} - g_m\frac{1}{s\left(C_{gs} + C_{gd}\right)}$$

At f_T, the above ratio is one. Therefore:

$$\frac{i_{out}}{i_{in}} = \frac{C_{gd}}{\left(C_{gs} + C_{gd}\right)} - g_m\frac{1}{j2\pi f_T\left(C_{gs} + C_{gd}\right)} = 1$$
$$\left|\frac{C_{gd}}{\left(C_{gs} + C_{gd}\right)} - g_m\frac{1}{j2\pi f_T\left(C_{gs} + C_{gd}\right)}\right|^2 = 1^2$$
$$\left[\frac{C_{gd}}{\left(C_{gs} + C_{gd}\right)}\right]^2 + \left[\frac{g_m}{2\pi f_T\left(C_{gs} + C_{gd}\right)}\right]^2 = 1 \tag{4.12}$$
$$f_T^{\,2} = \left(\frac{1}{2\pi}\right)^2\left[\frac{g_m^{\,2}}{\left(C_{gs} + C_{gd}\right)^2 - C_{gd}^{\,2}}\right]$$

$$f_T = \left(\frac{1}{2\pi}\right)\left[\frac{g_m}{\sqrt{\left(C_{gs} + C_{gd}\right)^2 - C_{gd}^{\,2}}}\right] \tag{4.13}$$

The denominator is in the form of $a^2 - b^2$, which can be written as $(a - b)(a + b)$ so the above equation can be reduced further as:

$$f_T = \left(\frac{1}{2\pi}\right)\left[\frac{g_m}{\sqrt{C_{gs}(C_{gs} + 2C_{gd})}}\right]$$

$$f_T = \left(\frac{1}{2\pi}\right)\left[\frac{g_m}{C_{gs}\sqrt{\left(1 + \frac{2C_{gd}}{C_{gs}}\right)}}\right] \tag{4.14}$$

If $C_{gd} << C_{gs}$, which is usually the case, then C_{gd}^2 in Eq. (4.13) can be ignored and

$$f_T \approx \frac{1}{2\pi}\left(\frac{g_m}{C_{gs} + C_{gd}}\right) \tag{4.15}$$

As can be seen, in order for a transistor to be able to operate at higher frequencies, g_m should be increased, but that is limited by our ability to increase I_D and/or decrease V_{OV}.

Additionally, from a device technology point-of-view, decreasing the size of the device decreases its internal capacitances C_{gd} and C_{gs}, thus increasing its f_T. This is the main reason why every newer technology is smaller than the previous one. The phenomenon is called technology scaling and is generally governed by a trend known as Moore's law.

4.2.2.2 MOSFET Small-Signal Model: Triode Region

In the triode region, and as can be seen in Fig. 4.6, r_{ds} is the inverse of the derivative of i_D with respect to $|v_{DS}|$ at the DC operating point which is usually taken when $|v_{DS}|$ is small, that is when $|v_{DS}|^2$ is too small and can be ignored.

$$\begin{aligned}
\frac{1}{r_{ds}} &= \left.\frac{\partial i_D}{\partial |v_{DS}|}\right|_{\text{at DC point}} \\
&= \left.\frac{\partial\left\{k'\left(\frac{W}{L}\right)\left[(|v_{GS}| - |V_{th}|)\,|v_{DS}| - \frac{1}{2}|v_{DS}|^2\right]\right\}}{\partial |v_{DS}|}\right|_{\text{at DC point}} \\
&\approx \left.\frac{\partial\left\{k'\left(\frac{W}{L}\right)\left[(|v_{GS}| - |V_{th}|)\,|v_{DS}|\right]\right\}}{\partial |v_{DS}|}\right|_{\text{at DC point}} \\
&= \left.k'\left(\frac{W}{L}\right)(|v_{GS}| - |V_{th}|)\right|_{\text{at DC point}} = k'\left(\frac{W}{L}\right)|V_{OV}|
\end{aligned} \tag{4.16}$$

As a result of this

Fig. 4.10 MOSFET small-signal triode region model

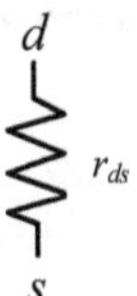

Table 4.6 BJT terminal names

Generic terminal name	BJT terminal name
X	Collector (C)
Y	Base (B)
Z	Emitter (E)

Table 4.7 BJT region names

Generic region name	BJT region name
$2a$	Saturation
$2b$	Forward active (linear)

Table 4.8 BJT symbols

NPN	PNP

$$r_{ds} \approx \frac{1}{k'\left(\frac{W}{L}\right)|V_{OV}|} \tag{4.17}$$

Thus, the MOSFET small-signal triode region model is a simple transistor as in Fig. 4.10.

4.3   Transistor Example 2: BJT

In this section, we will describe the Bipolar Junction Transistor (BJT) family.

The BJT is used mainly for high-frequency and high-power applications. The n-type BJT is called the NPN and the p-type is called the PNP.

The terminal names of BJTs are as in Table 4.6:

The region names of the BJTs are as in Table 4.7.

The symbols used for BJTs are shown in Table 4.8.

Table 4.9 Critical voltages

Generic voltage	BJT voltage
$|v_{XZ}|_{\text{critical}}$	$|V_{CE\text{-}EDGE}|$
$|v_{YZ}|_{\text{critical}}$	$|V_{BE\text{-}ON}|$

The arrows inside the symbols show the direction of the current. This current comes out of the emitter for the NPN and into the emitter for the PNP.

For a BJT, the critical voltages are as in Table 4.9.

For BJTs, we will concentrate on the behavior exclusively in the Forward Active region.

4.3.1 BJT Large-Signal Model: Forward Active Region

Regarding the voltage-current relations of a BJT, these are going to be presented against the backdrop of the general transistor discussions presented previously.

In the Forward Active region, the current is related to the voltages as follows. Note that the relation between i_C and v_{BE} is exponential.

$$i_C = I_s e^{\left(\frac{|v_{BE}|}{V_T}\right)}\left(1 + \lambda|v_{CE}|\right) \tag{4.18}$$

The term $(1 + \lambda|v_{CE}|)$ is included to account for the slope that i_C has when plotted versus $|v_{CE}|$. This effect is called the Early effect.

I_s is the saturation or scale current. It varies with the transistor's area. The larger the area, the higher is the I_s. Its value is much smaller than i_C.

V_T is the thermal voltage. Its value is around 25 mV at room temperature which is around 20 °C (293 K).

λ is the Early effect parameter (1/V) and its inverse is the Early voltage V_A.

$$\lambda = \frac{1}{|V_A|} \tag{4.19}$$

As a result, Figs. 4.3 and 4.4 for BJTs are as in Figs. 4.11 and 4.12.

A particularity of BJTs is that some current goes through the base terminal leading to the fact that $i_C \neq i_E$. In fact i_B is much smaller than i_C, which itself is slightly smaller than i_E and the directions of the current are as in Table 4.10.

More accurately, the currents have the following relation to each other:

$$i_E = i_C + i_B \tag{4.20}$$

$$i_C = \beta \ i_B \tag{4.21}$$

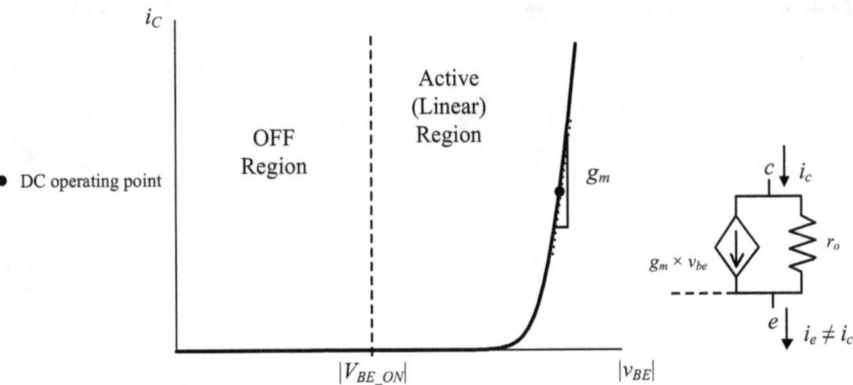

Fig. 4.11 BJT i_C versus $|v_{BE}|$

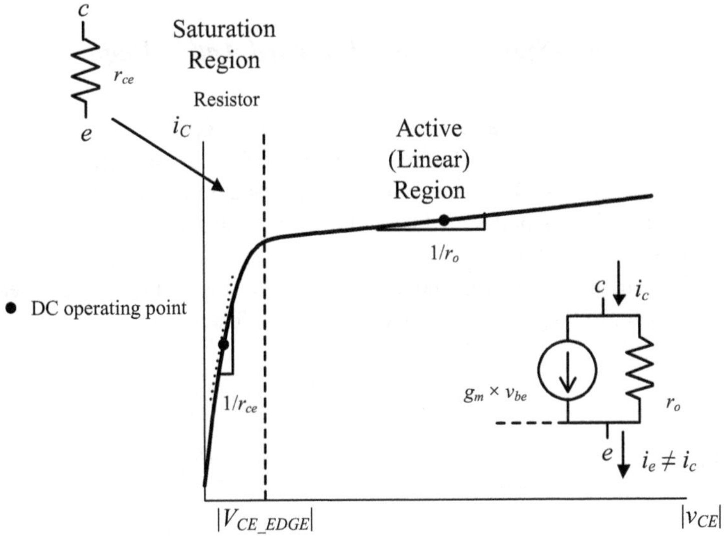

Fig. 4.12 BJT i_C versus $|v_{CE}|$

$$i_C = \alpha \ i_E \tag{4.22}$$

β is a parameter with a large and not well-specified value.
α is a parameter which is slightly less than one.
As a result, we can deduce the following

$$\alpha = \frac{\beta}{\beta + 1} \tag{4.23}$$

Table 4.10 BJT current direction

While concentrating only on the Forward Active region, as before, we will derive the expressions for g_m and r_o.

4.3.2   BJT Small-Signal Model: Forward Active Region

In Figs. 4.11 and 4.12, if v_{BE} and v_{CE} swing around a DC operating point and are kept small enough in amplitude, we can model the behavior as a straight line tangent to the original curve at that DC point. This is shown as a dotted line in Figs. 4.11 and 4.12. This is called the small-signal behavior of the transistor.

Now, we will derive the expressions for g_m and r_o.

In the Forward Active region, and as can be seen in Fig. 4.11, g_m is the derivative of i_C with respect to $|v_{BE}|$ at the DC operating point. Therefore:

$$
\begin{aligned}
g_m &= \left.\frac{\partial i_C}{\partial |v_{BE}|}\right|_{\text{at DC point}} = \left.\frac{\partial\left\{ I_s e^{\left(\frac{|v_{BE}|}{V_T}\right)}(1 + \lambda|v_{CE}|)\right\}}{\partial |v_{BE}|}\right|_{\text{at DC point}} \\[2mm]
&= \left.\frac{I_S}{V_T} e^{\left(\frac{|v_{BE}|}{V_T}\right)}(1 + \lambda|v_{CE}|)\right|_{\text{at DC point}} = \frac{I_S}{V_T} e^{\left(\frac{|V_{BE}|}{V_T}\right)}(1 + \lambda|V_{CE}|) \\[2mm]
&= \frac{I_C}{V_T}
\end{aligned}
\tag{4.24}
$$

Also in the Forward Active region, and as can be seen in Fig. 4.12, r_o is the inverse of the derivative of i_C with respect to $|v_{CE}|$ at the DC operating point. Therefore:

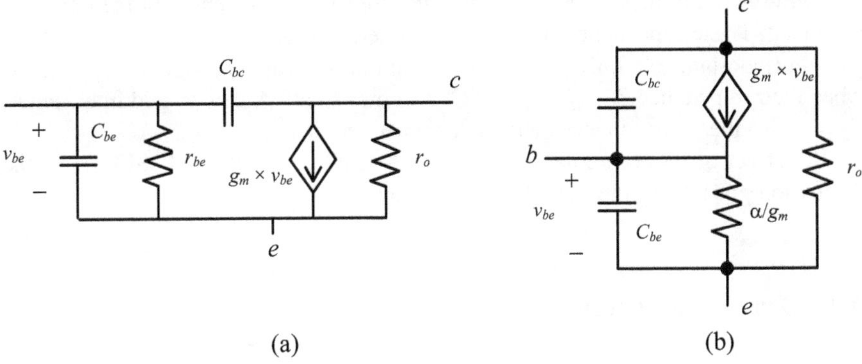

Fig. 4.13 BJT small-signal saturation region PI-model (**a**) and T-model (**b**)

$$\frac{1}{r_o} = \left.\frac{\partial i_C}{\partial |v_{CE}|}\right|_{\text{at DC point}} = \left.\frac{\partial\left\{I_{se}^{\left(\frac{|v_{BE}|}{V_T}\right)}(1+\lambda|v_{CE}|)\right\}}{\partial|v_{CE}|}\right|_{\text{at DC point}}$$

$$= \left.\lambda\times I_{se}^{\left(\frac{|v_{BE}|}{V_T}\right)}(1+\lambda|v_{CE}|)\right|_{\text{at DC point}} \qquad = \lambda\times I_{se}^{\left(\frac{|v_{BE}|}{V_T}\right)}(1+\lambda|V_{CE}|)$$

$$= \lambda\times I_C' \tag{4.25}$$

where I_C' is the DC current without the Early effect. Many times, this is approximated as I_C itself and as a result. This approximation becomes more accurate if λ is small (or conversely if V_A is large):

$$r_o = \frac{1}{\lambda\times|I_C'|} = \frac{|V_A|}{|I_C'|} \approx \frac{|V_A|}{|I_C|} \tag{4.26}$$

Alternatively to the analytical equations above, we can get more accurate approximations of the small-signal parameters g_m and r_o using simulations. This can be done by starting from Figs. 4.11 and 4.12 as was discussed in the case of MOSFETS in Sect. 4.2.2.1.

As a result of this, when operating around a DC point and keeping the amplitude small, the transistor in the Forward Active region can be modeled as a voltage-controlled current source in parallel with a resistor as in the PI-Model in Fig. 4.13a or equivalently the T-Model in Fig. 4.13b. Note that both small-signal models are valid without modifications for NPN and PNP transistors.

In the T-model, the resistance at the emitter whose value is α/gm is commonly known as r_e.

Additionally, in these models, the internal parasitic capacitances of the transistor are added. These capacitances are due to the internal structure of the transistor. They play an important role only when dealing with high-frequency signals since this is when their impedance $Z_C = 1 / (j\,2\pi f\,C)$ becomes small. At low to mid frequencies, $Z_C \approx \infty$ so these capacitances will act as open circuits.

The effect of these internal capacitances is similar to that in MOSFETs, so the discussion goes along the same lines as in Sect. 4.2.2.1.

4.4 Transistor Characterization

When dealing with a circuit, particularly during the design process, it is very important first to get to know the transistors that will be used very well. This involves knowing the voltage range that the transistors can take as well as the parameters of the transistors. Here we will limit the discussion to MOSFETs although a very similar procedure can be used for BJTs.

It all starts with knowing the maximum voltages v_{GS} and v_{DS} that the transistor can handle. This is given in the documentation of the transistors used. As for the parameters, we have seen two sets of them for each type of transistors: the large-signal parameters and the small-signal parameters.

4.4.1 Large-Signal Parameters

The large-signal parameters for MOSFETs are $|V_{th}|$, λ, and k. We can get these either from a datasheet or by characterizing the transistor ourselves using simulations.

If we are using discrete components, all these parameters are already fixed when we choose a certain transistor. However, if we are doing an integrated circuit design, we choose the transistor width W, and then figure out all the other parameters.

We will outline here a method in order to get these parameters using simulations. The circuits to construct are as shown in Fig. 4.14. In these circuits, using simulations, the voltages are swept within their allowable range and the current is plotted versus each one of these voltages.

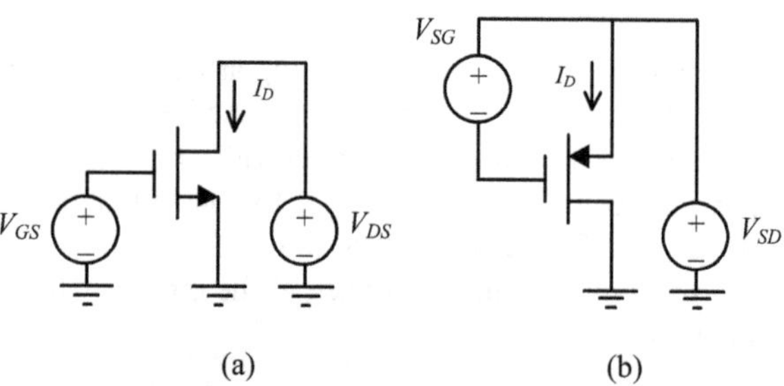

Fig. 4.14 Transistor characterization circuit for NMOS (**a**) and PMOS (**b**)

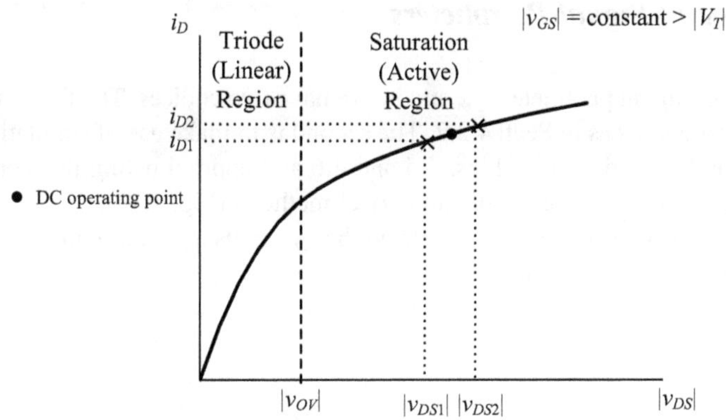

Fig. 4.15 i_D versus $|v_{DS}|$ to get λ

First, we need to get $|V_{th}|$. Looking into Fig. 4.5, $|V_{th}|$ is the voltage beyond which the current stats to increase quickly. Having the current axis with a logarithmic scale might help in this regards,

Second, we need to get λ. For this, we plot i_D versus $|v_{DS}|$ as in Fig. 4.15.

By taking the two points in the saturation region in the vicinity of the DC operating point, as indicated in Fig. 4.15, we can see that

$$i_{D1} = k(|v_{GS}| - |V_{th}|)^2(1 + \lambda|v_{DS1}|)$$
$$i_{D2} = k(|v_{GS}| - |V_{th}|)^2(1 + \lambda|v_{DS2}|)$$
$$\frac{i_{D1}}{i_{D2}} = \frac{1 + \lambda|v_{DS1}|}{1 + \lambda|v_{DS2}|} \tag{4.27}$$
$$i_{D1}(1 + \lambda|v_{DS2}|) = i_{D2}(1 + \lambda|v_{DS1}|)$$
$$i_{D1} + i_{D1}\lambda|v_{DS2}| = i_{D2} + i_{D2}\lambda|v_{DS1}|$$
$$\lambda(i_{D1}|v_{DS2}| - i_{D2}|v_{DS1}|) = i_{D2} - i_{D1}$$

$$\lambda = \frac{i_{D2} - i_{D1}}{i_{D1}|v_{DS2}| - i_{D2}|v_{DS1}|} \tag{4.28}$$

Third, we need to get k and this can simply be done using either of the two points above

$$k = \frac{i_{D1}}{(|v_{GS}| - |V_{th}|)^2(1 + \lambda|v_{DS1}|)} \tag{4.29}$$

4.4.2 Small-Signal Parameters

For the small-signal parameters g_m and r_o, we have two choices. The first is to use the analytical equations as in Sect. 4.2.2. The second is to make use of simulations. This can be done by starting from Figs. 4.5 and 4.6 and approximating the slopes at the DC operating points using simple numerical methods (Fig. 4.16).

For g_m, we can designate two points on the i_D versus $|v_{GS}|$ curve in the vicinity of the DC operating point as in

In this case, we get:

$$g_m = \frac{i_{D2} - i_{D1}}{|v_{DS2}| - |v_{DS1}|} \tag{4.30}$$

Regarding r_o, we can designate two points on the i_D versus $|v_{DS}|$ curve in the vicinity of the DC operating point in the saturation region as in Fig. 4.17.

In this case, we get:

$$r_o = \frac{|v_{DS2}| - |v_{DS1}|}{i_{D2} - i_{D1}} \tag{4.31}$$

This method is more accurate than the analytical one, since it makes use of very little approximations.

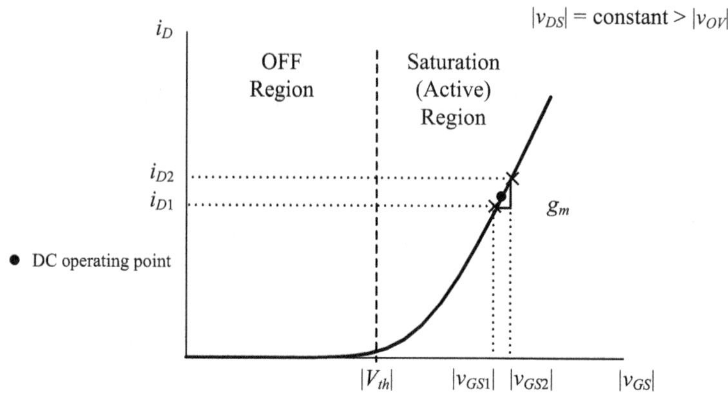

Fig. 4.16 Getting g_m using simulations

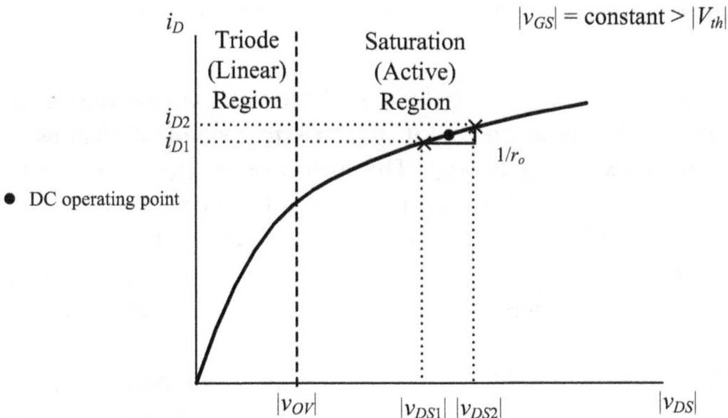

Fig. 4.17 Getting r_o using simulations

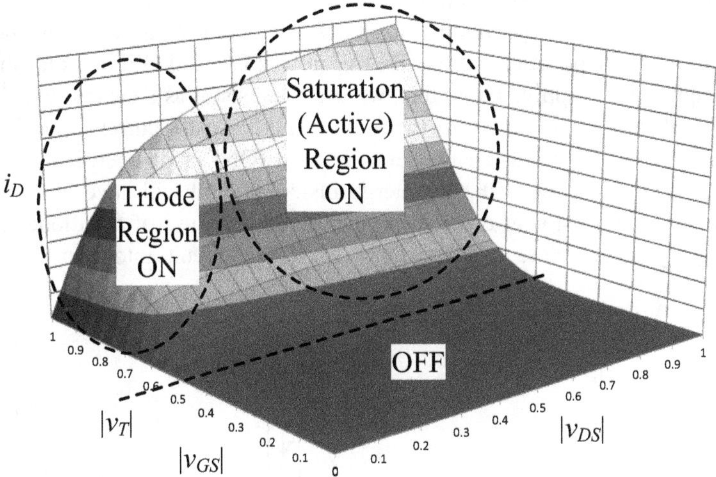

Fig. 4.18 MOSFET i_D versus |v| as a switch

4.5 Basic Circuits and Applications

The transistor's behavior in Figs. 4.2, 4.3, and 4.4 lends itself to many applications.
In this section we will highlight some of the most important ones. Without any loss
of generality, we will describe these applications using MOSFETs.

4.5.1 Switches

As can be seen in in Fig. 4.18, if $|v_{GS}|$ is varied between extreme values, the transistor would either pass the i_D, or prevent it from passing, essentially acting as a switch. The threshold of switching is $|V_{th}|$. This behavior is used extensively in digital electronics where switching is the cornerstone of logic design.

Compared to a diode, the MOSFET mimics a switch in a better way, since, unlike in a diode, the controlling voltage $|v_{GS}|$ is applied at the Gate terminal which is different from the output terminal. In a diode, the input and output terminals are the same.

When the transistor is OFF, it acts as an almost infinite resistance. When the transistor is ON, it has some internal resistance R to it. The value of this resistance depends on whether the transistor ends up in the triode or the saturation region. If in the triode region, then R is r_{ds}. If in the saturation region, then R is r_o. As a result, the simplest switch model is shown in Fig. 4.19 for an NMOS and a PMOS.

An example of an application for a transistor acting as a switch is shown in Fig. 4.20. In this application, we would like to turn ON and OFF a motor that runs from a high power supply. To do this, we use a transistor as a switch, which, in its turn, controls a relay that controls the motor. The transistor current i_D only passes through the coil of the relay, either energizing it (ON) or de-energizing it (OFF). When the relay switches OFF, the energy inside the coil will have to be diverted away from the transistor in order to protect it, thus the use of the diode, called a freewheeling diode. Note that the relay splits the circuit into two domains: a

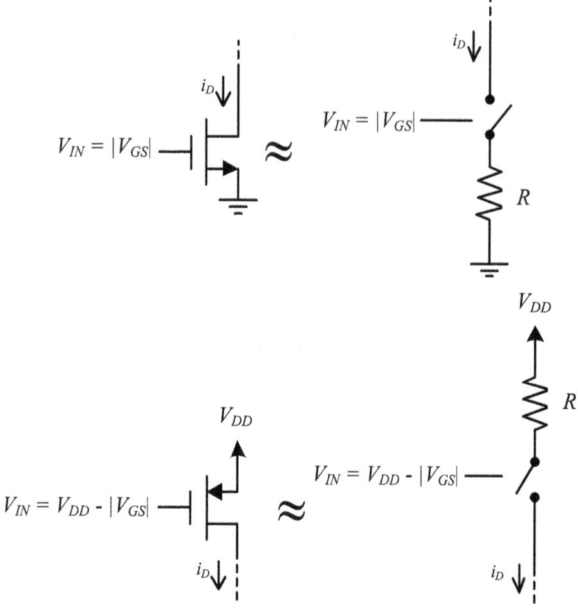

Fig. 4.19 Simple model of a transistor as a switch

Fig. 4.20 MOSFET driving
a relay

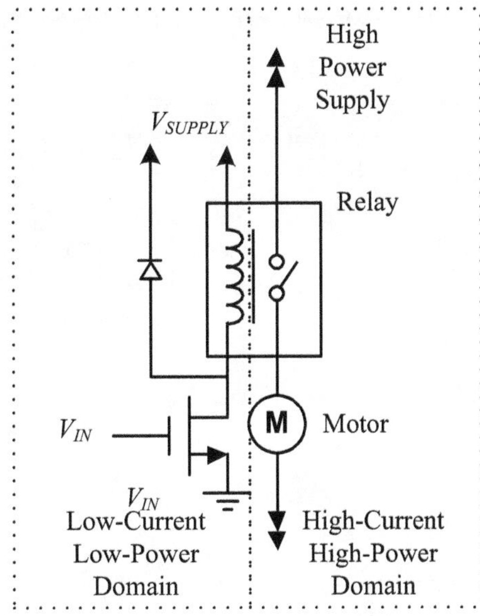

Fig. 4.21 Inverter truth
table and symbol

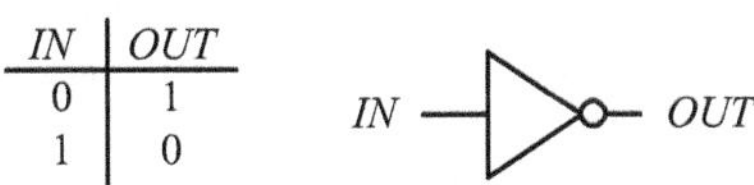

IN	OUT
0	1
1	0

low-current low-power domain to the left, and a high-current high-power domain to
the right.

4.5.2   Logic Gates

Following the usage of a transistor as a switch, the idea can be applied to create logic
gates, the simplest of which is the inverter whose truth table and symbol are shown in
Fig. 4.21.

In order to have this functionality, we will put a PMOS on top of NMOS,
effectively acting as switches working in opposite directions as in Fig. 4.22.

A sample plot depicting v_{IN} and v_{OUT} versus time is shown in Fig. 4.23.

When v_{IN} is equal to V_{SUPPLY}, it is high so the NMOS is ON and the PMOS is
OFF and v_{OUT} is connected to ground, thus low. When v_{IN} is equal to ground, it is
low so the NMOS is OFF and the PMOS is ON and v_{OUT} is connected to V_{SUPPLY},
thus high.

Note that both the input and output signals have a DC as well as several AC
components.

Fig. 4.22 CMOS inverter

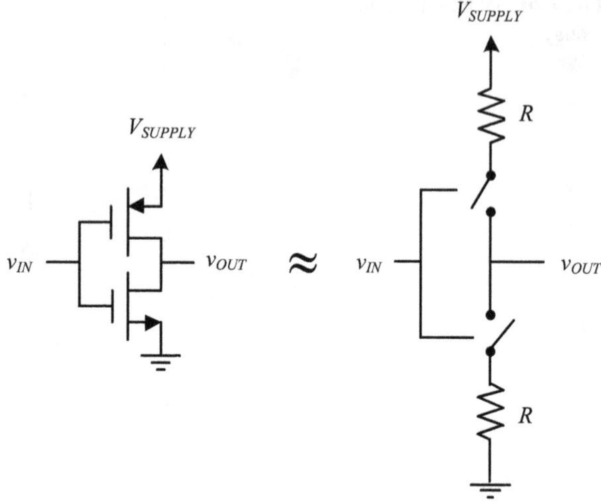

Fig. 4.23 Inverter input and output signals

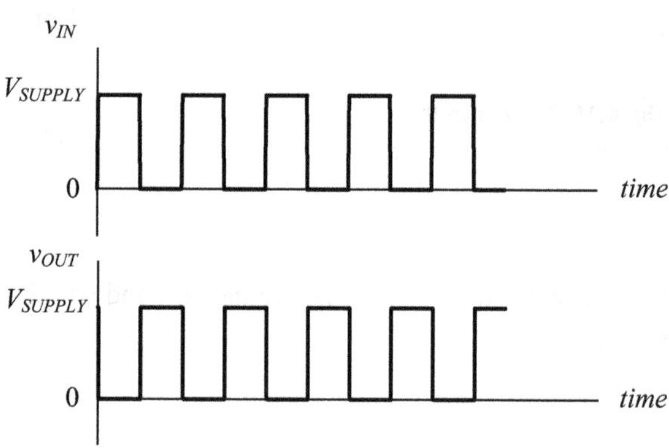

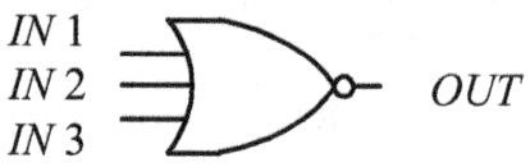

Fig. 4.24 3-input NOR truth table and symbol

IN 1	IN 2	IN 3	OUT
0	0	0	1
0	0	1	0
0	1	0	0
0	1	1	0
1	0	0	0
1	0	1	0
1	1	0	0
1	1	1	0

Another interesting logic gate is the NOR gate, which can implement all kinds of logic if combined probably with other NOR gates. Figure 4.24 depicts the truth table of a 3-input NOR gate along with its symbol.

The CMOS implementation is shown in Fig. 4.25.

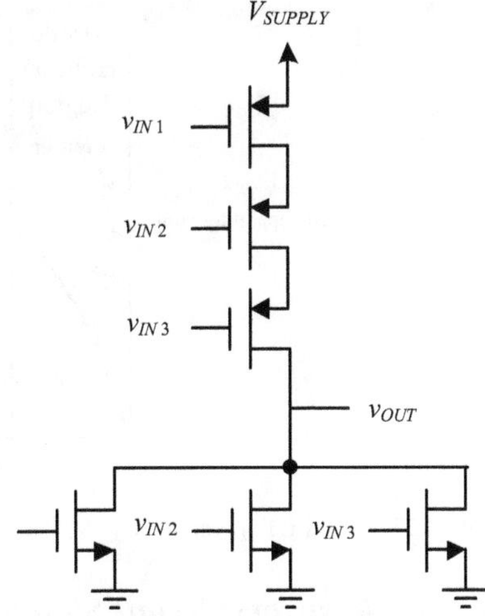

Fig. 4.25 CMOS 3-input NOR gate

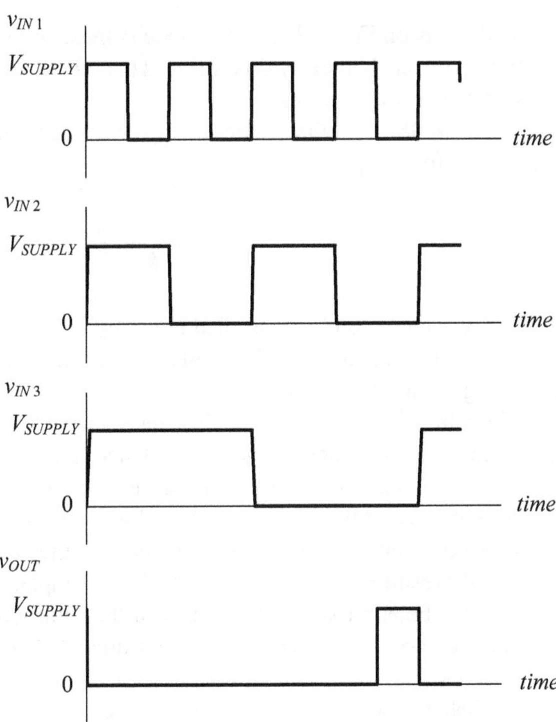

Fig. 4.26 NOR gate input and output signals

A sample plot depicting the inputs and output of such a NOR gate versus time is shown in Fig. 4.26.

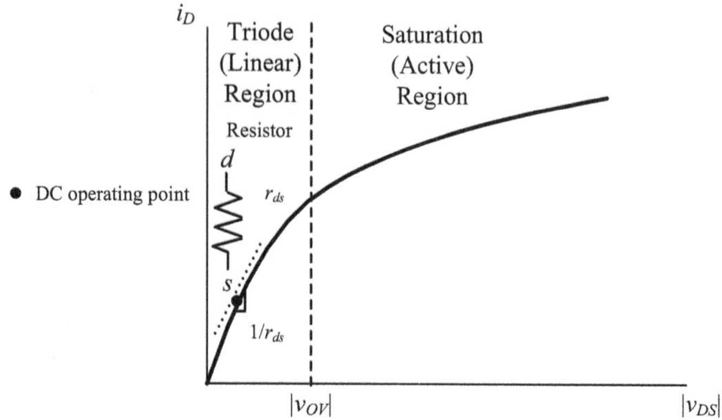

Fig. 4.27 MOSFET i_D versus $|v_{DS}|$

4.5.3 Resistors: Small Signal

We will focus on Fig. 4.27. In the triode (Linear or Ohmic) region, the small-signal relation between i_D and $|v_{DS}|$ is linear. This means that the transistor is acting as a resistor whose value is r_{ds}.

For a MOSFET, the value of r_{ds} was derived in (4.16) and (4.17) and repeated here:

$$r_{ds} \approx \frac{1}{k'\left(\frac{W}{L}\right)|V_{OV}|} \tag{4.32}$$

This resistance can be controlled either by selecting a different transistor with an appropriate aspect ratio (W/L), mainly by choosing a different W in IC design, or by designing for a different $|V_{OV}|$.

Wider transistors have a smaller resistance for a certain $|V_{OV}|$. Additionally, increasing $|V_{OV}|$ also decreases the resistance. However, increasing $|V_{OV}|$ is becoming more difficult particularly in IC design due to the scaling down of the power supply voltage and the $|V_{th}|$ not scaling down with it.

Notice that having a linear relation between the current and the voltage does not necessarily result in a resistance effect. For example, in Fig. 4.5, the i_D versus $|v_{GS}|$ relation can be approximated as linear in the saturation region; however, since the current does not pass through the two terminals where the voltage is taken, namely the Gate and the Source, this results in a voltage-controlled current source behavior, not a resistance behavior, as we are going to see next.

4.5.4 *Current Sources*

Looking at Figs. 4.28 and 4.29, we can see that in the saturation region, the behavior gives rise to a $|v_{GS}|$-controlled i_D in Fig. 4.28, where i_D slightly increases with $|v_{DS}|$ in Fig. 4.29. As a result, we can build a voltage-controlled current source using the transistor as such, where i_D is the current and $|v_{GS}|$ is the voltage controlling it. The small-signal relation between i_D and $|v_{GS}|$ is linear as was derived in (4.7), resulting in the following:

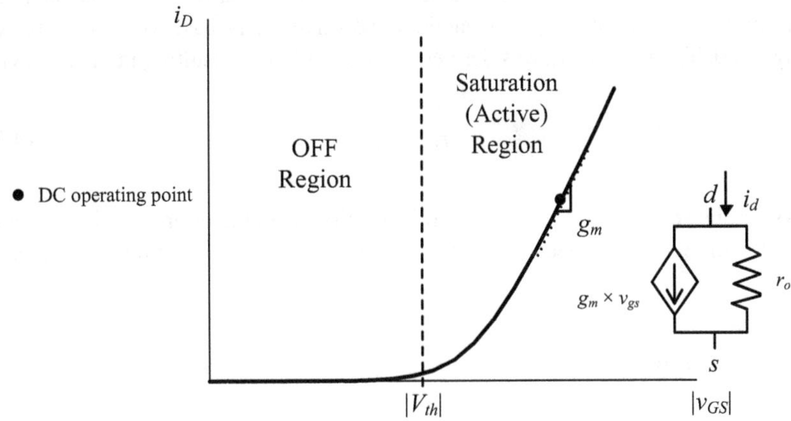

Fig. 4.28 MOSFET i_D versus $|v_{GS}|$

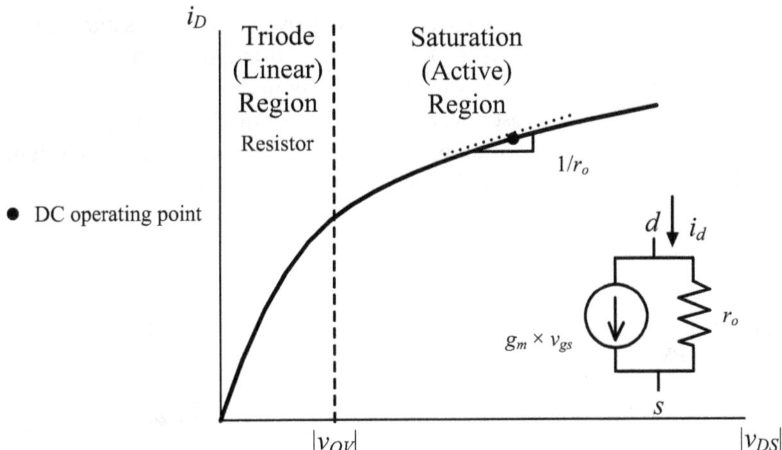

Fig. 4.29 MOSFET i_D versus $|v_{DS}|$

$$g_m = \frac{2I_D}{|V_{OV}|} \tag{4.33}$$

As can be seen, if we would like to increase the sensitivity of i_D with respect to $|v_{GS}|$, which is what g_m is, then we need to either increase the DC current I_D, or decrease $|V_{OV}|$. Increasing I_D results in higher power consumption, which is related generally to I_D and the power supply voltage, and decreasing $|V_{OV}|$ has its limits as was discussed in 4.5.3.

An interesting special case arises when $|v_{GS}|$ is kept constant (DC). This will result in an independent current source, which is how we implement them, primarily in IC design.

Continuing with the observations, in any case, the i_D generated is slightly dependent on $|v_{DS}|$, meaning that the current source has an output resistance and is not infinite. This output resistance, again in the small-signal case, is r_o as can be seen in Fig. 4.29. The value of r_o was derived in (4.8) and (4.9) resulting in the following:

$$r_o = \frac{|V_A|}{|I_D|} \tag{4.34}$$

As can be seen, increasing r_o, which is usually desirable in order to have a better current source, requires decreasing $|I_D|$ since $|V_A|$ is fixed for a certain transistor.

4.5.5 Diodes

If the gate terminal of a MOSFET is connected to its drain terminal, the resulting circuit will behave similarly to a diode, thus the name diode-connected transistor as in Fig. 4.30.

If it is ON, the diode-connected MOSFET will always be in the saturation region since $|v_{GS}| = |v_{DS}|$, so $|v_{DS}| > |v_{GS} - V_{th}|$. It differs from a diode in that the relation between i_D and $|v_{GS}|$ is quadratic instead of exponential.

This circuit enables us to use transistors instead of diodes in many applications. Also, we can use it as a level-shifter where we can lower the DC value of a signal by making it pass through this circuit instead of using a resistor.

Fig. 4.30 Diode-connected
NMOS (a) and PMOS (b)

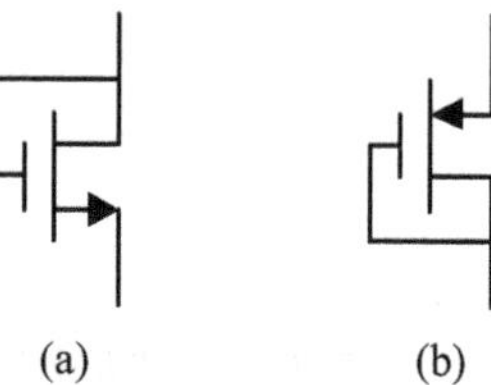

4.5.6   Amplifiers: Small Signal

Given that a MOSFET transistor acts as a voltage-controlled current source, the input is a voltage and the output is a current. This current can then be converted into a voltage by passing it through a resistor. The gain is then defined as the relation between the output voltage and the input voltage. The procedure is conceptually shown in Fig. 4.31.

There are more than one circuit topology that can implement this scheme. One simple way to do so is shown in Fig. 4.32.

The circuits surrounding the transistor are modeled as follows:

- The previous stage (signal source) is modeled using its Thevenin equivalent model. This signal source contains a DC component and a small-signal AC component. Also, the source has an output resistance R_{SRC}.
- The next stage (load) is modeled using its input resistance, indicated as R_{LD}.
- The current source, which, in combination with the DC component of the signal source, creates the DC operating point around which the transistor operates, is modeled using its Norton model with an output resistance R_{CT}.

Fig. 4.31 Using a transistor to build an amplifier

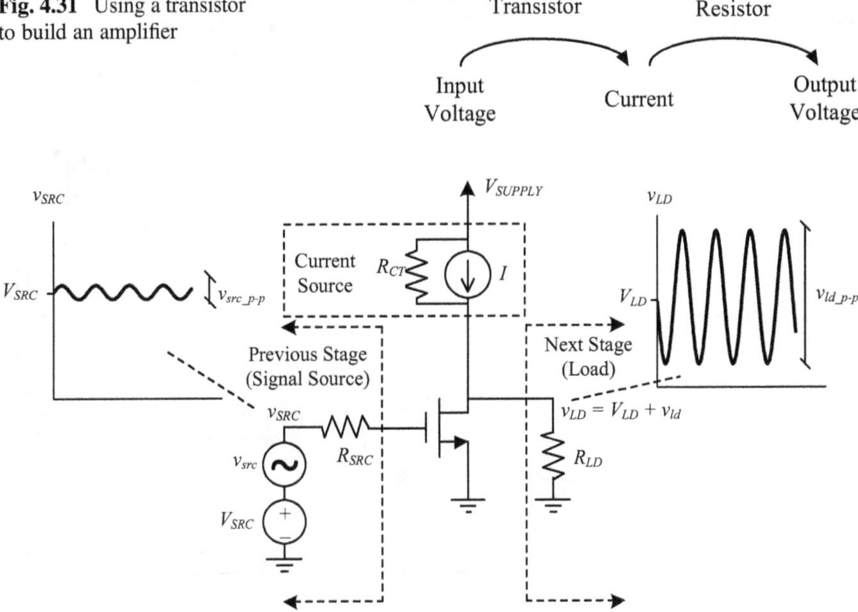

Fig. 4.32 Simple amplifier circuit topology

4.5.7 Capacitors

MOSFETs can be used as capacitors. The main idea is to connect the Source and Drain terminals together giving us one of the capacitor's connections, while the other connection is that to the Gate terminal as in Fig. 4.33. The resulting capacitor is called a MOS Cap.

There are three modes of operation:

- Accumulation mode: $|v_{GS}| < |V_{th}|$.
- Weak inversion mode: $|v_{GS}| \approx |V_{th}|$.
- Strong inversion mode: $|v_{GS}| > |V_{th}|$.

In all of these modes capacitance is approximately proportional to $W \times L$.

A sample NMOS capacitance versus voltage behavior is shown in Fig. 4.34.

As can be seen, in the accumulation and strong inversion modes, the capacitance is fairly constant with the value being higher in the strong inversion mode. Therefore, in these modes, the transistor acts like a fixed capacitor.

In the weak inversion mode, the capacitance is a function of the applied voltage. In this mode, the transistor acts as a variable voltage-controlled capacitor.

4.5.8 Active Inductors

Using transistors and capacitors, we can create an inductor. This is called an active inductor since it makes use of transistors and of course some power.

Fig. 4.33 MOS Caps using NMOS (**a**) and PMOS (**b**)

(a)   (b)

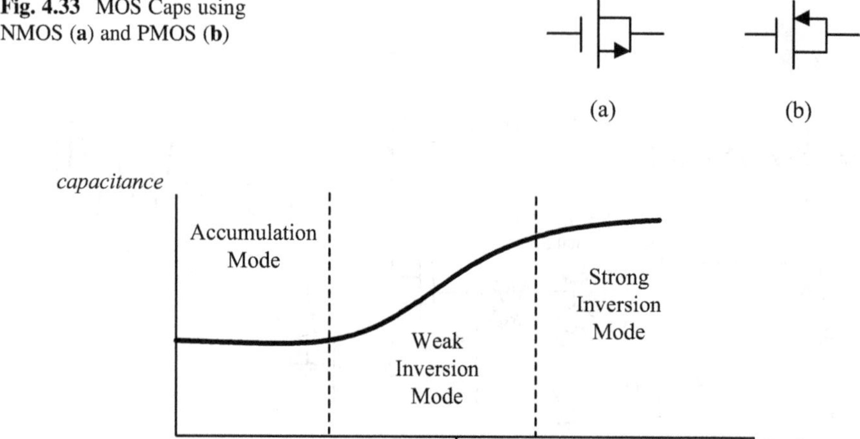

Fig. 4.34 NMOS Capacitance versus voltage

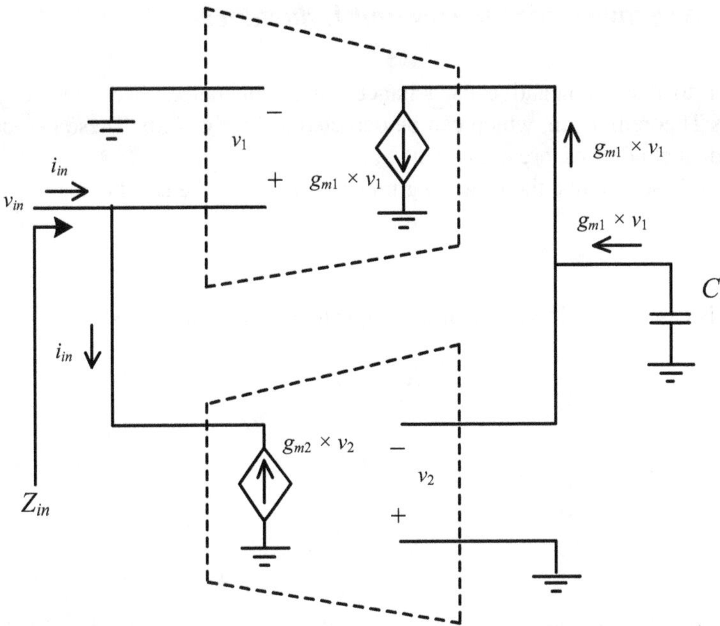

Fig. 4.35 Gyrator-C circuit implementing an active inductor

A popular way to do so is using a Gyrator-C circuit. This circuit is shown in Fig. 4.35. In its simplest implementation, it makes use of two transistors that give rise to two transconductances g_{m1} and g_{m2}. What follows is a simplified analysis of this ideal circuit that shows the desired effect as long as the transistors are operating in the appropriate region and are behaving correctly.

$$i_{in} = -g_{m2} \times v_2$$
$$v_2 = (-g_{m1} \times v_1) \times Z_C = \frac{-g_{m1} \times v_1}{sC}$$
$$v_1 = v_{in}$$
$$\Rightarrow i_{in} = \frac{g_{m2} \times g_{m1} \times v_{in}}{sC} \tag{4.35}$$
$$\Rightarrow Z_{in} = \frac{v_{in}}{i_{in}} = s\left(\frac{C}{g_{m1} \times g_{m2}}\right) = sL_{active}$$

Therefore, the input impedance Z_{in} acts as an active inductance, namely

$$L_{active} = \frac{C}{g_{m1} \times g_{m2}} \tag{4.36}$$

4.5.9 *Negative Capacitances and Inductances*

In order to discuss negative capacitances and inductances, we need to present Miller's Theorem again, which can be depicted as in Fig. 4.36. These two circuits, under ideal conditions, are equivalent.

Given these circuits, the following precondition must be satisfied:

$$v_2 = A \times v_1 \tag{4.37}$$

In this case, the following relations are satisfied according to Miller:

$$Z_1 = \frac{Z}{1 - A} \tag{4.38}$$

and

$$Z_2 = \frac{Z}{1 - \frac{1}{A}} \tag{4.39}$$

Knowing that the input impedance Z_{in} of the circuit is Z_1 itself, we will choose a couple of scenarios for Z to show how the circuit to the left might lead to negative capacitances and inductances.

If Z is a capacitor, then:

$$Z_{in} = Z_1 = \frac{Z}{1 - A} = \frac{1/sC}{1 - A} = \frac{1}{sC(1 - A)} = \frac{1}{sC_{in}} \tag{4.40}$$
$$\text{where } C_{in} = C(1 - A)$$

As a result, we have a couple of resulting situations:

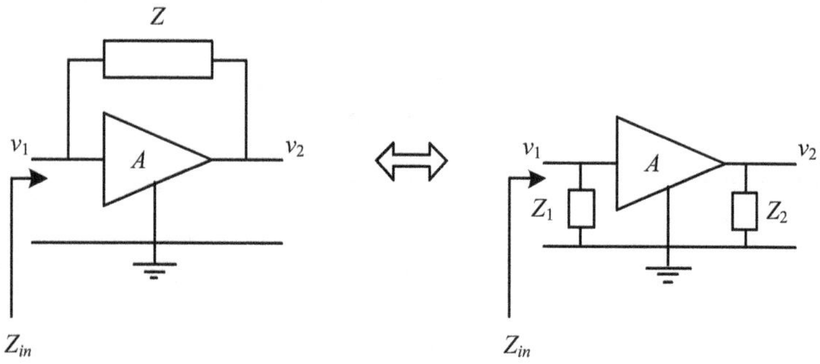

Fig. 4.36 Equivalent circuits Miller's theorem

$$A < 1 \Rightarrow C_{in} \text{ is positive}$$
$$A > 1 \Rightarrow C_{in} \text{ is negative}$$

(4.41)

In the second case, the circuit behaves as a negative capacitor from the source point of view.

As for the other scenario, if Z is an inductor, then:

$$Z_{in} = Z_1 = \frac{Z}{1-A} = \frac{sL}{1-A} = s\left(\frac{L}{1-A}\right) = sL_{in}$$
$$\text{where } L_{in} = \frac{L}{1-A}$$

(4.42)

As a result, we have a couple of resulting situations:

$$A < 1 \Rightarrow L_{in} \text{ is positive}$$
$$A > 1 \Rightarrow L_{in} \text{ is negative}$$

(4.43)

In the second case, the circuit behaves as a negative inductor from the source point of view.

Chapter 5
Technology and Limitations

Technology is constantly improving. However, it always has some limitations. In order to understand the reasons behind this, this chapter covers the physical aspects of technology starting at the system level down to the device level. The chapter discusses the materials used, the fabrication process, and the limitations including process variations, parasitics, and electromigration. Additionally, reliability and aging are also discussed along with heat management. This chapter concludes with an overview of past, present, and future trends in technology advancement.

5.1 System Overview

Circuits in modern electronic equipment are implemented in various ways and assembled using different techniques. If we remove the equipment cover we typically see the electronic circuit and possibly other components such as a battery.

The electronic circuit itself resides on a printed circuit board (PCB). There can be either one or many boards. In case there are many, the main one, typically the biggest one, is called the motherboard. The other boards, typically the smaller ones that connect to the motherboard, are called the daughter boards. The PCB serves to connect the different components that reside on it. The PCB itself does not need to take the shape of the equipment since many times cutouts are needed to fit other components such as a battery or a camera for example.

The components that the PCB hold can be either discrete such as single capacitors and resistors or complete integrated circuits (ICs), also known as dies or chips, residing typically in packages.

Figure 5.1 shows a simple setup with a single main board and two packages containing their ICs.

The PCB consists of a sandwich structure of alternating conductive, typically copper, layers and insulating, typically glass epoxy, layers. The layers, particularly

© The Editor(s) (if applicable) and The Author(s), under exclusive license to
Springer Nature Switzerland AG 2024
J. G. Atallah, M. Ismail, *Integrated Electronic Circuits*,
https://doi.org/10.1007/978-3-031-62707-1_5

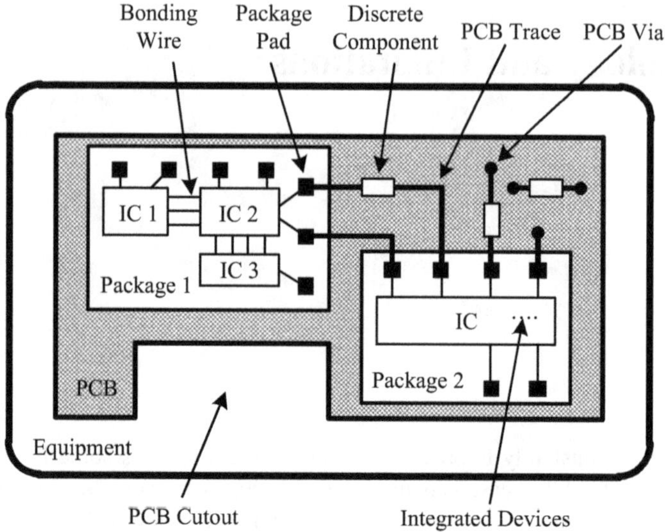

Fig. 5.1 The electronics in a modern equipment

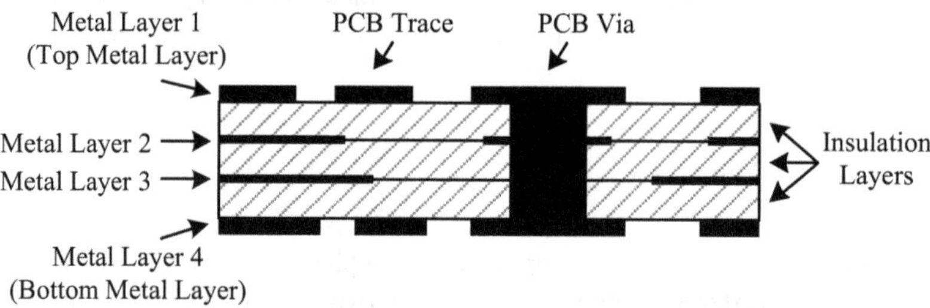

Fig. 5.2 Stack-up of a 4-layer PCB

the insulating layers, do not have to be of the same thickness. Also, the insulating layers can be made of different materials. The conductive layers are used to form traces, which are the connections. Traces on different layers can be connected together using vertical connections called vias. A PCB can have several conductive layers. A 4-layer PCB stack-up is shown in Fig. 5.2.

The components and packages are mounted on the PCB using the top and/or the bottom metal layers. The packaging technology plays an important role in this regard. Some components require through-hole mounting, which means that their connections have to go through the PCB and soldered on the reverse side. In this case, no component can be mounted on their reverse side. Other components and packages are surface-mount, which means that they are soldered on the side of the component only, thus allowing for other components to be mounted on their opposite side. The difference between the two is shown in Fig. 5.3.

Fig. 5.3 Through-hole and surface-mount packages

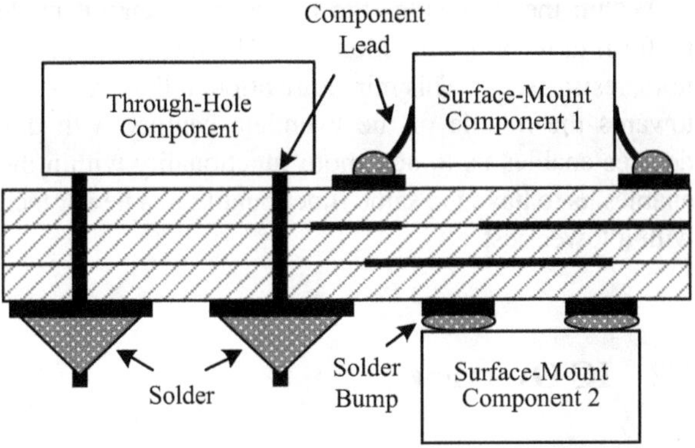

Fig. 5.4 Cross-section of a SiP using flip-chip ICs

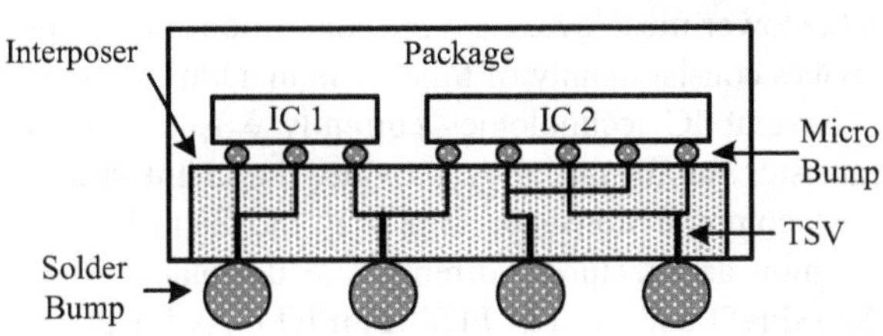

In case the component is a packaged IC, the package would contain either one or more ICs depending on the chosen system partition. The ICs are typically wire-bonded to the package using gold connections. If part of the system is implemented using a certain technology node and another uses another technology node, then two or more ICs, each using its own technology, can be combined in one package resulting in a System in Package (SiP) as in Package 1 of Fig. 5.1. On the other hand, if the whole circuit is implemented on a single IC, then the package will contain only this IC. If this IC implements most of the needed features, then the result would be a System-on-Chip (SoC) as in Package 2 of Fig. 5.1. Some of these packaged ICs can be proprietary such as microprocessors and communication transceivers while others can be standard off-the-shelf ones such as operational amplifiers and voltage regulators.

Figure 5.4 shows an alternative SiP technology where the ICs are flipped inside the package. It makes use of an interposer that serves as a tiny PCB inside the package, connecting the ICs to the solder bumps of the package using Through-Silicon Vias (TSVs). These ICs are sometimes called chiplets. In this case, the ICs make use of micro-bumps instead of bonding wires for their external connections.

Another alternative is to place the IC directly on the PCB without a package. This is called Chip-On-Board (COB) and can be used in applications where the package is undesirable such as when severe cost reduction is needed, when the IC does not lend itself naturally to packaging such as in LED applications, or when the resistances, capacitances, and inductances added by the package prevent the performance results from meeting the specifications.

Within the IC resides the miniaturized circuit made of integrated devices that perform most of the functionality. The integrated devices can be transistors, capacitors, resistors, etc. This miniaturization of the circuit in the form of an IC that started towards the middle of the twentieth century with the ever-shrinking size of the devices enables us to add more functionality within the same area, or, conversely, enables us to have the same functionality in a newer technology using a smaller area with a lower cost.

5.2 IC Overview

Similar to a PCB, an IC is made of several layers, but with the semiconductor devices in the lower-most layers and the connections in the upper layers. The semiconductor devices consist mainly of transistors in addition to resistors, capacitors, etc.

Several IC technologies currently exist. They mainly differ in terms of the transistor family that they provide, the architecture, and the materials used. The most common transistor family is the Field-Effect Transistor (FET). The most common architectures currently are the planar Metal-Oxide-Semiconductor FET (MOSFET) and the Fin FET (FinFET). As for the materials, the most commonly used material is silicon. A sample cross-section of a planar 6-metal-layer MOSFET technology is shown in Fig. 5.5.

The cross-section can be divided into two parts:

- Front End of Line (FEOL), which consists of the lower layers where the devices exist next to each other.
- Back End of Line (BEOL), which consists of the upper layers where the metal interconnects exist.

The FEOL is made of a substrate and all the necessary materials needed to create integrated devices. Notice how the substrate, which is typically made of p-type material, holds the whole circuit, so noise and interference can travel from one part of the circuit to another, creating the undesirable phenomenon of substrate-coupling. In order to avoid this, some technology processes use an insulator as a substrate resulting in what is known as Silicon-on-Insulator (SOI), which performs very well especially if the signals include high-frequency components.

The BEOL is primarily a sandwich of insulation and metal layers, usually copper. The metal layers can be connected together using vias. The top-most metal layer, Metal 6 in Fig. 5.5, is thicker than the lower ones. It is used to connect devices that are far away from each other. By making this layer thicker, the connection will have lower resistance. All these connections have unwanted resistance, capacitance, and inductance to them, called parasitics. These parasitics should be reduced and taken care of by the designer.

The BEOL can contain two special devices. The first can be seen in the structure made of the intermediate Capacitor-Top-Metal (CTM) layer and Metal 5. The combination along with the insulator between them creates a Metal-Insulator-

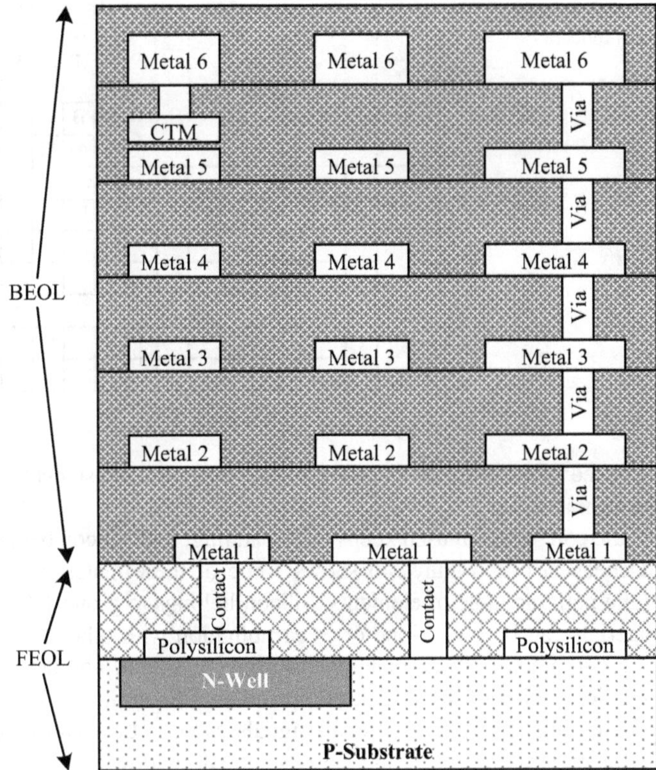

Fig. 5.5 Cross-section of a 6-metal-layer MOSFET technology

Metal (MIM) type of capacitor. The second device consists of a spiral inductor that can be built out of the top-most metal layer, Metal 6. Taking advantage of the increased thickness of this layer, the inductor will have less resistance, thus higher quality, compared to using other layers. These two types of components, especially the inductor, take up a lot of space on the IC. Therefore inductors should be avoided and MIM capacitors should be used sparingly.

5.3 Integrated Devices

Devices such as transistors are made of semiconductor materials. They are formed in the FEOL next to each other. The material used consists of a core material such as silicon (Si), which is modified, resulting in either n-type semiconductor or p-type semiconductor. These semiconductor materials, in addition to other ones including silicon dioxide, which is used as an insulator, polysilicon, and copper which is used as a conductor, form the core of modern technologies. In some technologies, silicon

Fig. 5.6 Some popular core materials used in modern technologies

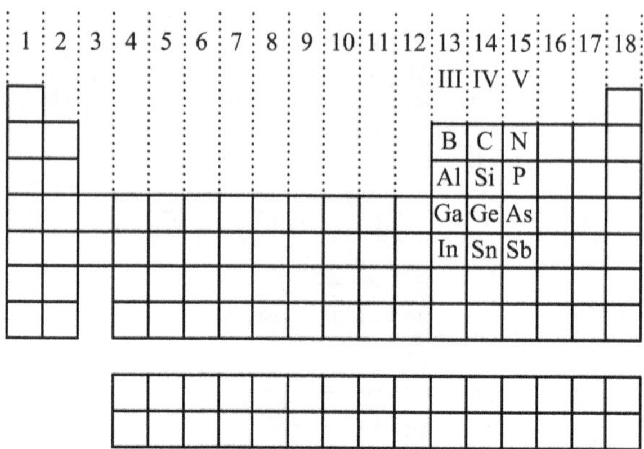

dioxide is replaced with materials that have higher dielectric constants (high-k), for improved performance.

The list of the core material used keeps on expanding, with silicon being the most popular and cheapest one. Figure 5.6 shows some of these materials as they are classified in the periodic table of elements. Silicon belongs to group IV.

These materials can be used alone or in combination with others. For example silicon can be combined with germanium resulting in silicon germanium (SiGe), which induces less noise in the circuit compared to silicon alone. This makes it suitable in circuits that process very small signals such as Low-Noise Amplifiers (LNAs) used in wireless receivers and Low-Noise Block downconverters (LNBs) used to convert a wireless signal into a wired one. Also, silicon can be combined with carbon resulting in silicon carbide (SiC) which can handle extreme temperatures making it suitable for space applications. Another interesting combination is the III-V compound semiconductor such as gallium nitride (GaN), which can handle higher power than silicon. This makes it suitable in circuits such as fast power chargers used for phones, power amplifiers used in wireless transmitters, and LED applications.

The modification of the material is called doping and consists of adding some impurities. This results in a doped material that can either be of an n-type where there are excess electrons or p-type where there are excess holes, depending on the impurities added. NMOS transistors rely on the electrons for their operation, thus their current-voltage relation makes use of electron mobility μ_n. On the other hand, PMOS transistors rely on the holes for their operation, thus their current-voltage relation makes use of hole mobility μ_p.

Several simplified views of a typical planar bulk NMOS structure are shown in Fig. 5.7.

L and W are the transistor channel length and width respectively. P-active is needed to create the Bulk terminal, which is the connection to the substrate. P-active is sometimes called p-implant or p-tap. A contact is used to connect the Drain (D), Source (S), and Bulk (B) from the active region directly to the first metal layer. As for

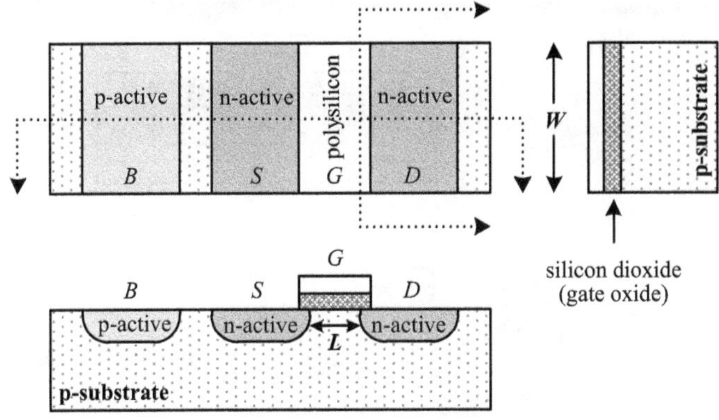

Fig. 5.7 Simplified views of a planar bulk NMOS structure

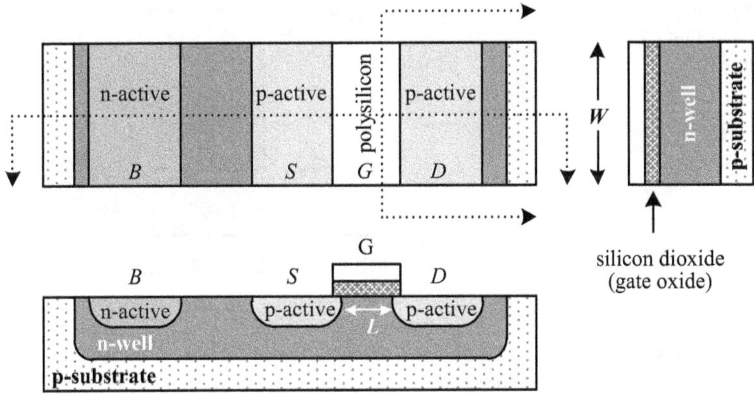

Fig. 5.8 Simplified views of a planar bulk PMOS structure

the Gate (G), a contact from the polysilicon to the first metal layer is used to connect it.

Observe how an NMOS is fundamentally symmetrical. However, some layout methods for wider devices will make the transistor asymmetrical. For this reason, it is important to keep track of the positions of the source and the drain at all times.

Similarly, several simplified views of a typical planar bulk PMOS structure are shown in Fig. 5.8.

Since the substrate is made of p-type material, an n-well is created for the PMOS. As a result, a PMOS is more complicated to create than an NMOS. Apart from this, the connections are similar to those in an NMOS.

When the transistor is ON, the current travels through the channel between the drain and the source. This is controlled by the gate terminal. With newer technologies, the channel length is getting smaller making it more difficult to fully turn OFF a transistor due to the effect of the electric field lines from the source and the drain. This is one example of the short-channel effects that arise in modern technologies.

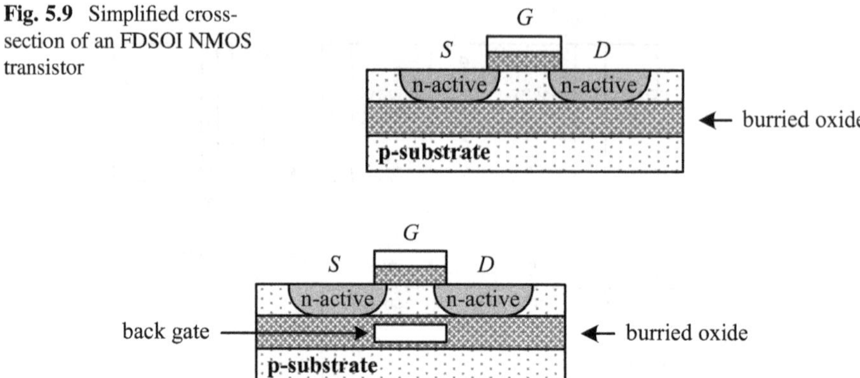

Fig. 5.9 Simplified cross-section of an FDSOI NMOS transistor

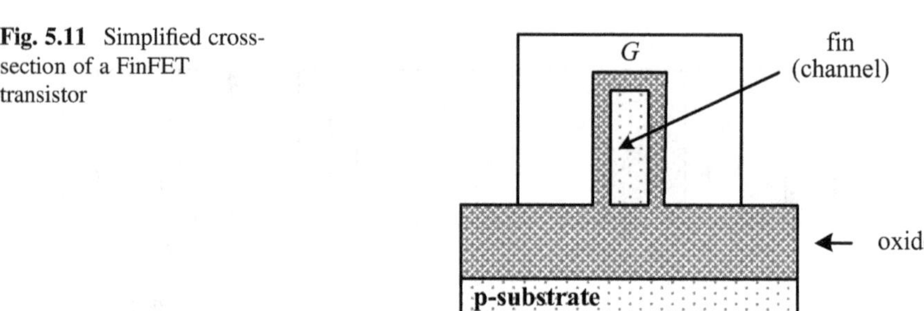

Fig. 5.10 Simplified cross-section of a planar FDSOI NMOS transistor with a back gate

Fig. 5.11 Simplified cross-section of a FinFET transistor

This can be reduced by increasing the doping concentration of the channel region, but this has its limits too. Additionally, in bulk processes such as in Figs. 5.7 and 5.8, the current has the tendency to escape the channel through the substrate. Add to this that every unintentional *p-n* junction such as the ones at the bottom of the source and drain terminals exhibits some junction capacitance which slows down the transistor, then we can see the several possibilities for potential needed improvements.

One such improvement is to incorporate a buried oxide below the transistor resulting in a Fully Depleted Silicon on Insulator (FDSOI) as shown in Fig. 5.9.

The drawback in this case is that to reduce the short-channel effects, the buried oxide has to be thin, which increases the junction capacitances,

A solution to this is to add another gate in the buried oxide, resulting in a double-gate structure: the main gate, and the back gate as shown in Fig. 5.10.

With this back gate, a better control of the channel is achieved. However, if we continue with the same reasoning, it would be even better to have the gate on not only two, but three sides of the channel. This gives rise to the FinFET, which, when combined with SOI, has the cross-section seen in Fig. 5.11.

In this case, the drain would be connected to the fin in the foreground and the source would be connected to the fin in the background, so the current will go through the figure.

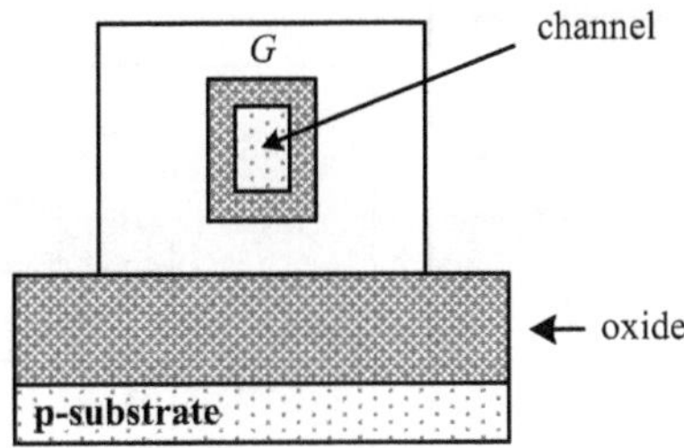

Fig. 5.12 Simplified cross-section of a GAAFET transistor

Yet another step is to have the gate completely surround the channel as in a Gate-All-Around FET (GAAFET) transistor, a cross-section of which is shown in Fig. 5.12.

These improvements help in having a better control of the gate and in achieving even smaller feature sizes. Note that a certain technology will have only one of these types implemented. This broadens the choices available to select the appropriate technology for a certain application. However, this comes at a price since the advanced technologies, particularly the FinFET and the GAAFET are considerably more expensive than planar bulk and planar SOI technologies.

Additionally, when designing using FinFET and GAAFET, the designer does not have control over the dimensions of the transistor. Thus, if a wider transistor is needed for example, then several transistors should be placed in parallel, similarly to board-level discrete designs. This reduces the possible size options and is in contrast to the freedom that the designer has when using planar FET technologies where the appropriate transistor width and length can be chosen.

BJTs can be implemented in planar CMOS processes for special applications. Two simplified views of a typical NPN BJT structure are shown in Fig. 5.13. Terminals C, B, and E are the Collector, Base, and Emitter respectively. Observe how the BJT is fundamentally asymmetrical with the collector being the largest terminal.

Additionally, two simplified views of a typical PNP BJT structure are shown in Fig. 5.14.

5.4   Semiconductor Manufacturing and the Foundry Ecosystem

Integrated circuits are manufactured in a semiconductor fabrication plant, also known as foundry or fab. The fab consists of several facilities that deal with the different stages of IC design. These facilities are collectively called the foundry ecosystem. The ICs are manufactured on a wafer in a process called a fabrication run. After fabrication, ICs get packaged in a packaging house, then head into a PCB facility where they get mounted. At every step of the way, there are several quality checks and testing procedures that take place.

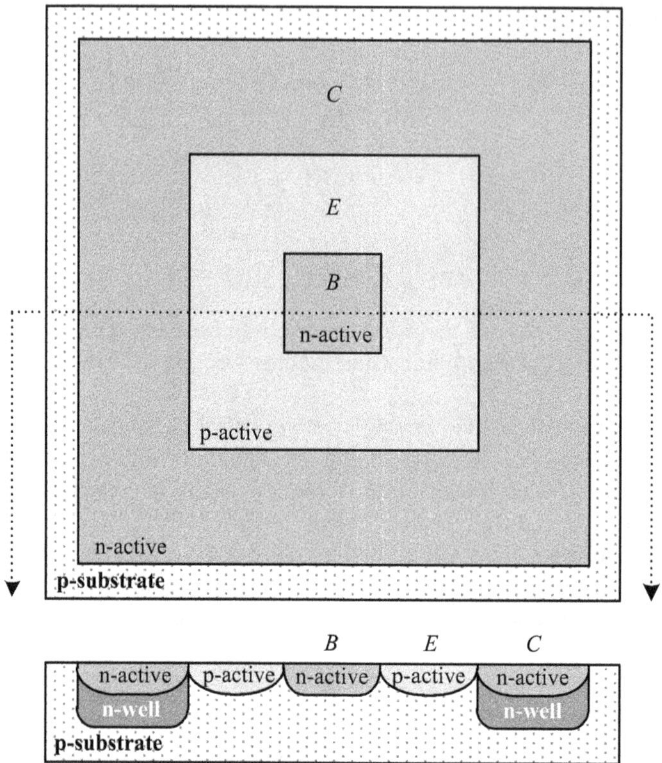

Fig. 5.13 Simplified views of an NPN BJT structure

The starting point consists of the design files that the designer sends to the foundry. These design files are typically in the Graphic Design System II (GDSII) stream format. They contain geometric shapes and other information derived from the IC layout. They, along with uniformly distributed test structures inserted by the fab, are used in order to generate the photomask set that will be used in the manufacturing process. Each photomask set is made of several glass photomask plates, which contain opaque and transparent patterns, thus blocking or allowing the light to shine through. These patterns correspond to the geometries that will be fabricated. Each wafer requires its own photomask set to fabricate the different layers of the ICs contained therein.

The designer might share a wafer with other designers in what is known as a Multi-Project Wafer (MPW) run. This is common for designs that are meant for research, prototyping, or small volume applications. Alternatively, the designer can book the whole wafer, which is typically the case for large-scale products.

Semiconductor manufacturing consists of several operations and the flow itself is subject to several variations. Notwithstanding this, the following steps describe the general operations that are undertaken:

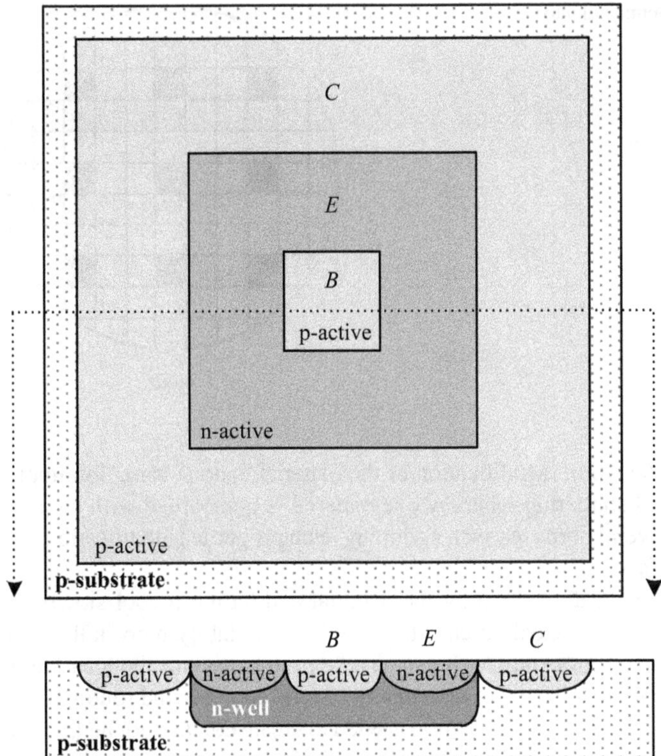

Fig. 5.14 Simplified views of a PNP BJT structure

1. Wafer manufacturing: A wafer is a thin circular disc made of silicon. It is fabricated by heating up the sand into a high-purity liquid, then solidifying it into an ingot. This ingot is then sliced into a disc, which is then polished to remove the defects.
2. Oxidation: A uniform oxide layer is formed on the wafer to make it semiconductive.
3. Photolithography: This process, also known as lithography, is used to pattern a region on a semiconductor material. A light-sensitive chemical called a photoresist (or simply resist) is deposited on top of the semiconductor assembly. A photomask is placed on top of the photoresist and light is shone to transfer the pattern on the photomask to the photoresist.
4. Etching: The areas where the light hits the photoresist should be removed. This is done using either a liquid etching technique (wet etching) or a gas etching technique (dry etching).
5. Deposition: Deposition consists of adding a new layer of material by coating a thin film. This can be done using either Physical Vapor Deposition (PVD) which makes use of sputtering or Chemical Vapor Deposition (CVD) which makes use of gases flowing over the wafer.

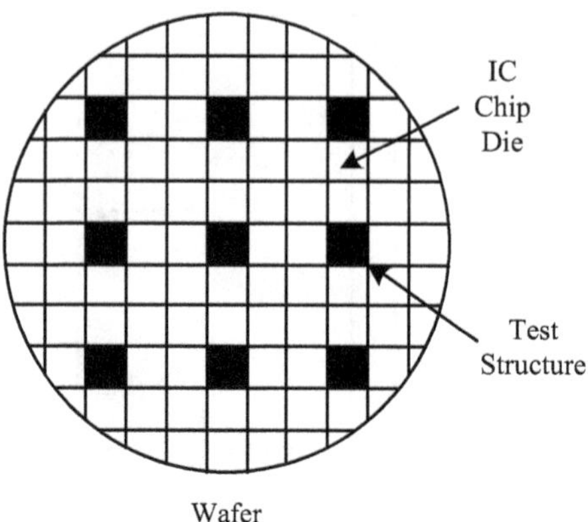

Fig. 5.15 Semiconductor wafer

6. Ion implantation: Modification of the material is done using ion implantation by the use of sputtering whereby one material is bombarded with another. This can be for several purposes such as doping, etch properties modification, and cleaning of a surface.

7. Thermal annealing: Heating the material and letting it cool slowly in a process called thermal annealing can stabilize the material by recrystallizing it after ion implantation for example. It can also be used to control the location of dopants.

8. Metallization and contact information: Metal contacts and routing are required in an IC to carry the signal. This is needed to create contacts to the transistor without impacting its performance. Metal silicides, which are compounds made of metal and silicon, are used for the drain and source contacts and to contact with the polysilicon of the gate. As for the several interconnect layers above the transistors, they are typically made of copper along with the vias which serve to connect one layer to another. The first few layers are called the local interconnects. They have a very fine pitch and are used for local connections. The upper layers are called the global interconnects. They are relatively thick and are used to create the long-haul connections across the IC.

9. Quality check: After the IC is fabricated, checking the quantity of the elements present in a semiconductor as well as the thickness values is a crucial step in ensuring the quality of the results. This is done by doing measurements using either a scanning electron microscope (SEM) and/or a transmission electron microscope (TEM) and analyzing the results using energy dispersive spectroscopy (EDS).

Figure 5.15 shows a sketch of a semiconductor wafer along with the ICs therein. Note the test structures inserted by the fab in order to test for defects, indicating whether the whole wafer meets the fab's quality requirements.

Many of these operations such as photolithography, etching, deposition, ion implantation, and metallization are repeated several times to build up the required IC layers.

Table 5.1 OBM/ODM/OEM

Type of company	Design	Manufacture	Brand
OBM	✓	✓	✓
ODM	✓	✓	
OEM		✓	

After fabrication, the fab tests the structures that it has inserted in order to check whether the whole wafer meets the limits of defects acceptable for the fab. If the wafer does not pass this test, the whole wafer is rejected.

If the wafer passes the test, the ICs therein are functionally tested. On-wafer testing is important because the ICs that do not meet their intended functional specifications should not move to the next stage. Determining which ICs are defective early-on in the process is preferable so that time and money is not spent on them in subsequent stages. This testing process should be as comprehensive as possible, while still being as fast as possible so that it does not delay the fabrication line.

After on-wafer testing, the wafer is diced, which is the process of separating the individual ICs. The defective ones are discarded and the remaining ones are sent to a packaging facility.

In the packaging facility, the ICs are placed inside a package and tested again before they get shipped to the PCB facility.

In the PCB facility, the ICs are mounted along with the other needed components and the whole PCB gets tested.

The above steps, IC fabrication, packaging, and testing are typically handled at different facilities belonging potentially to different service providers. Eventually, the mounted PCB is placed inside the equipment casing.

The company that oversees all these steps across the different service providers can be one of several types:

- Original Brand Manufacturer (OBM): This type of company designs, produces, and brands the product under its own brand name.
- Original Device Manufacturer (ODM): This type of company designs and manufactures the product while another company brands it under its own name.
- Original Equipment Manufacturer (OEM): This type of company only manufactures a product, while another company does the design and brands it under its own name.

A summary of the above is shown in Table 5.1.

Looking at Table 5.1, we can see that the companies that are complementary to ODMs only do the branding and its associated activities such as marketing. Additionally, the companies that are complementary to OEMs do the design and branding. These are sometimes known as fabless companies. Additionally, there are companies that only do the design, and these are called IP companies or chipless companies.

5.5 Process Variations and Corners

The IC fabrication process is a subtractive bulk process. It is subtractive since material is first deposited across the wafer and eventually part of it is etched out. This is in contrast to an additive process where only the needed amount of material is applied in the first place. Additionally, it is a bulk process since it deals with the carriers (electrons and holes) in bulk, not individually. Therefore, the distribution of these carriers within a certain region is not fully controlled even though considerable effort is put in order to have an even distribution. Given that additive quantum-level fabrication processes are not available, the subtractive bulk process results in quite some variations in the ICs across the wafer. Therefore, process variations are an inevitable challenge in the fabrication of ICs.

These variations may cause the IC not to meet the specifications resulting in a yield loss where the yield can be defined as

$$\text{yield} = \frac{\text{number of ICs that meet the specifications}}{\text{total number of fabricated ICs}} \qquad (5.1)$$

This yield loss increases the cost per IC sold. Fabrication-wise, these variations need to be minimized, and design-wise, they should be taken into account to make sure that the yield is acceptable.

Process variations are statistical by nature and cause Intrinsic Parameter Fluctuations (IPF) due to the discreteness of charge and matter. This is due to the continuing scaling-down of the devices from one generation to the next. Below the 45 nm technology node, this variability presents a major challenge, and as such the move to FinFETs and other advanced technologies became imminent. IPF results in the following:

- Random Dopant Fluctuations (RDF).
- Line Edge Roughness (LER).
- Polysilicon Granularity (PSG).
- Oxide Thickness Fluctuations (OTF).
- Metal Grain Granularity (MGG).

IPF affect the parameters of transistors across a wafer. Quality check can determine those ICs that are outside the specifications, but even the ones that are within the specifications contain some variations which the designer should take into account in the design phase. These variations increase with newer technologies owing to their smaller feature sizes, thus putting a limitation on the speed and even feasibility of scaling things down.

5.6   Parasitics, Electromigration, and IR Drop

All the materials used for metal connections have a finite conductivity. Therefore, every connection has its own unwanted resistance. Additionally, connections have unwanted capacitances with respect to ground and with respect to neighboring connections. Also, if the connection length is within a certain range of values, then the connection will also have unwanted inductance. Additionally, the substrate itself has its own unwanted resistances and capacitances that allow a signal from one side of the IC to couple to the other side in a phenomenon called substrate coupling. All of the unwanted resistances, inductances, and capacitances are called parasitics. They limit the performance of the IC by adding power consumption due to the parasitic resistances, by inducing corner frequencies thus lowering the speed, and by creating coupling between neighboring connections and between different parts of the IC through the substrate. On the power supply side, the excess parasitic resistances will reduce the quality, namely the stability, of the supply voltage as well as that of ground. These parasitics should be taken care of during the design and implementation of the IC, especially when the signals involved have high-frequency components such as in RF applications and high-speed digital applications, including clock distribution. An example of two neighboring metal connections and their RC parasitics is shown in Fig. 5.16.

If the current density (that is the current per unit area) in a certain metal connection, including vias, gets too high when the IC is used, the connection will fail. The mechanism that causes this failure is called Electromigration (EM). This mechanism is getting worse with newer and thus smaller feature sizes where the current densities are increasing. As a result, it is the designer's responsibility to make sure that the connections can handle high currents.

High current densities cause the metal connection to overheat due to the excess collision of the electrons with the metal atoms. With time, the metal atoms will

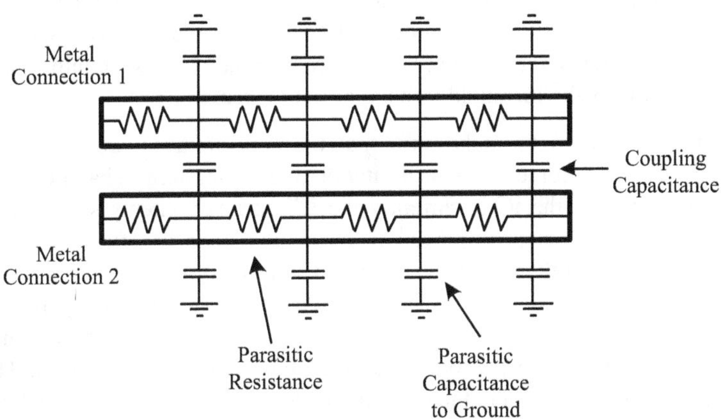

Fig. 5.16 Two neighboring metal connections and their RC parasitics

dislocate from the metal connection leading to either an open circuit in the connection and/or a short circuit to the neighboring connection.

This effect is accompanied initially by a substantial voltage (IR) drop across the connection which also causes degradation in the performance, eventually leading to IC malfunction. The solution usually is to make these connections wider in order to handle the required currents.

As a result, EM/IR effects also put a limitation on the feature sizes given that a certain amount of current needs to pass through them.

5.7 Reliability and Aging

With electronics finding their way into critical applications and infrastructure, their reliability in the long run is at question. Several mechanisms affect the device performance in the long run. These can be summarized as follows:

- Hot Carrier Injection (HCI): When the transistor gate length gets shorter, the transistor might get damaged by the hot electrons attracted to the gate, which can cause oxide damage.
- Negative Bias Temperature Instability (NBTI): This phenomenon occurs in PMOS transistors when the gate-source voltage is negative. This causes electrical stress which traps carriers (holes) at the interface between the oxide and silicon or within the oxide layer itself. This causes degradation in the performance of the PMOS over time.
- Positive Bias Temperature Instability (PBTI): This phenomenon occurs in NMOS transistors when the gate-source voltage is positive and the temperature is high. Unlike NBTI, interface traps are not created, so full recovery is possible when the electrical stress is removed (that is the transistor is turned off) and/or when the temperature is lowered.
- Self-Heating Effects (SHE): In semiconductor devices, the phenomenon of impact ionization, which contributes to HCI, can lead to increased self-heating. Remarkably, higher self-heating tends to reduce the carriers' mobility and subsequently lowers the threshold voltage of the device.

The designer needs to make sure that these phenomena are under control, otherwise they will enforce a serious limitation on the long-term reliability of the IC.

Heat generated by the IC is currently the IC's primary enemy so much so that several schemes are being employed in order to reduce its effect on electronic circuits. This is called thermal management. The primary purpose of thermal management is to efficiently channel the heat outside the IC. This is done by using a heat sink, with an optional fan or liquid cooling on top of the package and by connecting the package to a thick metal layer inside the PCB to distribute the heat. Thermal paste between the package and the heat sink is used to have a good thermal contact. This setup is seen in Fig. 5.17.

Fig. 5.17 Heat management

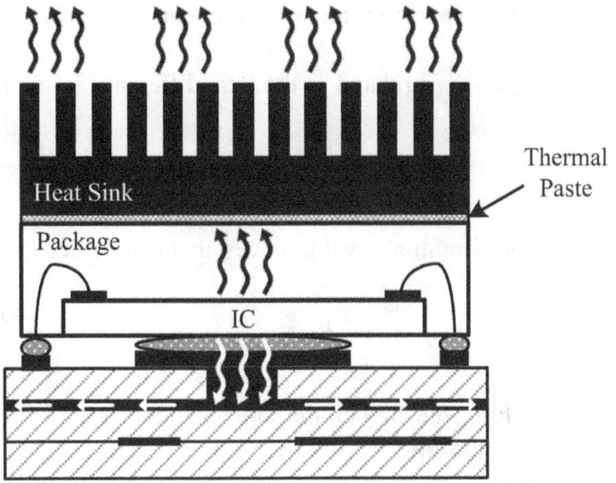

Table 5.2 Parameters used in thermal management calculations

Parameter	Explanation
T_{JMAX}	Maximum allowable junction temperature
T_J	Junction (IC) temperature
θ_{JC}	Thermal resistance between junction and package
T_C	Package (case) temperature
θ_{CS}	Thermal resistance between package and heat sink
T_S	Heat sink temperature
θ_{SA}	Thermal resistance between heat sink and the surrounding environment
T_A	Surrounding environment (ambient) temperature

With the IC having the highest temperature in the setup, the more efficiently the heat is transferred, the lower is the temperature difference between the adjoining parts, and the higher is the power dissipation the IC can handle, which translates to higher performance.

As an example of such an analysis, let us assume for simplicity that all the heat from the IC in Fig. 5.17 flows upwards. In this case, the power dissipated ($P_{DISSIPATED}$) is transformed into heat which will flow through the package, the heat sink, then spreads in the environment. The critical temperature is that of the junctions inside the transistors, here called T_J, which needs to stay below a critical value called T_{JMAX}. The temperature drops as the heat flows into the package whose temperature is T_C, where 'C' stands for case. It drops further as the heat flows into the heat sink whose temperature is T_S. It drops even further when it spreads in the environment whose temperature is T_A, where 'A' stands for ambient. As the heat crosses the boundary between two media, it encounters some resistance called the thermal resistance measured in degrees Celsius per Watt ($°C/W$). Therefore we have the thermal resistance θ_{JC} between the junction and a specific point on the package, known as the case, θ_{CS} between the case and the heat sink, and θ_{SA} between the heat sink and the surrounding environment. Table 5.2 presents a summary of these parameters.

$P_{DISSIPATED}$ is related to the total difference in temperature as follows

$$P_{DISSIPATED} = \frac{T_J - T_A}{\theta_{JC} + \theta_{CS} + \theta_{SA}} \tag{5.2}$$

The maximum allowable power that can be dissipated is $P_{DISSIPATEDMAX}$, where

$$P_{DISSIPATEDMAX} = \frac{T_{JMAX} - T_A}{\theta_{JC} + \theta_{CS} + \theta_{SA}} \tag{5.3}$$

Therefore, the higher the ambient temperature or the higher the total thermal resistance, the lower is the allowable $P_{DISSIPATEDMAX}$. We say that the dissipated power has to be de-rated in case of a high ambient temperature and the de-rating factor is the total thermal resistance.

For optimal performance, the primary focus is to reduce this total thermal resistance so that we do not need to heavily de-rate the dissipated power, thus keeping the IC operational at its maximum capabilities, which translates into high clock speeds for digital applications for example.

5.8 Past, Present, and Future Trends

Cramming more components onto ICs has been the norm since more than 60 years. This has been done by reducing the feature size of every newer technology, better known as scaling, as well as increasing the size of the IC itself. This was applied first to digital applications, then to RF/analog/mixed-signal ones, with digital applications remaining the main target.

The main driver behind this trend is reduced cost per device such as transistors. However, there is a point when, if too many devices are integrated on a single IC, the yield will suffer, thus cancelling the benefit of the cost reduction. According to Moore's Law, which is actually an industrial trend not a physical law, the number of devices per IC is expected to double every year. That was based on the industrial trend until the year 1965 and extrapolated until the year 1975.

Another driver behind this trend is that by reducing the feature size, the power needed to drive all the unwanted, that is parasitic, capacitances such as those in the metal connections is also decreased. Add to that the effect of higher speed when these parasitics are reduced and there is a win-win situation.

All of this applies nicely to digital applications, which is what Moore's Law primarily targets. However, at any point in the technology development, the yield of digital ICs is much higher than that of RF/analog/mixed-signal ICs.

As a result, and in order to optimize the implementation, the total cost has to be minimized. This might result in either integrating all the circuitry in one IC, or

Fig. 5.18 Stacked ICs in a single package

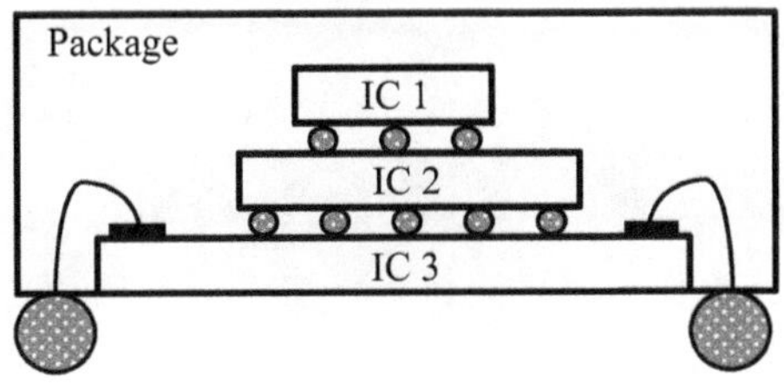

different ICs, but in one package, or in different packages altogether. As for the physical space that these ICs might occupy in a single package, stacking ICs on top of each other has become the norm where each IC might be using a different technology as in Fig. 5.18. Of course, in such a situation great care should be taken regarding the SHE, since the IC 2, for example, does not have a clear path to release the heat.

Additionally, in order to keep the yield high, new techniques and transistor structures have been devised such as SOI, FinFETs, and GAAFETs that allow better control over the yield.

Chapter 6
IC Design Flow

A custom IC design flow is presented in this chapter with all the major steps that need to be taken. This includes the theoretical analysis, schematic entry, simulations, layout, and post-layout simulations. This design flow is meant to be applicable to any IC design suite since it targets the fundamentals without going into vendor-specific features.

6.1 Design Flow

The design of integrated circuits has been made both more complicated and more accelerated during the last few decades. The complications stem from the complexity in modeling the behavior of electronic devices, including the secondary effects, individually and collectively, in addition to the complexity of the features needed in order to meet the marketing requirements. On the other hand, acceleration has been possible by the advent of initially simple, then advanced, Electronic Design Automation (EDA) tools that automate some tasks, make others faster, and more importantly allow the designer to use more accurate although complex device models in order to better-predict the actual performance of the circuit.

Despite all the complications, the custom IC design flow, although not unique, can be summarized as seen in Fig. 6.1.

The flow can be split into front-end tasks and back-end tasks. The white and gray blocks constitute the front-end tasks, whereas the black and gray ones constitute the back-end tasks. The gray blocks, being common for the two, consist of the simulation steps in general.

The number sequence is the suggested full IC design flow. The dashed arrows show the allowed flow as per current EDA tools. The solid lines show the inputs required for every stage. The following is a brief description of every step:

© The Editor(s) (if applicable) and The Author(s), under exclusive license to
Springer Nature Switzerland AG 2024
J. G. Atallah, M. Ismail, *Integrated Electronic Circuits*,
https://doi.org/10.1007/978-3-031-62707-1_6

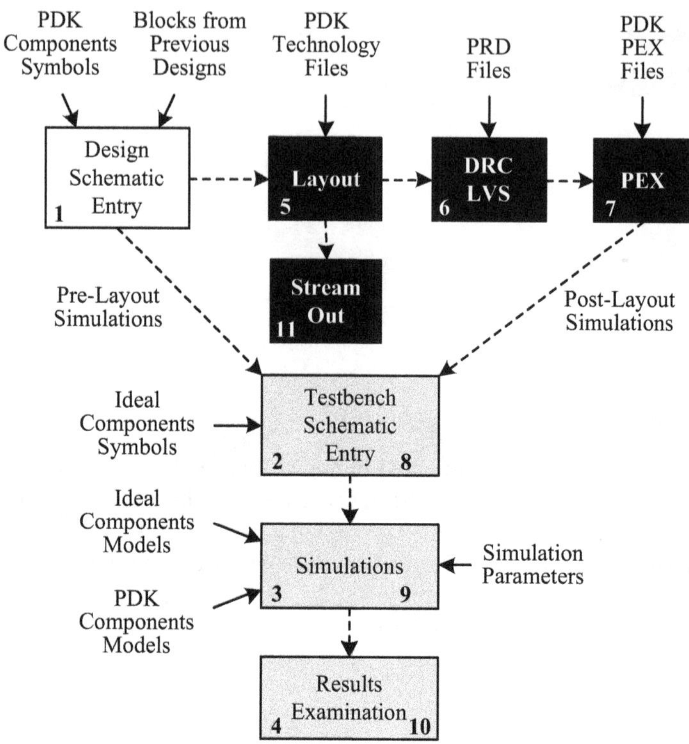

Fig. 6.1 IC design flow overview

1. Circuit Schematic Entry: The designer draws the schematic of the circuit using
   the schematic symbols provided by the foundry, possibly in addition to any
   previous circuit block that the designer has. The designer then creates a symbol,
   namely a black-box with interface ports, for this circuit.
2. Testbench Schematic Entry: The designer creates a testbench, which is simply
   another schematic, where the previously created symbol is used as a Design
   Under Test (DUT). An example of such a testbench is shown in Fig. 6.2. This
   testbench schematic requires modeling the source(s) and load(s) using ideal
   components such as VSUPPLY, vSRC, RSRC, and RLD in Fig. 6.2 at the
   schematic or behavioral level.
3. Simulations: The testbench subsequently undergoes a series of simulation
   analyses. These simulation analyses use the supplied models for the components, but the connections between them are treated as ideal. These are known as
   pre-layout simulations. Internally, the simulator does not deal with the schematic directly. Instead, it converts it into a netlist, which is a textual description
   of the circuit that the user can read if needed. The netlist format depends on
   the simulator, but, in general, it contains the name of each device in the circuit,
   the nodes to which the device is attached, the parameters of the device, etc. The

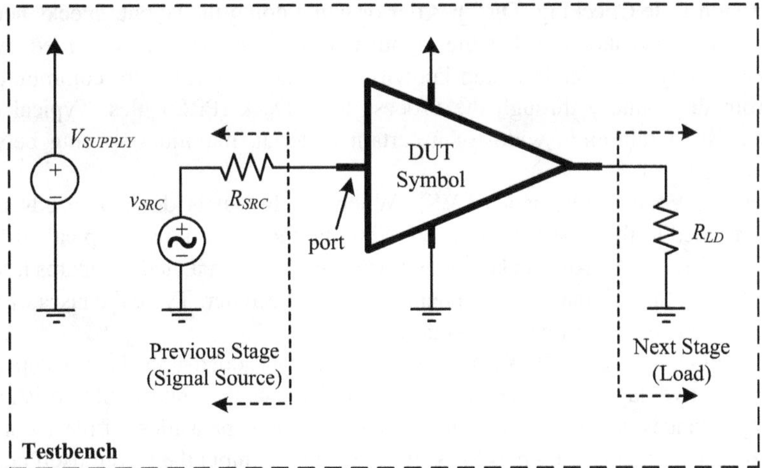

Fig. 6.2 Testbench example

simulations also require parameters such as the analysis type, duration, and outputs. Some of the most important simulation analysis types are as follows:

- DC Analysis: This analysis finds the DC operating point of the DUT given that the input is a single-point value. The result for every node is a single value.
- Transient Analysis: This is a time-domain analysis. It computes the response of the DUT over an interval of time. The result for every node is a wave versus time.
- AC Analysis: This analysis first linearizes the circuit about a DC operating point. Then it computes the response of the DUT to a sinusoidal stimulus while sweeping a parameter such as frequency, in which case the result of every node is a wave versus frequency.

4. Results Examination: Basically, the results consist of table of values that are commonly plotted and analyzed.
5. Layout: Layout is the first back-end task. It consists of transforming the DUT from a schematic to a collection of geometries that describe the constituencies of every component in every layer of the IC. The components are also connected using the conductive layers. As a result, for the first time in the design flow, the connections between the components are not ideal anymore, but consist of actual conductive paths. The layout is done looking at the IC from the top with each layer consisting of a certain color and hash pattern depending on the Process Design Kit (PDK) technology files. All the components in the DUT are from the PDK, so they all have pre-built layouts that are parameterized. Therefore, in most situations, the designer only needs to decide where to place the components (floorplan) and how to route them.

6. Design Rule Checking (DRC): After or while doing the layout, checks need to be done to make sure that the layout conforms to the rules provided by the technology provider in a step known as DRC. These rules are communicated from the foundry through the Process Rule Deck (PRD) files. Typical rules consist of minimum width of a certain material, minimum spacing between materials, etc.

7. Layout Versus Schematic (LVS): When the layout is done, it needs to be compared to the schematic to make sure that they match, in a step called LVS. Therefore, LVS takes the layout, extracts a netlist from it, and compares it to the netlist of the original circuit schematic of the designer. Typical causes of LVS failure include improper connection of devices.

8. Parasitic EXtraction (PEX): With the connections added at the layout step using conductive layers, we can now extract the usually undesirable resistances, capacitances, and even inductances, also known as parasitics, of these connections. This step is known as PEX. It requires as an input the layout itself and the PDK PEX files that describe these parasitics and the parasitics of the substrate on-top of which the design sits, now that we know the exact position of every component inside the IC. The output of this step consists of another netlist and optionally its schematic that includes the original one in addition to all the parasitics. This is known as the extracted netlist or the extracted schematic view.

9. Testbench Schematic Entry: The extracted schematic view is now a more complete model of the DUT, called DUTextracted.

10. Simulations: The DUTextracted is tested in this step to make sure that the performance is still within the specifications even in the presence of the parasitics.

11. Results Analysis: The DUTextracted results analysis in this step constitutes the closest view the designer can have to the proper measurement results before actually sending the IC for fabrication.

12. Stream Out: The designer streams the layout into an industry-standard file format such as GDSII to be sent to the foundry for fabrication.

13. In practice, the steps above are iterated in order to reach the target results. The important overall aim is to do so within the minimum amount of time. Therefore, any iteration that encompasses several steps is not desired, which means the sooner a problem is uncovered, the better.

6.2   Transistor Models

Several models exist for each of the devices in the IC. Here we will concentrate on the transistor models owing to the primary importance of particularly MOSFET transistors in current ICs. Models are a set of equations used to capture the actual behavior of the device. Therefore, the closer the equation results are to actual measurements, the more accurate is the model.

Transistor models range from simple ones that are used for hand calculations to complex ones that are used for simulations. Transistor models are also sometimes known as Simulation Program with Integrated Circuit Emphasis (SPICE) models owing to the first standardized set of models used in SPICE simulators. This naming is a carry-over from the past and is not very accurate since several of the models currently in use are not related to the original SPICE models. Transistor models are charted by the Compact Model Coalition (CMC) working group, which is affiliated with the Silicon Integration Initiative (Si2), a consortium of companies involved in the IC ecosystem.

MOSFET models are usually arranged in levels where the numbering depends on the EDA vendor, especially for the higher levels. Level 0 and level 1 models are the ones used for hand calculations and are rarely used in IC simulations. As the level increases, the models start to include more effects in order to give more accurate results. Also, some of these models are designed for specific FET architectures such as FinFET and SOI. Transistor models are developed by research groups or individuals whose names they carry. For example, the level 1 model is called the Schichman-Hodges model, with the level 0 model being its simplified version. Higher-level models include Berkeley Short-channel IGFET Model (BSIM) which is quite popular, Enz, Krummenacher, Vittoz (EKV), Hiroshima-university STARC IGFET Model (HiSIM), and Philips.

As for BJTs, several models exist for these too such as Ebers-Moll and Berkeley-Spice Gummel-Poon.

Lower-level models are not accurate for IC design. They are used for hand calculations in order to provide the following:

- An insight into the dynamics of the circuit: This helps the designer to understand how a certain circuit works and to come up with ideas for new circuits. Also, in case the simulation results do not meet the performance specifications, this insight into the dynamics will help the designer to know which component parameter value to change and in which direction (increase or decrease). All of this can be achieved by doing a symbolic analysis of the circuit using these models, with no numbers included.
- The approximate numerical values of the circuit component parameters: With these numerical values, the designer can start the simulations and, based on the simulation results, can refine the values as well as the circuit in order to achieve the desired performance.

Complex models are more accurate and are used for simulations using EDA tools. However, the more complex the model is, the more time and possibly resources the simulator will need to finish the simulation. There are always limits no matter what machines are used for the simulations. Therefore, if the design is large, it is customary to model one part of it at a time using these detailed circuit-level models, while the other parts can be modeled using behavioral hardware descriptive languages such as VHDL (IEEE standard 1076) or Verilog (IEEE standard 1800) or any of their dialects that provide approximate results compared to circuit-level models, but can be simulated quickly.

6.3   Corner and Statistical Simulations

The IC fabrication process involves variability resulting in device parameters that are different from the typical ones. These are called process variations. The process variations result in faster or slower than normal MOSFET transistors. Therefore, if we assign the first letter to the NMOS and the second letter to the PMOS, we can say that we might have in any given IC the following combinations: TT, SS, SF, FS, and FF, where T stands for typical, S stands for slow, and F stands for Fast.

In addition to this, while the circuit is operational, changes can take place to its supply voltage and temperature. Therefore, the operational voltage has a range and the operational temperature has a range. The voltage range depends on the power supply implementation and the temperature range depends on the product's market. For example, consumer products are meant to operate within a narrower temperature range than industrial products.

These three variations are combined in what is called Process-Voltage-Temperature (PVT) variations that have to be considered by the designer to ensure the proper operation of the IC in realistic situations.

Each combination of process variation, voltage value, and temperature value is called a corner. The designer decides on the specific combinations which will constitute the corners to be simulated. In today's ICs, the results from all the corners have to meet the specifications, even though these corners vary in their probability of occurrence. For example, it might be that a specific corner has a 50% probability to occur in reality while another corner has a 5% probability. By making sure that all the corner simulation results meet the specifications, we would be overdesigning the IC. Alternatively, the designer might include only those corners with high probability of occurrence, in which case the design effort would be more reasonable.

Another way to deal with the process variation part of the PVT is to let the tools sample the combination of transistors on its own based on a certain distribution provided by the fab. This is called a Monte Carlo simulation whereby the tool selects the combinations of transistors with high probability of occurrence more often than the ones with low probability of occurrence.

6.4   Implementation Considerations

The back-end tasks, particularly the layout, play a crucial role in attaining a quality IC implementation. The optimizations at this implementation stage are all about parasitics, process variations, heat, and area. We will present some of the most important techniques used in order to address these aspects.

6.4.1 *Dealing with Connection Parasitics*

Connection parasitics are of primary concern when implementing an IC especially in the presence of signals with high-frequency components such as in RF and high-speed digital applications. They are also of utmost concern when implementing power supply and ground connections.

A connection can be modeled in any of the following ways:

- A single resistor (R).
- A single resistor and a single capacitor (lumped RC).
- A string of resistors and capacitors (distributed RC).
- A string of resistors, inductors and capacitors (distributed RLC).

The resistance in the connection is due to the nonzero resistivity of the material, and is a function of the connection width. As a result the power supply and ground connections, which handle large currents, are usually made wide in order to reduce the IR drop across them. Alternatively, more than one connection in parallel can be used to achieve the same effect. Additionally, vias are also arrayed in order to reduce the IR drop across them. The top-most metal layers in an IC are thicker than the lower ones. For this reason, global connections make use of these top-most layers and local connections make use of the lower-most layers.

Another undesirable effect due to this resistance is that the IR drop across the connection will create self-heating, thus increasing the temperature of the IC. This will accelerate the degradation of the IC, thus reducing its reliability.

Every connection can handle a maximum average current density denoted as J_{max}. The average current density J, measured in A/m^2, is given as

$$J = \frac{I}{S} \tag{6.1}$$

where I is the current and S is the cross-sectional area of the connection. If too much current is pushed into the connection, exceeding J_{max}, the material in the connection will start to migrate and the connection will fail after some time. This is called electromigration, and is a serious effect in the long run.

The capacitance of the connection is due to its proximity to the substrate, which is grounded, resulting in capacitances to ground, and to its proximity to other connections, resulting in coupling capacitances between them.

Each one of the models above has its benefits and drawbacks. The simpler lumped models are fast to simulate but do not capture the behavior accurately, whereas the distributed models are slower to simulate, but are more accurate. These models are automatically extracted by the EDA tool in the parasitic extraction step 7 of Fig. 6.1.

In practice, a good compromise is the distributed RC model since it provides a good balance between speed and accuracy. For example, a cross-section view of a 4-section distributed model of a connection on top of the substrate is shown in Fig. 6.3.

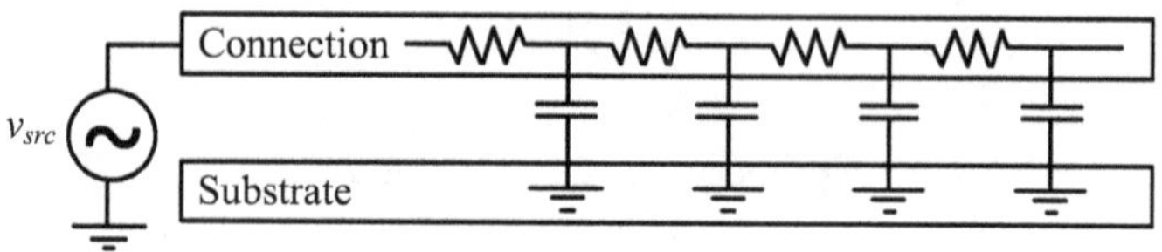

Fig. 6.3 Distributed RC model of a connection on top of the substrate

Fig. 6.4 One element from a distributed model

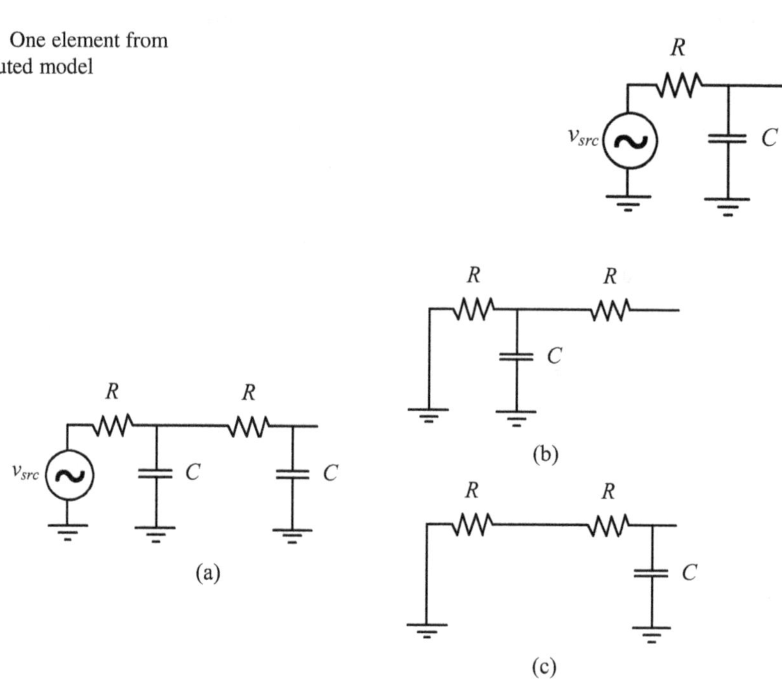

Fig. 6.5 Two elements from a distributed model

How do we calculate the bandwidth of a connection modeled using the distributed RC model? We will analyze this using mathematical induction. Let us take one element from an N-element distributed model as shown in Fig. 6.4.

This element has a time constant equal to RC, and consequently a 3-dB bandwidth given as

$$f_{3dB\_element} = \frac{1}{2\pi RC} \tag{6.2}$$

Now let us take two elements from the model as shown in Fig. 6.5a.

Using the open-circuit time constant method, the time constant related to the first capacitor can be calculated by disabling the voltage source and replacing the second capacitor with an open circuit as in Fig. 6.5b. Therefore, the first time constant is

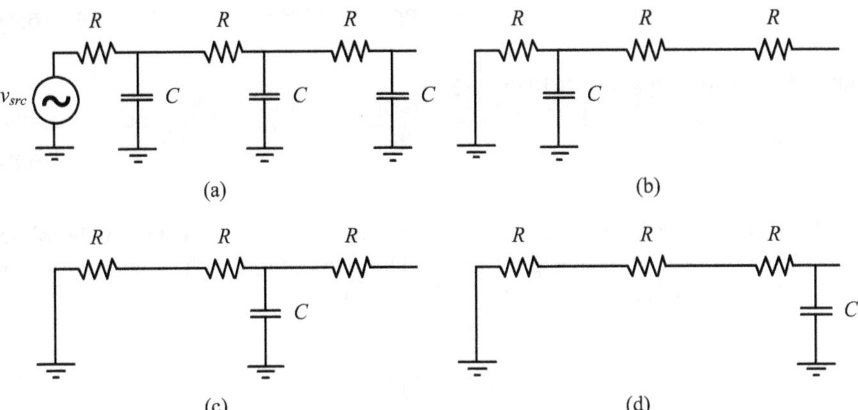

Fig. 6.6 Three elements from a distributed model

$$\tau_1 = RC \tag{6.3}$$

and the corresponding corner frequency is

$$f_1 = \frac{1}{2\pi RC} \tag{6.4}$$

The time constant related to the second capacitor can be calculated by disabling the voltage source and replacing the first capacitor with an open circuit as in Fig. 6.5c. Therefore, the second time constant is

$$f_2 = \frac{1}{2\pi(2RC)} \tag{6.5}$$

Combining f_1 and f_2, we can get a good estimate of the 3-dB corner frequency of the model in Fig. 6.5a:

$$\frac{1}{f_{3dB\_2}} \approx \frac{1}{f_1} + \frac{1}{f_2} \Rightarrow f_{3dB\_2} \approx \frac{1}{\dfrac{1}{f_1} + \dfrac{1}{f_2}} \tag{6.6}$$

$$\Rightarrow f_{3dB\_2} \approx \frac{1}{2\pi RC + 2\pi(2RC)}$$

Now let us take three elements from the model as shown in Fig. 6.6a.

Using the open-circuit time constant method, the time constant related to the first capacitor can be calculated by disabling the voltage source and replacing the second and third capacitors with an open circuit as in Fig. 6.6b. Therefore, the first time constant is

$$\tau_1 = RC \tag{6.7}$$

and the corresponding corner frequency is

$$f_1 = \frac{1}{2\pi RC} \tag{6.8}$$

The time constant related to the second capacitor can be calculated by disabling the voltage source and replacing the first and third capacitors with an open circuit as in Fig. 6.6c. Therefore, the second time constant is

$$f_2 = \frac{1}{2\pi(2RC)} \tag{6.9}$$

The time constant related to the third capacitor can be calculated by disabling the voltage source and replacing the first and second capacitors with an open circuit as in Figure 6.6d. Therefore, the third time constant is

$$f_3 = \frac{1}{2\pi(3RC)} \tag{6.10}$$

Combining f_1, f_2, and f_3, we can get a good estimate of the 3-dB corner frequency of the model in Fig. 6.6a:

$$\frac{1}{f_{3dB\_3}} \approx \frac{1}{f_1} + \frac{1}{f_2} + \frac{1}{f_3} \Rightarrow f_{3dB\_3} \approx \frac{1}{\frac{1}{f_1} + \frac{1}{f_2} + \frac{1}{f_2}}$$

$$\Rightarrow f_{3dB\_3} \approx \frac{1}{2\pi RC + 2\pi(2RC) + 2\pi(3RC)} \tag{6.11}$$

Comparing (6.2), (6.6), and (6.11), we can see the following pattern:

$$f_{3dB\_N} \approx \frac{1}{2\pi RC \sum_{1}^{N} x} = f_{3dB\_element} \left(\frac{1}{\sum_{1}^{N} x} \right) = f_{3dB\_element} \left(\frac{2}{N(N+1)} \right) \tag{6.12}$$

The calculated equation above corresponds very well with the simulation results as can be seen in Fig. 6.7.

What we observe is that if we take a connection and divide it into squares, where each square is modeled using a lumped model as in Fig. 6.8, then quadrupling the number of squares, which corresponds to quadrupling the length of the connection, leads to a ten-times reduction in the 3-dB bandwidth according to Fig. 6.7. This is

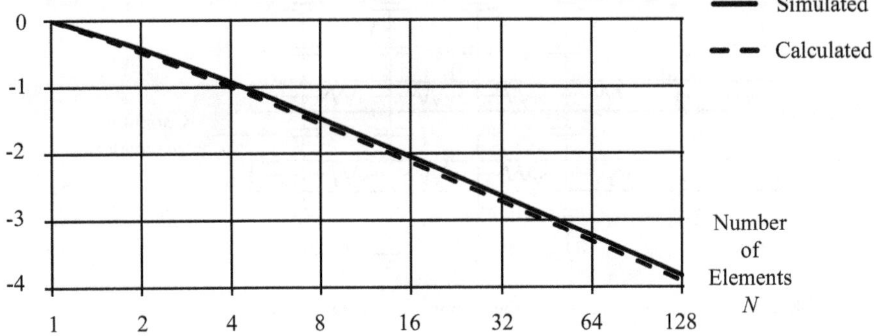

Fig. 6.7 Simulated and calculated normalized 3-dB bandwidth versus number of elements

Fig. 6.8 Connection
divided into N squares

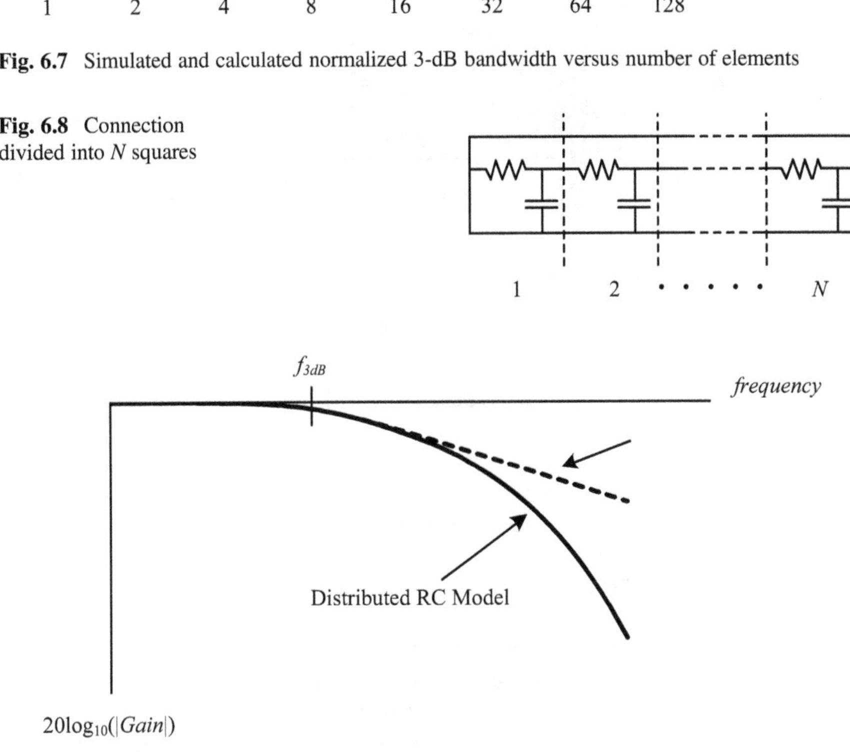

Fig. 6.9 Lumped RC model compared to distributed RC model

significant and should be taken into account when routing lines carrying signals with high-frequency components.

Another question to answer is what exactly is the difference between modeling a connection using a lumped RC model and using a distributed RC model? If we take a connection and build these two models for it, the results will be as in Fig. 6.9.

The *Gain* in this case is the voltage at the output of the connection divided by the voltage at its input, which is naturally less than unity. As can be seen, the correspondence can be very good at low frequencies within the 3-dB bandwidth, but at

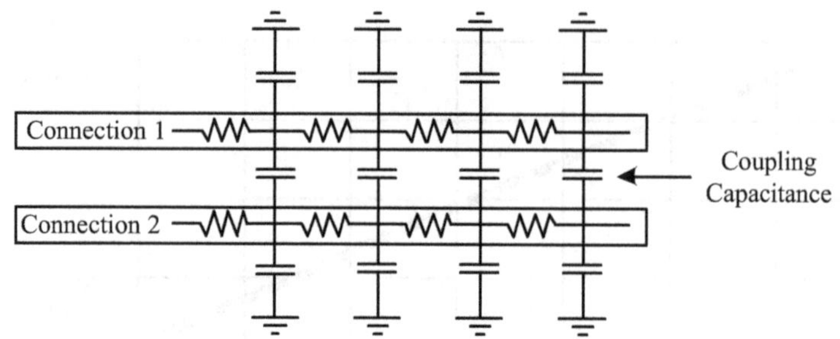

Fig. 6.10 Coupling capacitance due to the proximity of connections

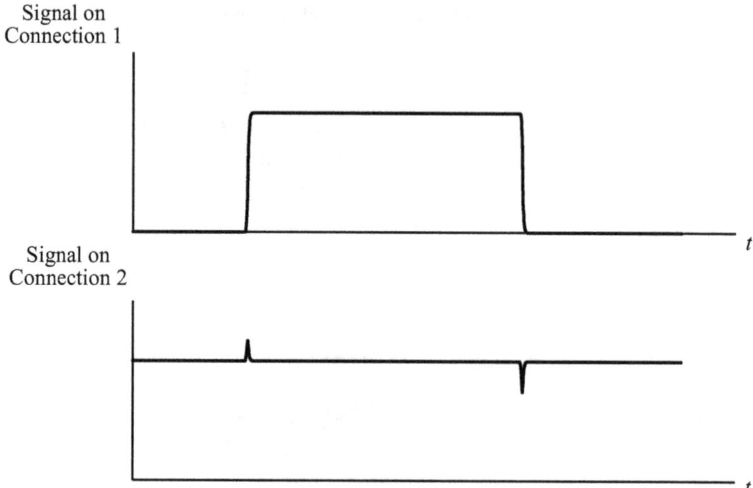

Fig. 6.11 Effect of coupling capacitance between connection 1 and connection 2

high frequencies, the results diverge significantly, with the distributed RC model being the more accurate one. This is why the distributed RC model should be used whenever signals with high-frequency components are involved.

Another example of connection parasitics consists of a connection in proximity to another one. This creates coupling capacitance in addition to the grounded capacitance as shown in the top view of the model in Fig. 6.10.

This coupling capacitance is a function of the connections' lengths and their proximity. If the connections are too close, they will be subject to similar noise contamination, which is desirable when routing differential signals. However, if the connections carry different data such as in a bus, then the coupling capacitance will allow a signal in one connection to affect the other.

For example, if the signal in connection 1 changes and that of connection 2 is supposed to stay the same, the result due to the coupling capacitance will be as shown in Fig. 6.11. The glitches on connection 2 are called interference due to the instantaneous high frequencies on connection 1 finding their way through the coupling capacitance. One way to mitigate this is to distance the connections from each other, thus reducing the coupling capacitance. However, if the bus carries several connections, this strategy will take too much space. Therefore, a better strategy is to reduce the number of connections in a bus and have the connections run at higher speeds, thus to replace parallel connections with high-speed serial connections.

6.4.2 Splitting a Transistor into Fingers

A MOSFET transistor is fundamentally symmetrical with its width typically much greater than its length. This results in an awkward layout shape as can be seen for the NMOS in Fig. 6.12a. The small white squares are vias. The difficulty with this transistor is to find an optimum place to connect the terminals. For example, if we connect the drain from the middle, and due to the large width of the channel, the extremities of the drain might not be at the same voltage potential. Additionally, process variations will result in variability between the extreme points of this transistor.

A better way to implement a wide transistor is shown in Fig. 6.12b. Here, the transistor is split into N fingers (three in this case) in such a way that the aspect ratio

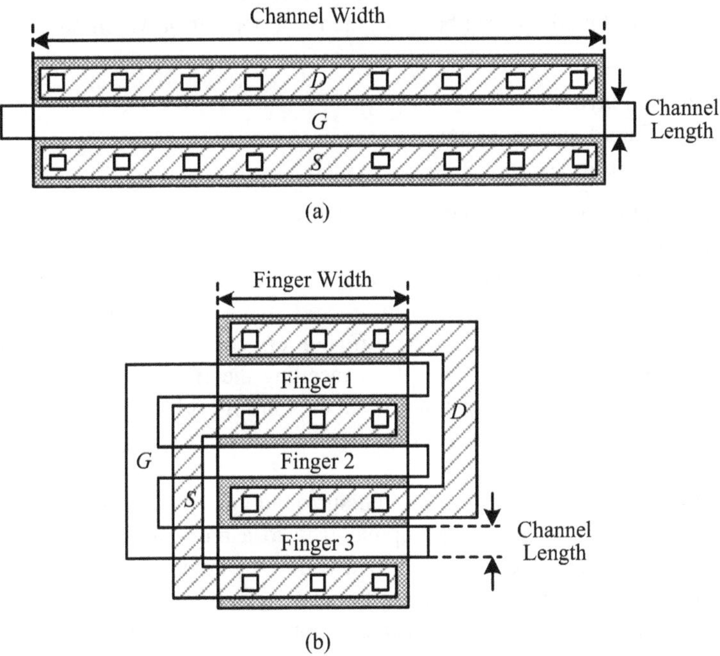

Fig. 6.12 Wide transistor layout using one finger (a) and using three fingers (b)

of the whole transistor looks like that of a square. The resulting channel length is the same as that in Fig. 6.12a and so is the transistor width, which can be calculated as follows:

$$\text{Channel Width} = N \times \text{Finger Width} \tag{6.13}$$

6.4.3 Matching Devices

Many circuits require matching between the components. For example, a differential amplifier requires the two branches to be matched. Matching is also needed in current mirrors. Also, data converters require matched components such as transistors and capacitors. This is a challenge with today's technologies where process variations are becoming more prominent in conjunction to layout-dependent effects. It is the extra effort that is put in the layout that ensures that components are as close in performance to each other as possible. As a result, perfect matching is not possible, but the more effort is exerted in the layout, the better matching we can get. Without loss of generalization, we will discuss the schemes used for matching transistors. Similar schemes can be used for other devices.

Let us take two transistors Q_1 and Q_2 that we need to match. The first step is to divide the transistors into fingers. In this example, we have four fingers per transistor. Several layout methods can be used, the most popular of which are as follows:

- Symmetry layout.
- Interdigitated layout.
- Symmetry interdigitated layout.
- Common centroid layout.

These layout methods, without the detailed connections for visibility reasons, are shown in Fig. 6.13.

Symmetry layout in Fig.e 6.13a requires the least amount of effort but performs the worst compared to the others. Interdigitated layout in Fig. 6.13b works best in order to minimize the length of oxide diffusion effects. As for symmetry interdigitated layout in Fig. 6.13c, it is good for cancelling the effect of the linear gradient differences. As for the common centroid layouts in Fig. 6.13d, e, they work best if the number of fingers or the number of components to be matched is large such as in data converter applications.

In addition to the schemes above, the surroundings of the matched transistors should be the same. Therefore, a popular way to achieve this is to add dummy devices around the matched transistors. These dummy devices should be connected to the appropriate voltages in order not to create forward-biased diodes leading to latchup. An example of dummy devices is shown in Fig. 6.14. The dummy devices are in white.

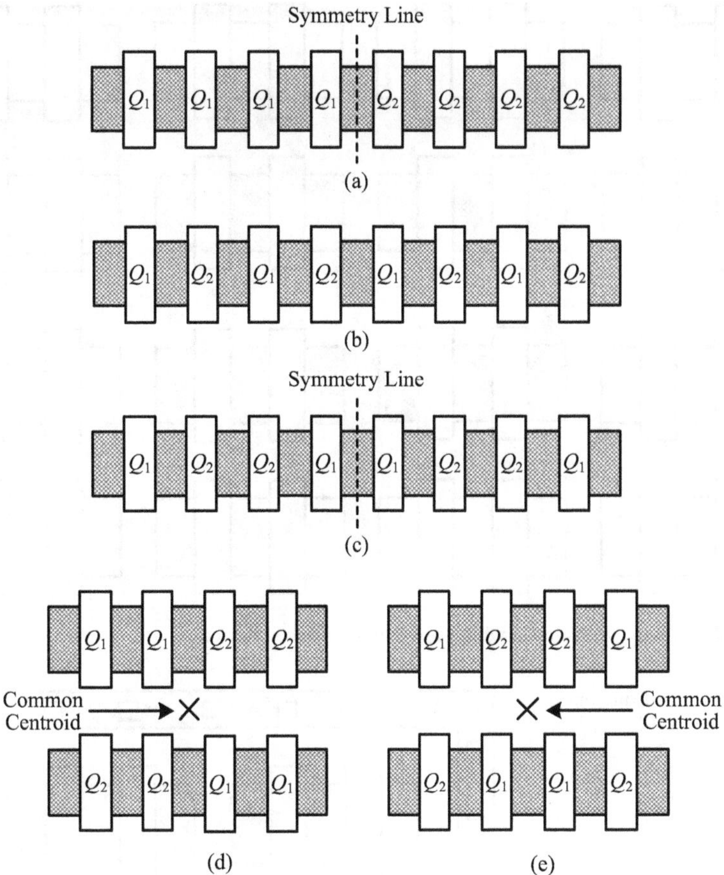

Fig. 6.13 Layout methods for matching: symmtery layout (**a**), interdigitated layout (**b**), symmetry interdigitated layout (**c**), common centroid layout variation 1 (**d**), and common centroid layout variation 2 (**e**)

6.4.4 Shielding and Guard Rings

Sensitive blocks in the layout such as RF and analog blocks dealing with small signals have to be shielded from noisy blocks such as digital blocks. The shielding is done by creating a guard ring around the sensitive blocks and by having separate supply and ground connections from those of the noisy blocks.

A guard ring consists of a stack of metal layers connected using vias. The ring is typically connected to ground and can be used simultaneously as a substrate contact for NMOS transistors. The guard ring can surround either one device or a group of devices. An example of a guard ring made of stacked metal layers surrounding four devices is shown in Fig. 6.15.

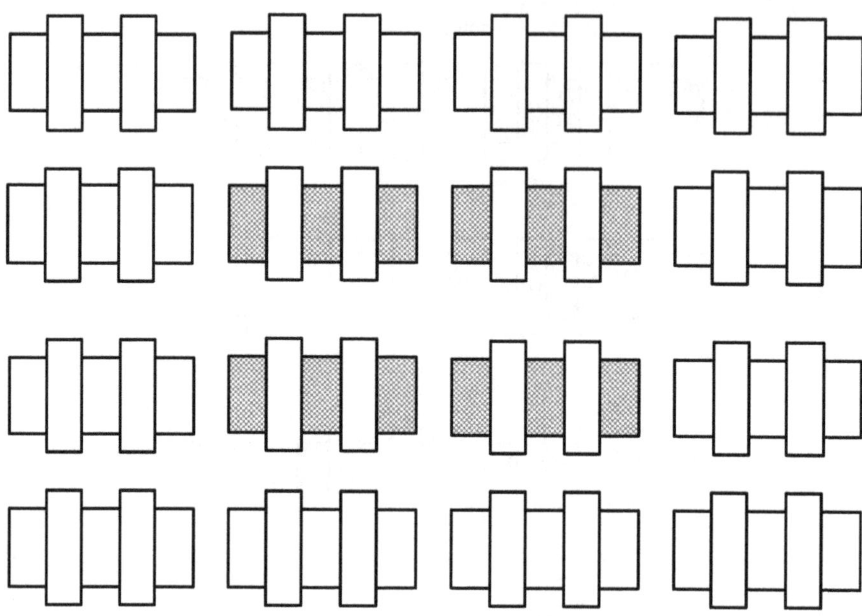

Fig. 6.14 Dummy structures in white surrounding matched devices

Fig. 6.15 Guard ring made
of stacked metal layers

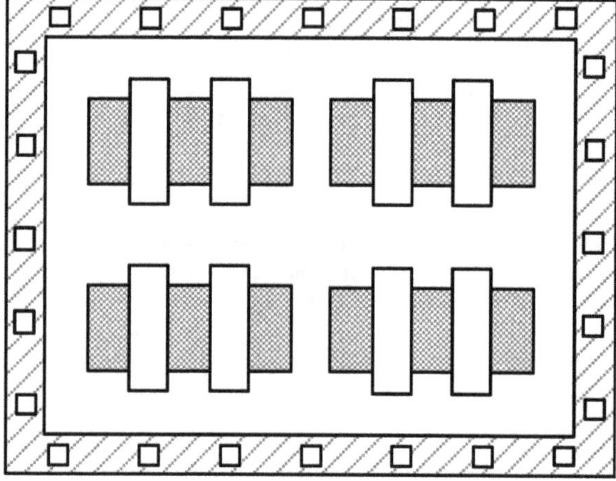

Chapter 7
Single-Ended Amplifiers

In this chapter, we will describe in details the different amplifier configurations that we can build using transistors. These amplifiers are single-ended, which means they have one input terminal and one output terminal, of course in addition to a power terminal and a ground terminal. Without any loss of generality, we will describe these applications using MOSFETs. These configurations can be implemented using any other technology including BJTs.

7.1 Introduction

Given that a MOSFET transistor acts as a voltage-controlled current source, the input is a voltage and the output is a current. This current can then be converted into a voltage by passing it through a resistor. The gain is then defined as the relation between the output voltage and the input voltage. The procedure is conceptually shown in Fig. 4.31 (Fig. 7.1).

The MOSFET transistor has only one current as an output candidate. However, it has several candidates for the input voltage. These candidates are shown in Table 7.1.

Since there cannot be a current at the Gate at medium to low frequencies, then the two configurations CS2 and CD2 are not valid. Another reason for this is that even if the input is at the other terminals, this will not result in a voltage at the Gate since the MOSFET is unidirectional.

As for configuration CG1, it does not result in any useful applications since the small-signal voltage gain itself is very low, while the current gain is unity.

As a result, we are left with three configurations that we are going to analyze and figure out which ones are useful in which situations. These are:

© The Editor(s) (if applicable) and The Author(s), under exclusive license to
Springer Nature Switzerland AG 2024

J. G. Atallah, M. Ismail, *Integrated Electronic Circuits*,
https://doi.org/10.1007/978-3-031-62707-1_7

Fig. 7.1 Using a transistor to build an amplifier

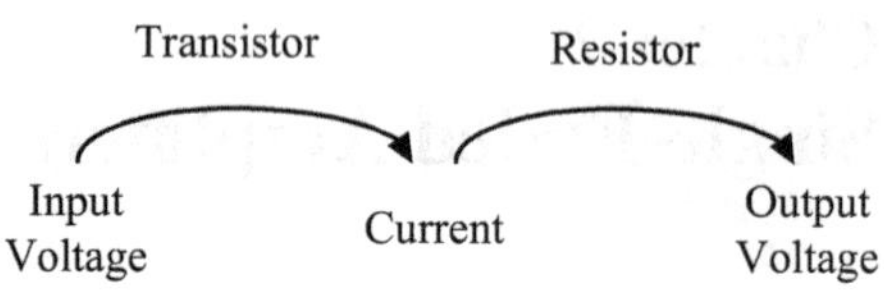

Table 7.1 All the transistor input and output configuration possibilities

Configuration name	Input	Output
Common-Source 1 (CS1)	Gate	Drain
Common-Drain 1 (CD1)	Gate	Source
Common-Source 2 (CS2)	Drain	Gate
Common-Gate 1 (CG1)	Drain	Source
Common-Drain 2 (CD2)	Source	Gate
Common-Gate 2 (CG2)	Source	Drain

Fig. 7.2 Analysis using superposition

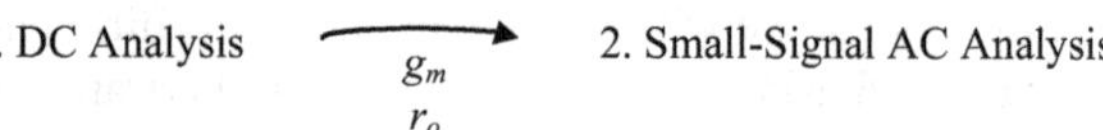

- Common-Source configuration (CS1).
- Common-Gate configuration (CG2).
- Common-Drain configuration (CD1).

Realizing that all the signals are the sum of a DC and AC components, and keeping the AC component small (thus small-signal), the analysis follows a two-step superposition approach as shown in Fig. 7.2:

1. The first step consists of determining the following:

   - Maximum output (v_{LD_MAX}).
   - Minimum output (v_{LD_MIN}).

2. The second step consists of determining the DC operating point. This is done by analyzing the behavior of the circuit with the DC inputs only. Out of this analysis, we can get the:

   - average power consumption (P_{av}),
   - the parameters needed (g_m and r_o) in order to do the small-signal analysis later.

3. The third step consists of doing the AC small-signal analysis by looking into the behavior of the circuit with the AC small-signal inputs only. In this analysis, we can gauge the following:

   - Input resistance (r_{in}).
   - Output resistance (r_{out}).
   - AC Small-signal voltage gain (v_{ld} / v_{src}).
   - Bandwidth (BW).

7.2   Common-Source (CS) Configuration

The common-source configuration is shown in Fig. 7.3. The circuits surrounding the transistor are modeled as follows:

- The previous stage (signal source) is modeled using its Thevenin equivalent model. This signal source contains a DC component and a small-signal AC component. Also, the source has an output resistance R_{SRC}.
- The next stage (load) is modeled using its input resistance, indicated as R_{LD}.
- The current source, which, in combination with the DC component of the signal source, creates the DC operating point around which the transistor operates, is modeled using its Norton model with an output resistance R_{CT}.

Typical v_{SRC} and v_{LD} waveforms of a common-source configuration are shown in Fig. 7.4.

Looking at the NMOS-based implementation in Fig. 7.3a, we can see that the maximum value the output voltage can take is governed by what is happening above the output node. Therefore:

$$v_{LD\_MAX} = V_{SUPPLY} - V_{CT\_REQ} \tag{7.1}$$

where V_{CT_REQ} is the smallest required voltage across the current source in order for it to operate properly.

The minimum value of the output voltage is governed by what is happening below the output node. Since R_{LD} is a passive component, it does not set a limit on the output voltage. However, the transistor needs to stay in the saturation region, so it requires some voltage to breathe. Therefore:

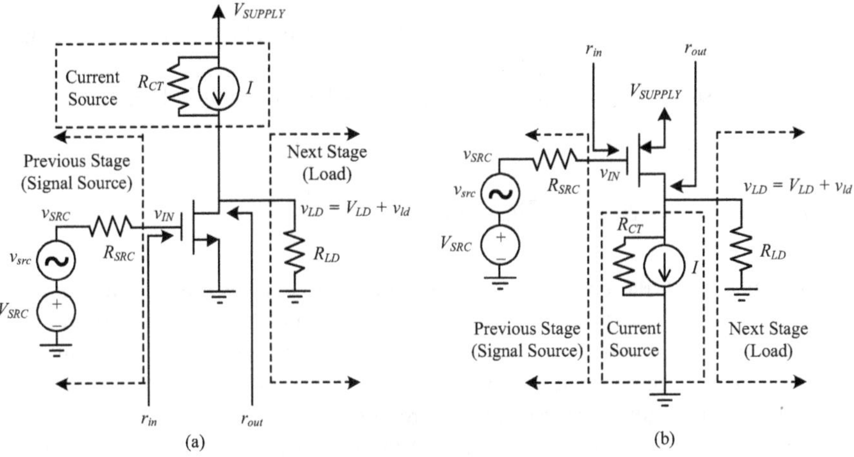

Fig. 7.3 Common-source configuration: (**a**) NMOS-based and (**b**) PMOS-based

Fig. 7.4 Typical (**a**) v_{SRC} and (**b**) v_{LD} waveforms of a common-source configuration

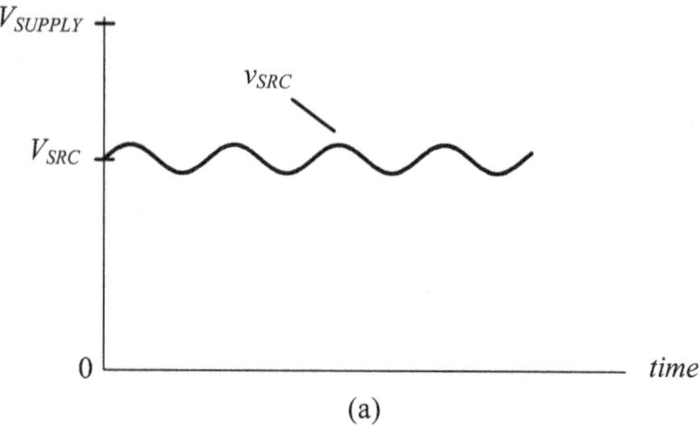

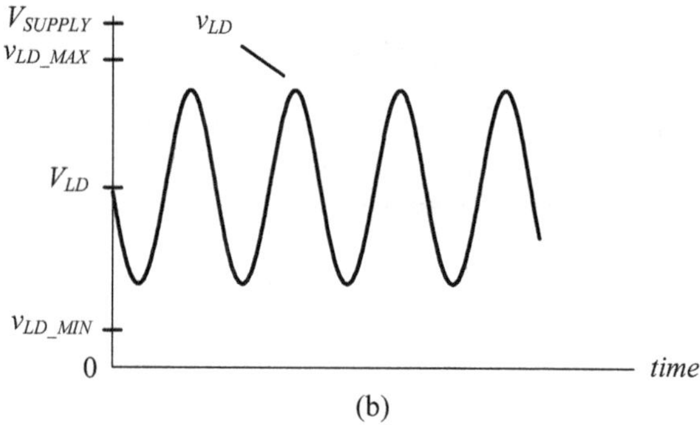

$$v_{LD\_MIN} = V_{OV\_REQ} \tag{7.2}$$

where V_{OV_REQ} is the minimum required overdrive voltage of the NMOS transistor. If v_{LD} goes below this value, the transistor will get out of the saturation region and into the triode region.

As a result, in order to get the maximum swing at the output, it is sensible to have the DC point at the output (V_{LD}) in the middle between v_{LD_MIN} and v_{LD_MAX}.

Similar reasoning can be applied to the PMOS-based circuit in Fig. 7.3b.

7.2.1 Common-Source DC Analysis

We will disable all the AC sources and keep the DC sources only. As a result, the small-signal voltage sources, if any, are replaced with a short circuit (zero voltage) and the small-signal current sources, if any, are replaced with an open circuit (zero current). The resulting circuit for DC analysis is shown in Fig. 7.5. This is what the DC components of all the signals in the circuit see.

There is no current through the gate of the transistor, so V_{SRC} and V_{IN} are the same, which means that R_{SRC} plays no role.

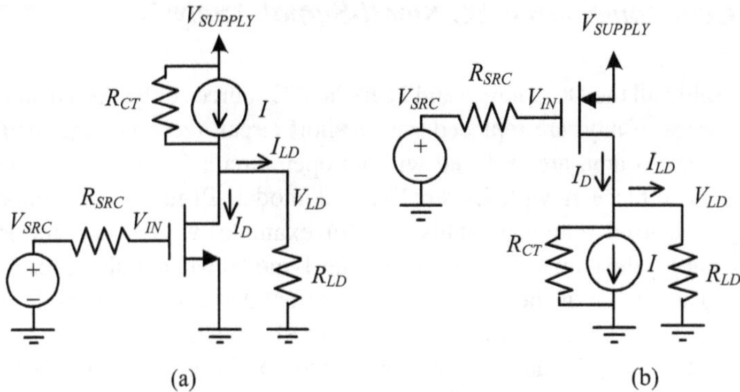

Fig. 7.5 DC model of the common-source configuration: (**a**) NMOS-based and (**b**) PMOS-based

Also, looking at the current source, usually this has a fairly large output resistance R_{CT}. Therefore, the current coming out of the current source is almost equal to I. Consequently, we can see that the average power consumption can be approximated as follows:

$$P_{av} \approx V_{SUPPLY} \times I \tag{7.3}$$

As a result, given that the power supply V_{SUPPLY} is fixed, the current I plays a crucial role in determining the power consumption of the circuit.

The current I is related to I_D and I_{LD}. It is in our interest to have most of the current I going into I_D in order to maximize g_m and thus the gain. Therefore, we usually aim for a high R_{LD}.

From a design point-of-view, we need to make sure that the V_{GS}, V_{DS}, and I_D with which we provide the transistor have to belong to the three-dimensional plot for that particular transistor. In other words, these values have to be compatible, and of course the transistor has to be in the saturation region following its large-signal model:

$$I_D = \frac{1}{2}k'\left(\frac{W}{L}\right)(|V_{GS}| - |V_{th}|)^2(1 + \lambda|V_{DS}|) \tag{7.4}$$

Now that I_D is known, we can get the small-signal AC model parameters as follows:

$$g_m = \frac{2I_D}{|V_{OV}|} \tag{7.5}$$

$$r_o \approx \frac{|V_A|}{|I_D|} \tag{7.6}$$

These parameters will be used next in the AC small-signal analysis.

7.2.2 Common-Source AC Small-Signal Analysis

We will disable all the DC sources and keep the AC sources only. As a result, the DC voltage sources, if any, are replaced with a short circuit (zero voltage) and the DC current sources, if any, are replaced with an open circuit (zero current). As for the transistor, we replace it with its small-signal model, PI-model or T-model. The models can be used interchangeably, so, for example, we can use the PI-model when looking for the gain of the circuit and the T-model when looking for the output resistance. The choice of the model does not affect the results, but might make the analysis a bit easier. In this case, we opt for the PI-model.

We will start with the analysis at low to medium frequencies. In this frequency range, the internal capacitances have very high impedances, so they will be replaced with open circuits. The resulting circuit for AC analysis is shown in Fig. 7.6. This is what the AC components of all the signals in the circuit see. Note that the small-signal model of the circuit is the same, whether it is NMOS-based or PMOS-based.

As can be seen in Fig. 7.6, the input of the circuit is an open circuit. As a result, the input resistance is:

$$r_{in} = \infty \tag{7.7}$$

This is ideal when the information is in the voltage, which is mandatory in this case.

In order to get the output resistance, we need to disable the input source, which is a voltage source in this case. Therefore, we will short-circuit it. Also, we will put our test source exactly where we would like to see the output resistance as shown in Fig. 7.7.

As a result, and since we have disabled the input voltage source,

$$v_{gs} = 0$$
$$r_{out} = \frac{v_{test}}{i_{test}} = r_o // R_{CT} \tag{7.8}$$

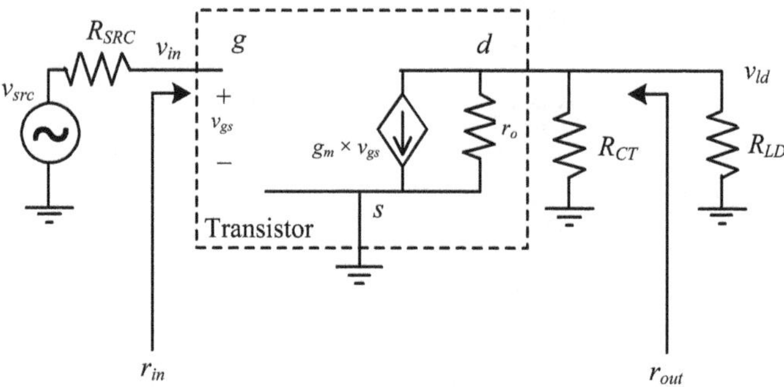

Fig. 7.6 AC small-signal model of the common-source configuration

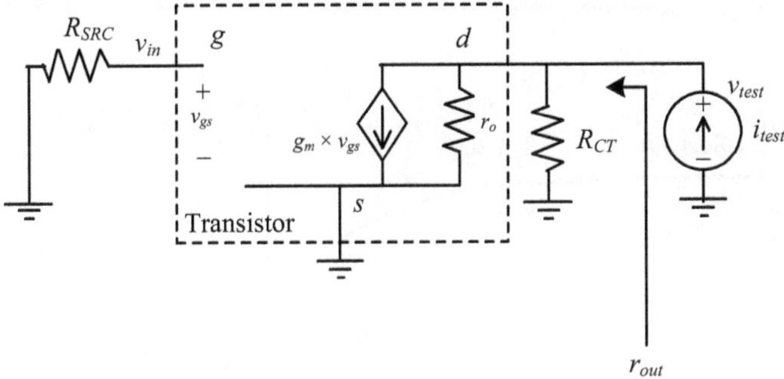

Fig. 7.7 AC small-signal model of the common-source configuration—output resistance

Next, we will get the voltage gain. Looking into Fig. 7.6, we see that

$$v_{ld} = -g_m \times v_{gs}(r_o//R_{CT}//R_{LD})$$
$$v_{gs} = v_{in} = v_{src} \qquad (7.9)$$
$$\text{gain} = \frac{v_{ld}}{v_{src}} = -g_m(r_o//R_{CT}//R_{LD})$$

This gain expression can be alternatively written as:

$$\text{gain} = \frac{v_{ld}}{v_{src}} = -\frac{r_o//R_{CT}//R_{LD}}{1/g_m} \qquad (7.10)$$

The gain is negative, which means that v_{ld} is 180 degrees out-of-phase with respect to v_{src} as seen in Fig. 7.4.

Therefore, the gain in the common-source configuration is the total resistance at the drain terminal of the transistor divided roughly by the total resistance at the source terminal of the transistor (think in terms of the T-model).

As a result, one way to decrease the gain is to add a resistor at the source terminal of the transistor in this configuration.

If the frequency of the input signal is increased beyond a certain value called the 3-dB frequency (f_{3db}), the output signal, and thus the gain, will start to drop. As a result, f_{3db} is the maximum frequency around which the amplifier will operate correctly. The zone of correct operation is called the bandwidth. These results are shown in Fig. 7.8.

In order to get the bandwidth, we will draw the high-frequency small-signal model of the configuration by replacing the transistor with the complete small-signal model that includes the internal capacitances. The procedure we will follow is called the open-circuit time constant method, which consists of getting:

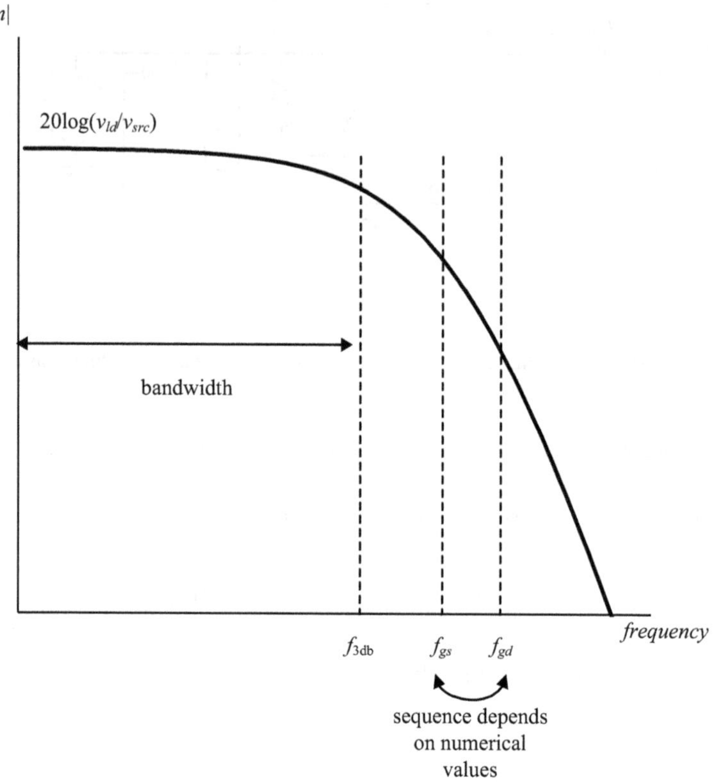

Fig. 7.8 Gain versus frequency

1. the resistance seen by each one of these capacitances while disabling the others by open-circuiting them.
2. the time-constant associated with each one of these capacitance and resistance combination.
3. the corner frequency that each time constant generates.
4. the combined corner frequency which we will call the bandwidth.

We will start with the high-frequency small-signal model as in Fig. 7.9.

Now, taking capacitance C_{gs} into consideration, we will disable the source and open-circuit the capacitance C_{gd} as in Fig. 7.10.

By inspection, we can put ourselves in the position of C_{gs} with one eye at the gate terminal and another at the source terminal. The resistance R_{gs} that we see is R_{SRC}. Therefore, the time constant associated with C_{gs} and the resistance it sees is

$$\tau_{gs} = C_{gs}R_{gs} = C_{gs}R_{SRC} \tag{7.11}$$

Consequently, this gives rise to the first corner frequency

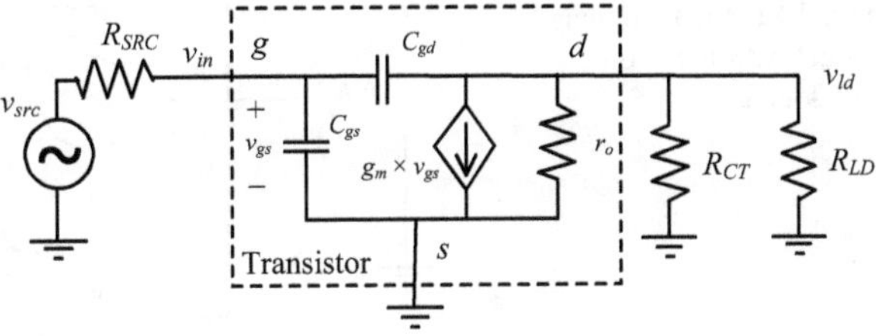

Fig. 7.9 High-frequency small-signal model of the common-source configuration

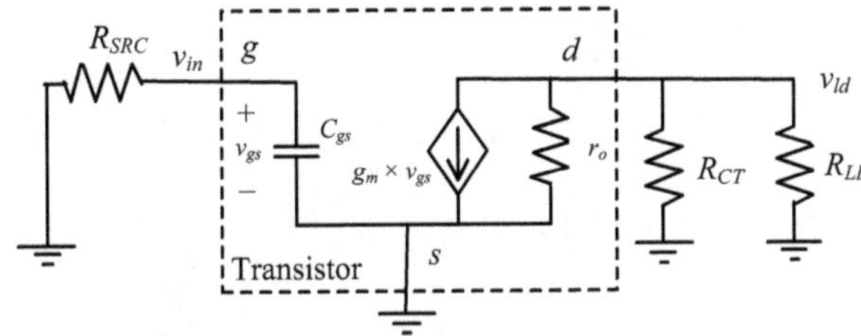

Fig. 7.10 High-frequency small-signal model with focus on C_{gs}

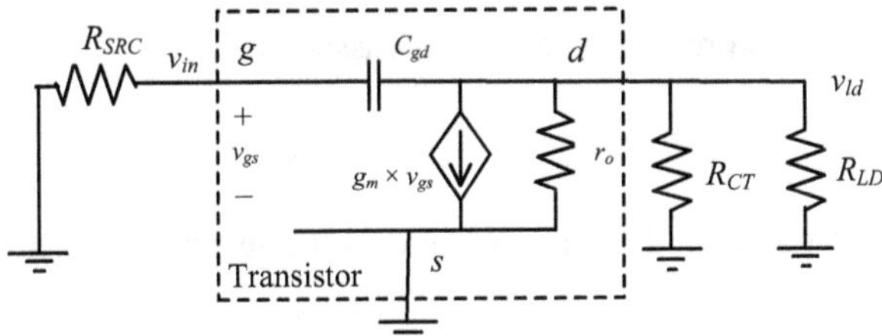

Fig. 7.11 High-frequency small-signal model with focus on C_{gd}

$$\omega_{gs} = \frac{1}{\tau_{gs}} = \frac{1}{C_{gs}R_{gs}}$$
$$f_{gs} = \frac{1}{2\pi \times C_{gs}R_{gs}} \tag{7.12}$$

Now, we will get the second time constant associated with C_{gd} and the resistance it sees. Redrawing the circuit again, now focusing of C_{gd}, we get Fig. 7.11.

We cannot get the resistance seen by C_{gd} using inspection since we have a dependent current source that affects the result, so we will have to replace C_{gd} with a test source as in Fig. 7.12.

First, let

$$R_{total} = r_o // R_{CT} // R_{LD} \tag{7.13}$$

Consequently

Fig. 7.12 Small-signal model with focus on C_{gd}—test source

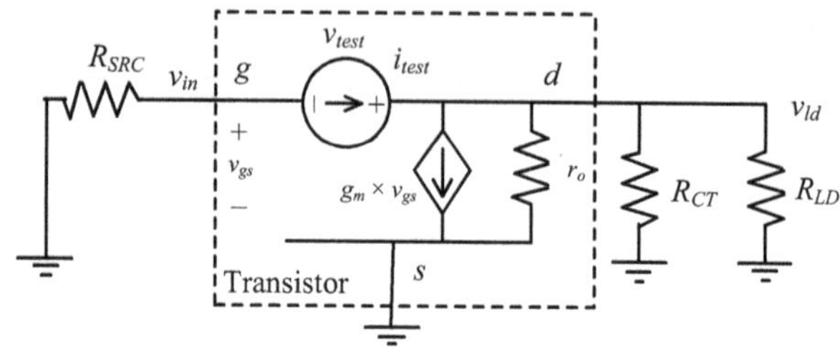

$$v_{gs} = -i_{test}R_{SRC}$$

$$i_{test} = g_m v_{gs} + \frac{v_{ld}}{R_{total}} = g_m(-i_{test}R_{SRC}) + \frac{v_{test} - i_{test}R_{SRC}}{R_{total}}$$

$$i_{test}\left(1 + g_m R_{SRC} + \frac{R_{SRC}}{R_{total}}\right) = \frac{v_{test}}{R_{total}}$$

$$R_{gd} = \frac{v_{test}}{i_{test}} = R_{SRC} + R_{total} + g_m R_{SRC} R_{total}$$

$$(7.14)$$

Therefore, the time constant associated with C_{gd} and the resistance it sees is

$$\tau_{gd} = C_{gd}R_{gd} \tag{7.15}$$

Consequently, this gives rise to the second corner frequency

$$\omega_{gd} = \frac{1}{\tau_{gd}} = \frac{1}{C_{gd}R_{gd}}$$

$$f_{gd} = \frac{1}{2\pi \times C_{gd}R_{gd}}$$

$$(7.16)$$

Whether f_{gs} is smaller than f_{gd} depends on their numerical values. Looking at the slope of the curve on a logarithmic scale as in Fig. 7.8, we can see that eventually, at high frequencies, the curve goes down by 40 dB/decade since the circuit contains two corner frequencies f_{gs} and f_{gd}.

We will now estimate f_{3db} by combining f_{gs} and f_{gd}. As can be seen in Fig. 7.8, f_{3db} is smaller than the smallest; therefore, any frequency below f_{3db} is within the bandwidth of the amplifier. The simplest way to combine them is

$$\frac{1}{f_{3db}} \approx \frac{1}{f_{gs}} + \frac{1}{f_{gd}} \tag{7.17}$$

7.3 Common-Gate (CG) Configuration

The common-gate configuration is shown in Fig. 7.13. The circuits surrounding the transistor are modeled as follows:

- The previous stage (signal source) is modeled using its Thevenin equivalent model. This signal source contains a DC component and a small-signal AC component. Also, the source has an output resistance R_{SRC}.
- The next stage (load) is modeled using its input resistance, indicated as R_{LD}.
- The current source, which, in combination with the DC component of the signal source, creates the DC operating point around which the transistor operates, is modeled using its Norton model with an output resistance R_{CT}.

Typical v_{SRC} and v_{LD} waveforms of a common-gate configuration are shown in Fig. 7.14.

Looking at the NMOS-based implementation in Fig. 7.13a, we can see that the maximum value the output voltage can take is governed by what is happening above the output node. Therefore:

$$v_{LD\_MAX} = V_{SUPPLY} - V_{CT\_REQ} \tag{7.18}$$

where V_{CT_REQ} is the smallest required voltage across the current source in order for it to operate properly.

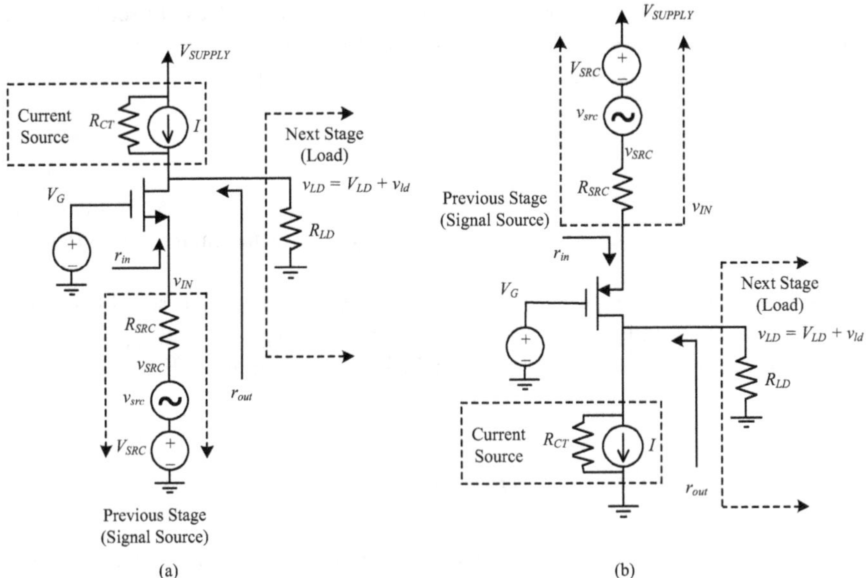

Fig. 7.13 Common-gate configuration: (a) NMOS-based and (b) PMOS-based

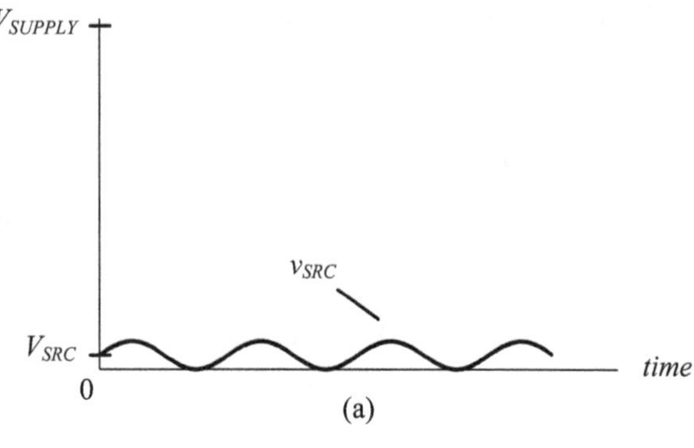

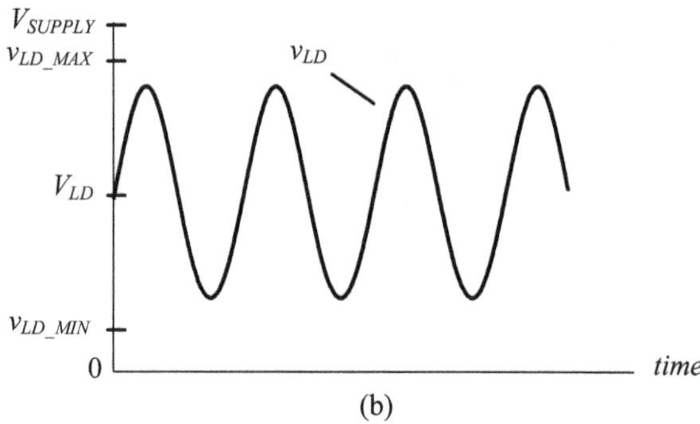

Fig. 7.14 Typical (**a**) v_{SRC} and (**b**) v_{LD} waveforms of a common-gate configuration

The minimum value of the output voltage is governed by what is happening below the output node. Since R_{LD} is a passive component, it does not set a limit on the output voltage. However, the transistor needs to stay in the saturation region, so it requires some voltage to breathe on top of v_{IN_MAX}. Therefore:

$$v_{LD\_MIN} = V_{OV\_REQ} + v_{IN\_MAX} \tag{7.19}$$

where V_{OV_REQ} is the minimum required overdrive voltage of the NMOS transistor. If v_{LD} goes below this value, the transistor will get out of the saturation region and into the triode region.

As a result, in order to get the maximum swing at the output, it is sensible to have the DC point at the output (V_{LD}) in the middle between v_{LD_MIN} and v_{LD_MAX}.

Similar reasoning can be applied to the PMOS-based circuit in Fig. 7.13b.

7.3.1 Common-Gate DC Analysis

We will disable all the AC sources and keep the DC sources only. As a result, the small-signal voltage sources, if any, are replaced with a short circuit (zero voltage) and the small-signal current sources, if any, are replaced with an open circuit (zero

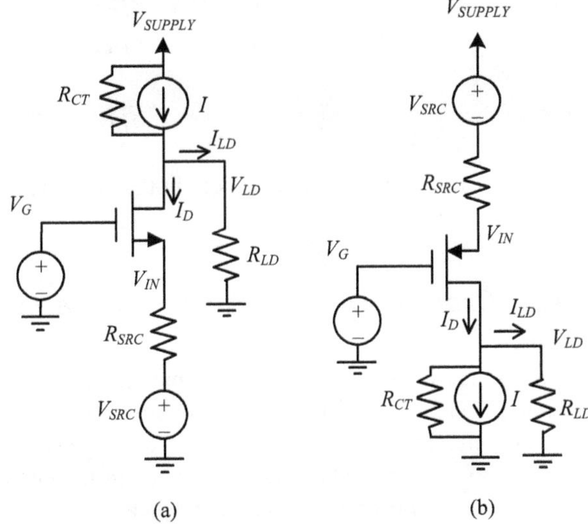

Fig. 7.15 DC model of the common-gate configuration: (**a**) NMOS-based and (**b**) PMOS-based

current). The resulting circuit for DC analysis is shown in Fig. 7.15. This is what the DC components of all the signals in the circuit see.

Looking at the current source, usually this has a fairly large output resistance R_{CT}. As a result, the current coming out of the current source is almost equal to I. Therefore, we can see that the average power consumption can be approximated as follows:

$$P_{av} \approx V_{SUPPLY} \times I \tag{7.20}$$

As a result, given that the power supply V_{SUPPLY} is fixed, the current I plays a crucial role in determining the power consumption of the circuit.

The current I is related to I_D and I_{LD}. It is in our interest to have most of the current I going into I_D in order to maximize g_m and thus the gain. Therefore, we usually aim for a high R_{LD}.

From a design point-of-view, we need to make sure that the V_{GS}, V_{DS}, and I_D with which we provide the transistor have to belong to the three-dimensional plot for that particular transistor. In other words, these values have to be compatible, and of course the transistor has to be in the saturation region following its large-signal model:

$$I_D = \frac{1}{2} k' \left(\frac{W}{L}\right) (\mid V_{GS} \mid - \mid V_{th} \mid)^2 (1 + \lambda \mid V_{DS} \mid) \tag{7.21}$$

Now that I_D is known, we can get the small-signal AC model parameters as follows:

$$g_m = \frac{2I_D}{|V_{OV}|} \tag{7.22}$$

$$r_o \approx \frac{|V_A|}{|I_D|} \tag{7.23}$$

These parameters will be used next in the AC small-signal analysis.

7.3.2 Common-Gate AC Small-Signal Analysis

We will start with the analysis at low to medium frequencies. In this frequency range, the internal capacitances have very high impedances, so they will be replaced with open circuits. The resulting circuit for AC analysis is shown in Fig. 7.16. This is what the AC components of all the signals in the circuit see. Note that the small-signal model of the circuit is the same, whether it is NMOS-based or PMOS-based.

In order to get the input resistance, we need to disable the input source, which is a voltage source in this case. Therefore, we will short-circuit it. Also, we will put our test source exactly where we would like to see the output resistance as shown in Fig. 7.17.

The analysis goes as follows:

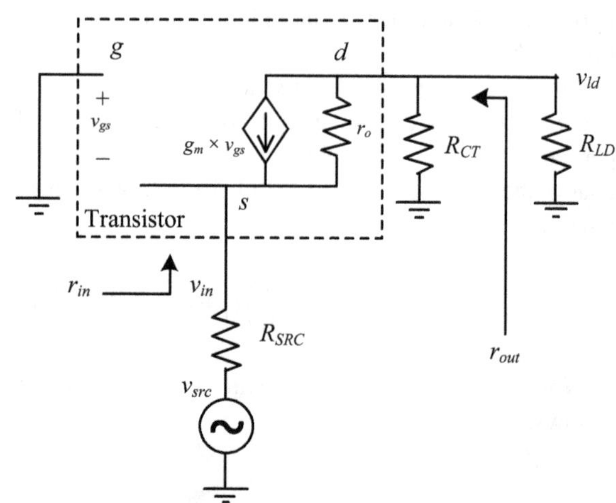

Fig. 7.16 AC small-signal model of the common-gate configuration

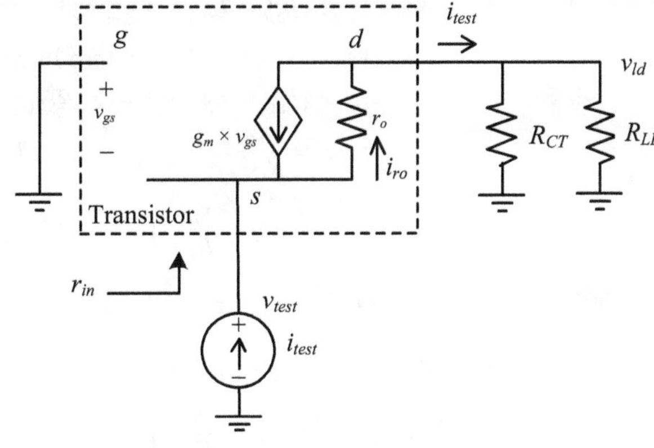

Fig. 7.17 AC small-signal model of the common-gate configuration—input resistance

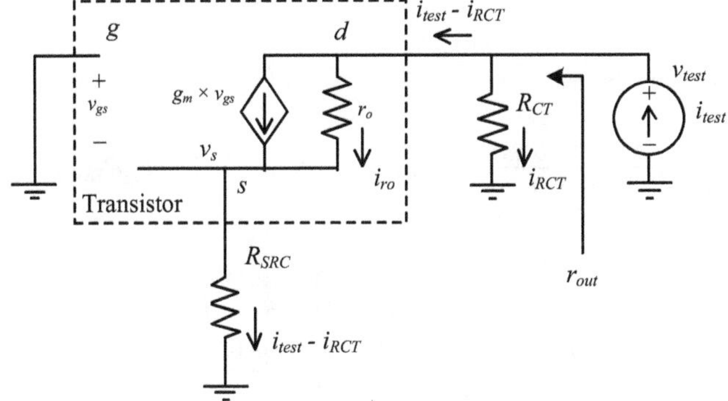

Fig. 7.18 AC small-signal model of the common-gate configuration—output resistance

$$
\begin{aligned}
v_{gs} &= -v_{test} \\
i_{ro} &= i_{test} + g_m v_{gs} = i_{test} - g_m v_{test} \\
&\quad - v_{test} + i_{ro} r_o + i_{test}(R_{CT}//R_{LD}) = 0 \\
&\quad - v_{test} + r_o(i_{test} - g_m v_{test}) + i_{test}(R_{CT}//R_{LD}) = 0
\end{aligned}
\tag{7.24}
$$

Therefore:

$$
r_{in} = \frac{v_{test}}{i_{test}} = \frac{r_o + (R_{CT}//R_{LD})}{1 + g_m r_o} = \frac{1 + \dfrac{R_{CT}//R_{LD}}{r_o}}{\dfrac{1}{r_o} + g_m}
\tag{7.25}
$$

As can be seen, the input resistance of a common-gate configuration is quite small.

The analysis for the output resistance follows Fig. 7.18.

$$i_{ro} = (i_{test} - i_{RCT}) - g_m v_{gs}$$

$$v_{gs} = -v_s = -(i_{test} - i_{RCT})R_{SRC}$$

$$i_{ro} = (i_{test} - i_{RCT})(1 + g_m R_{SRC})$$

$$i_{RCT} = \frac{v_{test}}{R_{CT}}$$

$$i_{ro} = \left(i_{test} - \frac{v_{test}}{R_{CT}} \right)(1 + g_m R_{SRC}) \tag{7.26}$$

Therefore, doing a KVL:

$$-(i_{test} - i_{RCT})R_{SRC} - i_{ro}r_o + v_{test} = 0$$

$$-\left(i_{test} - \frac{v_{test}}{R_{CT}} \right)R_{SRC} - \left[\left(i_{test} - \frac{v_{test}}{R_{CT}} \right)(1 + g_m R_{SRC}) \right] r_o + v_{test} = 0 \tag{7.27}$$

Separating v_{test} and i_{test}:

$$v_{test} \left[\frac{R_{SRC}}{R_{CT}} + \frac{(1 + g_m R_{SRC})r_o}{R_{CT}} + 1 \right] = i_{test}[R_{SRC} + (1 + g_m R_{SRC})r_o] \tag{7.28}$$

As a result:

$$\frac{v_{test}}{i_{test}} = \frac{R_{SRC} + (1 + g_m R_{SRC})r_o}{\dfrac{R_{SRC}}{R_{CT}} + \dfrac{(1 + g_m R_{SRC})r_o}{R_{CT}} + 1}$$

$$\frac{v_{test}}{i_{test}} = \frac{R_{SRC} + r_o + g_m R_{SRC} r_o}{\dfrac{R_{SRC}}{R_{CT}} + \dfrac{(1 + g_m R_{SRC})r_o}{R_{CT}} + 1} \tag{7.29}$$

$$r_{out} = \frac{v_{test}}{i_{test}} = \frac{\dfrac{R_{SRC}}{r_o} + 1 + g_m R_{SRC}}{\dfrac{R_{SRC}}{R_{CT} r_o} + \dfrac{(1 + g_m R_{SRC})}{R_{CT}} + \dfrac{1}{r_o}} = \frac{r_o + R_{SRC} + g_m r_o R_{SRC}}{\dfrac{R_{SRC}}{R_{CT}} + \dfrac{r_o(1 + g_m R_{SRC})}{R_{CT}} + 1} \tag{7.30}$$

Next, we will get the voltage gain. The focus will be on Fig. 7.19.
We will first get $v_{ld}\,/\,v_{in}$ and then $v_{in}\,/\,v_{src}$

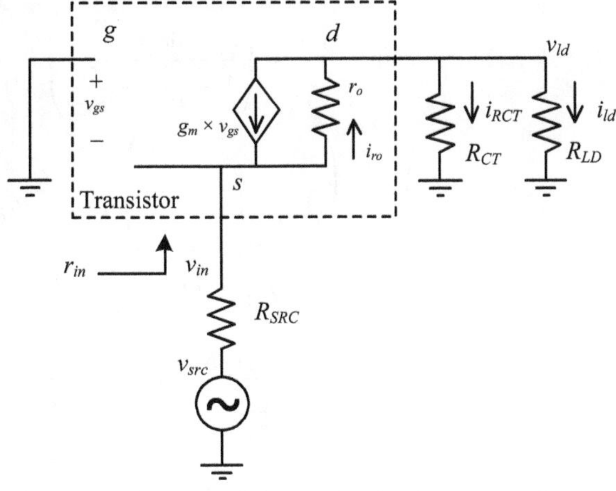

Fig. 7.19 AC small-signal model of the common-gate configuration—gain

$$v_{ld} = -i_{ro}r_o + v_{in}$$

$$i_{ro} = g_m v_{gs} + i_{ld} + i_{RCT}$$

$$v_{ld} = -(g_m v_{gs} + i_{ld} + i_{RCT})r_o + v_{in}$$

$$v_{gs} = -v_{in}$$

$$v_{ld} = -\left(-g_m v_{in} + \frac{v_{ld}}{R_{LD}} + \frac{v_{ld}}{R_{CT}}\right)r_o + v_{in} \tag{7.31}$$

$$v_{ld}\left(1 + \frac{r_o}{R_{LD}} + \frac{r_o}{R_{CT}}\right) = v_{in}(g_m r_o + 1)$$

$$\frac{v_{ld}}{v_{in}} = \frac{g_m r_o + 1}{1 + \dfrac{r_o}{R_{LD}} + \dfrac{r_o}{R_{CT}}}$$

In order to get $v_{in}\,/\,v_{src}$ and thus the total gain $v_{ld}\,/\,v_{src}$, we can just use the input resistance r_{in} that we found before:

$$\text{gain} = \frac{v_{ld}}{v_{src}} = \left(\frac{v_{ld}}{v_{in}}\right)\left(\frac{v_{in}}{v_{src}}\right) = \left(\frac{g_m r_o + 1}{1 + \frac{r_o}{R_{LD}} + \frac{r_o}{R_{CT}}}\right)\left(\frac{r_{in}}{r_{in} + R_{SRC}}\right) \tag{7.32}$$

In case $g_m r_o >> 1$, which is usually the situation, then:

$$\text{gain} = \frac{v_{ld}}{v_{src}} \approx \left(\frac{g_m r_o}{1 + \dfrac{r_o}{R_{LD}} + \dfrac{r_o}{R_{CT}}} \right) \left(\frac{r_{in}}{r_{in} + R_{SRC}} \right)$$

$$\frac{v_{ld}}{v_{src}} \approx \left(\frac{g_m}{\dfrac{1}{r_o} + \dfrac{1}{R_{LD}} + \dfrac{1}{R_{CT}}} \right) \left(\frac{r_{in}}{r_{in} + R_{SRC}} \right) \tag{7.33}$$

$$\frac{v_{ld}}{v_{src}} \approx g_m (r_o // R_{LD} // R_{CT}) \left(\frac{r_{in}}{r_{in} + R_{SRC}} \right)$$

$$\frac{v_{ld}}{v_{src}} \approx \frac{r_o // R_{LD} // R_{CT}}{\dfrac{1}{g_m}} \left(\frac{r_{in}}{r_{in} + R_{SRC}} \right)$$

This result looks similar to that of a common-source with two differences:

- the finite input resistance r_{in} affects the gain,
- the gain is positive, which means that v_{ld} is in phase with respect to v_{src} as seen in Fig. 7.14.

Similarly to a common-source configuration, and with similar gain plot as in Fig. 7.8, in order to get the bandwidth, we will draw the high-frequency small-signal model of the configuration by replacing the transistor with the complete small-signal model that includes the internal capacitances. The procedure we will follow is called the open-circuit time constant method, which consists of getting:

1. the resistance seen by each one of these capacitances while disabling the others by open-circuiting them.
2. the time-constant associated with each one of these capacitance/resistance combination.
3. the corner frequency that each time constant generates.
4. the combined corner frequency which we will call the bandwidth.

We will start with the high-frequency small-signal model as in Fig. 7.20.

Now, taking capacitance C_{gs} into consideration, we will disable the source and open-circuit the capacitance C_{gd} as in Fig. 7.21.

We will have to replace C_{gs} with a test source as in Fig. 7.22.

Starting with a KVL:

$$v_{test} + \left(g_m v_{gs} + i_{ro} - i_{test} \right) R_{SRC} = 0$$

$$v_{gs} = v_{test} \tag{7.34}$$

$$i_{ro} = \frac{-v_{test} + i_{test} R_{SRC} - g_m v_{test} R_{SRC}}{R_{SRC}}$$

Another KVL:

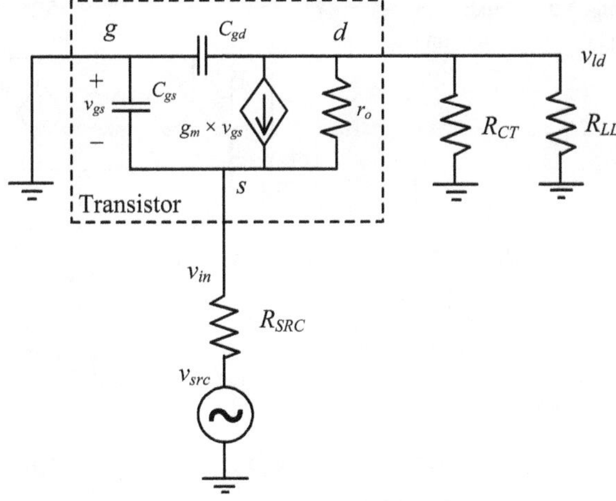

Fig. 7.20 Common-gate configuration high-frequency ac small-signal model

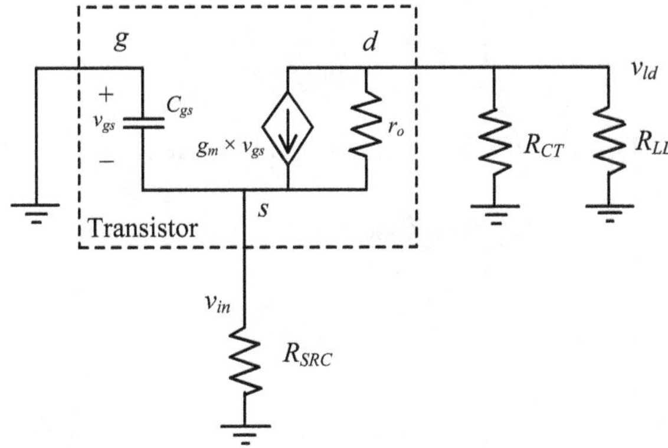

Fig. 7.21 Common-gate ac small-signal model with focus on C_{gs}

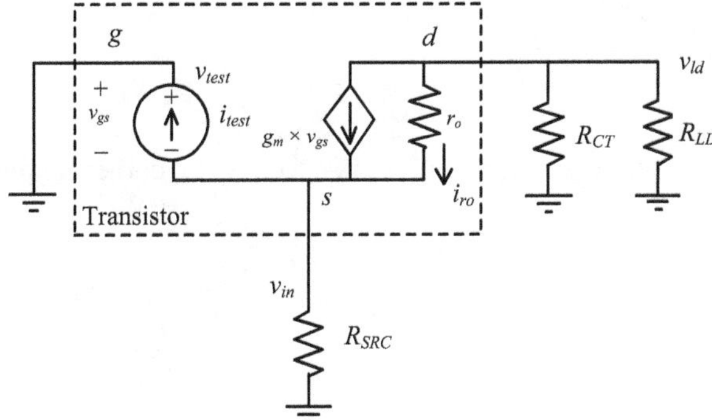

Fig. 7.22 Common-gate ac small-signal model with focus on C_{gs}—test source

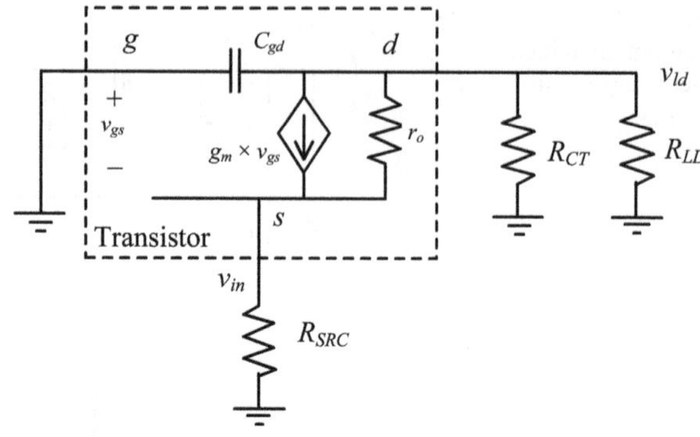

Fig. 7.23 Small-signal model with focus on C_{gd}

$$v_{test} - i_{ro}r_o - \left(g_m v_{gs} + i_{ro}\right)\left(R_{CT}//R_{LD}\right) = 0$$

$$v_{gs} = v_{test}$$

$$v_{test} - i_{ro}\left[r_o + \left(R_{CT}//R_{LD}\right)\right] - g_m v_{test}\left(R_{CT}//R_{LD}\right) = 0 \tag{7.35}$$

$$i_{ro} = \frac{v_{test} - g_m v_{test}\left(R_{CT}//R_{LD}\right)}{r_o + \left(R_{CT}//R_{LD}\right)}$$

Equating the above results to eliminate i_{ro}:

$$\frac{-v_{test} + i_{test}R_{SRC} - g_m v_{test}R_{SRC}}{R_{SRC}} = \frac{v_{test} - g_m v_{test}\left(R_{CT}//R_{LD}\right)}{r_o + \left(R_{CT}//R_{LD}\right)} \tag{7.36}$$

Isolating v_{test} and i_{test}:

$$R_{gs} = \frac{v_{test}}{i_{test}} = \frac{R_{SRC}\left[r_o + \left(R_{CT}//R_{LD}\right)\right]}{R_{SRC} + r_o + \left(R_{CT}//R_{LD}\right) + g_m R_{SRC}r_o} \tag{7.37}$$

Therefore:

$$\tau_{gs} = C_{gs}R_{gs} \tag{7.38}$$

Consequently, this gives rise to the first corner frequency

$$\omega_{gs} = \frac{1}{\tau_{gs}} = \frac{1}{C_{gs}R_{gs}}$$

$$f_{gs} = \frac{1}{2\pi \times C_{gs}R_{gs}} \tag{7.39}$$

Now, we will get the second time constant associated with C_{gd} and the resistance it sees. Redrawing the circuit again, now focusing of C_{gd}, we get the circuit in Fig. 7.23.

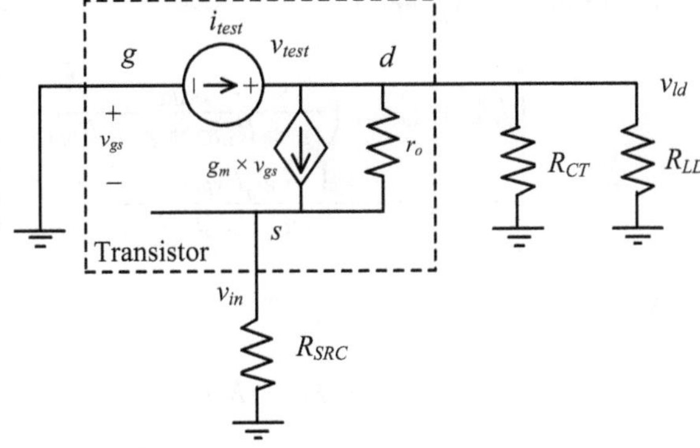

Fig. 7.24 Small-signal model with focus on C_{gd}—test source

We cannot get the resistance seen by C_{gd} using inspection since we have a current source, so we will have to replace C_{gd} with a test source as in Fig. 7.24.

Starting with the first KVL:

$$-v_{test} + i_{ro}r_o + (g_m v_{gs} + i_{ro})R_{SRC} = 0$$
$$-v_{test} + i_{ro}(r_o + R_{SRC}) + g_m v_{gs}R_{SRC} = 0 \qquad (7.40)$$
$$v_{gs} = \frac{v_{test} - i_{ro}(r_o + R_{SRC})}{g_m v_{gs}R_{SRC}}$$

A second KVL:

$$v_{gs} + (g_m v_{gs} + i_{ro})R_{SRC} = 0$$
$$v_{gs} = \frac{-i_{ro}R_{SRC}}{1 + g_m R_{SRC}} \qquad (7.41)$$

Equating the above results to eliminate v_{gs}:

$$\frac{-i_{ro}(r_o + R_{SRC}) + v_{test}}{g_m R_{SRC}} = \frac{-i_{ro}R_{SRC}}{1 + g_m R_{SRC}}$$
$$i_{ro} = v_{test}\left(\frac{1 + g_m R_{SRC}}{r_o + R_{SRC} + g_m r_o R_{SRC}}\right) \qquad (7.42)$$

Substituting in the second KVL:

$$v_{gs} = -v_{test}\left(\frac{R_{SRC}}{r_o + R_{SRC} + g_m r_o R_{SRC}}\right) \qquad (7.43)$$

A third KVL:

$$-v_{test} + (i_{test} - g_m v_{gs} - i_{ro})(R_{CT}//R_{LD}) = 0 \qquad (7.44)$$

Substituting for v_{gs} and i_{ro}:

$$- v_{test} + \left[\begin{array}{c} i_{test} + g_m v_{test} \left(\dfrac{R_{SRC}}{r_o + R_{SRC} + g_m r_o R_{SRC}} \right) \\[2em] - v_{test} \left(\dfrac{1 + g_m R_{SRC}}{r_o + R_{SRC} + g_m r_o R_{SRC}} \right) \end{array} \right] (R_{CT}//R_{LD}) = 0 \qquad (7.45)$$

As a result:

$$R_{gd} = \frac{v_{test}}{i_{test}} = R_{CT}//R_{LD}//(r_o + R_{SRC} + g_m r_o R_{SRC}) \qquad (7.46)$$

The smallest resistance governing R_{gd} is typically R_{LD}. Therefore R_{LD} usually dominates R_{gd}.

Therefore, the time constant associated with C_{gd} and the resistance it sees is

$$\tau_{gd} = C_{gd} R_{gd} \qquad (7.47)$$

Consequently, this gives rise to the second corner frequency

$$\omega_{gd} = \frac{1}{\tau_{gd}} = \frac{1}{C_{gd} R_{gd}}$$

$$f_{gd} = \frac{1}{2\pi \times C_{gd} R_{gd}} \qquad (7.48)$$

The same comments made above for the common-source configuration apply also here.

$$\frac{1}{f_{3db}} \approx \frac{1}{f_{gs}} + \frac{1}{f_{gd}} \qquad (7.49)$$

7.4 Common-Drain (CD) Configuration

The common-drain configuration is shown in Fig. 7.25. The circuits surrounding the transistor are modeled as follows:

- The previous stage (signal source) is modeled using its Thevenin equivalent model. This signal source contains a DC component and a small-signal AC component. Also, the source has an output resistance R_{SRC}.

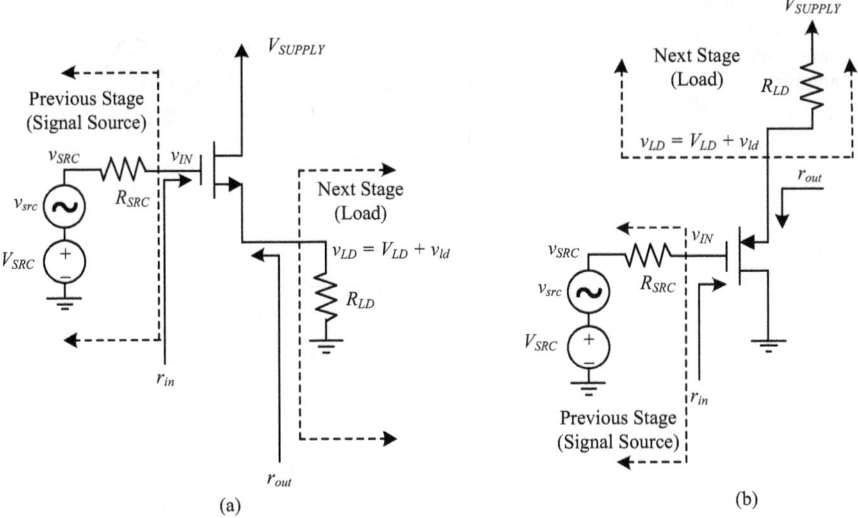

Fig. 7.25 Common-drain configuration: (**a**) NMOS-based and (**b**) PMOS-based

- The next stage (load) is modeled using its input resistance, indicated as R_{LD}.

Note that unlike the previous configurations, this one does not make use of a current source.

Typical v_{SRC} and v_{LD} waveforms of a common-drain configuration are shown in Fig. 7.26.

Looking at the NMOS-based implementation in Fig. 7.25a, we can see that the maximum value the output voltage can take is governed by what is happening above the output node. Therefore:

$$v_{LD\_MAX} = V_{SUPPLY} - V_{OV\_REQ} \tag{7.50}$$

where V_{OV_REQ} is the minimum overdrive voltage of the NMOS transistor. If v_{LD} goes above this value, the transistor will get out of the saturation region and into the triode region.

The minimum value of the output voltage is governed by what is happening below the output node. Since R_{LD} is a passive component, it does not set a limit on the output voltage. Therefore:

$$v_{LD\_MIN} = 0 \tag{7.51}$$

As a result, in order to get the maximum swing at the output, it is sensible to have the DC point at the output (V_{LD}) in the middle between v_{LD_MIN} and v_{LD_MAX}.

Similar reasoning can be applied to the PMOS-based circuit in Fig. 7.25b.

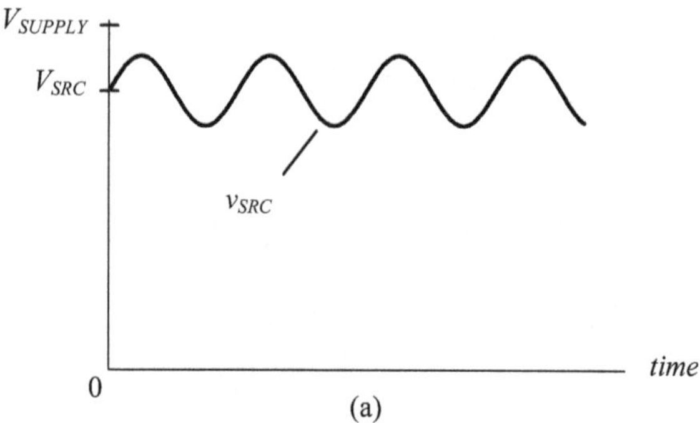

Fig. 7.26 Typical (**a**) v_{SRC} and (**b**) v_{LD} waveforms of a common-drain configuration

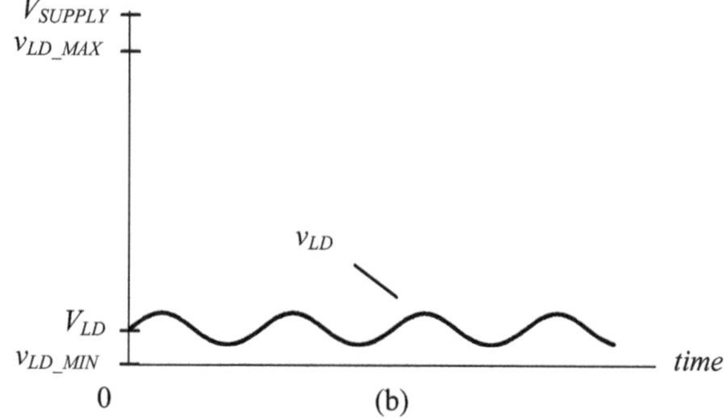

7.4.1 Common-Drain DC Analysis

We will disable all the AC sources and keep the DC sources only. As a result, the small-signal voltage sources, if any, are replaced with a short circuit (zero voltage) and the small-signal current sources, if any, are replaced with an open circuit (zero current). The resulting circuit for DC analysis is shown in Fig. 7.27. This is what the DC components of all the signals in the circuit see.

Concentrating on the NMOS-based implementation as in Fig. 7.27a, we need to get the current I_D.

We have:

$$I_D = k(V_{GS} - V_{th})^2(1 + \lambda V_{DS})$$

$$V_{GS} = \sqrt{\frac{I_D}{k(1 + \lambda V_{DS})}} + V_{th} \tag{7.52}$$

Continuing with a KVL:

$$-V_{SUPPLY} + V_{DS} + I_D R_{LD} = 0$$

$$V_{DS} = V_{SUPPLY} - I_D R_{LD} \tag{7.53}$$

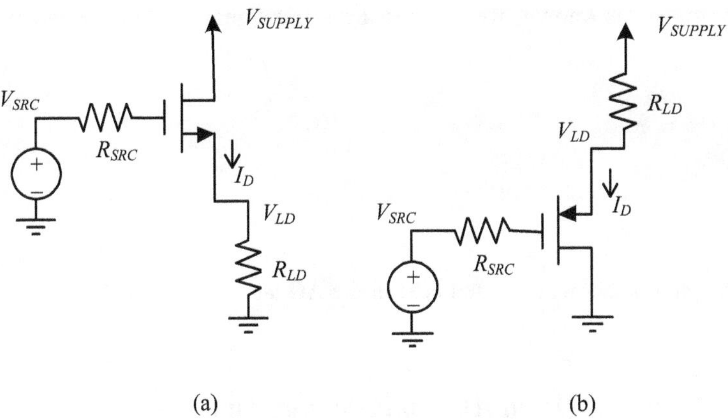

(a)   (b)

Fig. 7.27 DC model of the common-drain configuration: (**a**) NMOS-based and (**b**) PMOS-based

Substituting for V_{DS}:

$$V_{GS} = \sqrt{\frac{I_D}{k[1 + \lambda(V_{SUPPLY} - I_D R_{LD})]}} + V_{th} \tag{7.54}$$

Another KVL and noting that the current at the gate is zero:

$$-V_{SRC} + V_{GS} + I_D R_{LD} = 0$$
$$V_{GS} = V_{SRC} - I_D R_{LD} \tag{7.55}$$

Equating (7.54) and (7.55) to cancel V_{GS}:

$$\sqrt{\frac{I_D}{k[1 + \lambda(V_{SUPPLY} - I_D R_{LD})]}} + V_{th} = V_{SRC} - I_D R_{LD}$$
$$\frac{I_D}{k[1 + \lambda(V_{SUPPLY} - I_D R_{LD})]} = (V_{SRC} - I_D R_{LD} - V_{th})^2 \tag{7.56}$$
$$k[1 + \lambda(V_{SUPPLY} - I_D R_{LD})](V_{SRC} - I_D R_{LD} - V_{th})^2 - I_D = 0$$

With all the parameters at hand, this gives rise to a third-order equation in I_D which needs to be solved.

For hand calculations, a simplification can be done at this point by neglecting the effect of λ. This will result in a second-order equation.

After getting I_D:

$$P_{av} \approx V_{SUPPLY} \times I_D \tag{7.57}$$

Now that I_D is known, we can get the small-signal AC model parameters as follows:

$$g_m = \frac{2I_D}{|V_{OV}|} \tag{7.58}$$

$$r_o \approx \frac{|V_A|}{|I_D|} \tag{7.59}$$

These parameters will be used next in the AC small-signal analysis.

7.4.2 *Common-Drain AC Small-Signal Analysis*

We will start with the analysis at low to medium frequencies. In this frequency range, the internal capacitances have very high impedances, so they will be replaced with open circuits. The resulting circuit for AC analysis is shown in Fig. 7.16. This is what the AC components of all the signals in the circuit see. Note that the small-signal model of the circuit is the same, whether it is NMOS-based or PMOS-based (Fig. 7.28).

There is no current at the input of the circuit, which means it is open-circuited. As a result, the input resistance is:

$$r_{in} = \infty \tag{7.60}$$

In order to get the output resistance, we need to disable the input source, which is a voltage source in this case. Therefore, we will short-circuit it. Also, we will put our test source exactly where we would like to see the output resistance as shown in Fig. 7.29.

Since there is no current at the gate, then v_{in}, the voltage across R_{SRC}, is zero. As a result:

$$r_{out} = \frac{v_{test}}{i_{test}} = \frac{1}{g_m} // r_o \tag{7.61}$$

As can be seen, the output resistance of a common-drain configuration is quite small.

Next, we will get the voltage gain. The focus will be on Fig. 7.28.

Since there is no current at the gate, then v_{in}, the voltage across R_{SRC}, is zero. As a result, from v_{SRC} to v_{ld}, there is a voltage divider and the voltage gain:

Fig. 7.28 AC small-signal model of the common-drain configuration

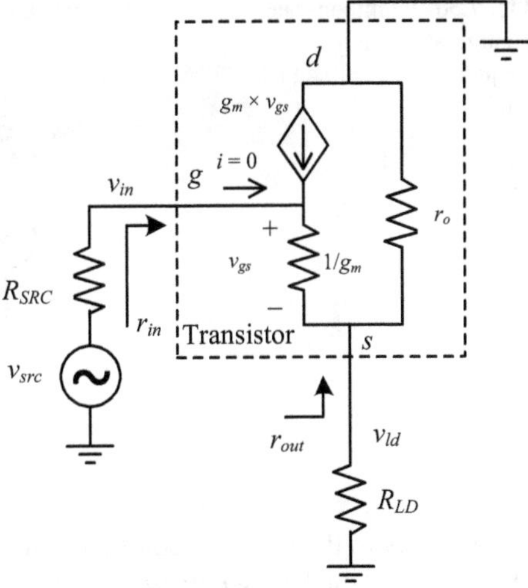

Fig. 7.29 AC small-signal model of the common-drain configuration—output resistance

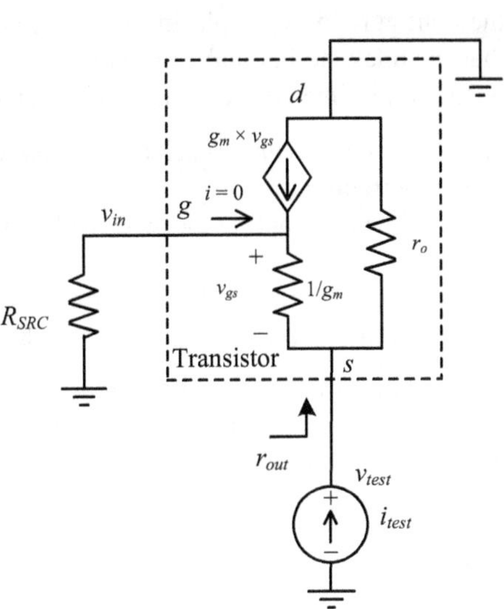

$$\text{gain} = \frac{v_{ld}}{v_{src}} = \frac{r_o//R_{LD}}{\frac{1}{g_m} + (r_o//R_{LD})} \tag{7.62}$$

As can be seen, the voltage gain has a magnitude less than one, which means that this is not a gain at all. In fact, the circuit acts as a voltage divider. Still, as we will see later, this amplifier is very useful as an output stage to drive large loads (small R_{LD}) since it amplifies the current due to its high input resistance and low output resistance.

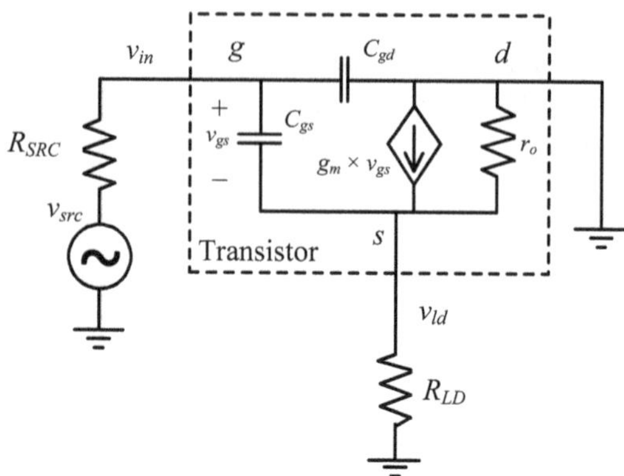

Fig. 7.30 Common-gate configuration high-frequency ac small-signal model

Similarly to the previous configurations, and with similar gain plot as in Fig. 7.8, in order to get the bandwidth, we will draw the high-frequency small-signal model of the configuration by replacing the transistor with the complete small-signal model that includes the internal capacitances. The procedure we will follow is called the open-circuit time constant method, which consists of getting:

1. the resistance seen by each one of these capacitances while disabling the others by open-circuiting them,
2. the time-constant associated with each one of these capacitance/resistance combination,
3. the corner frequency that each time constant generates,
4. the combined corner frequency which we will call the bandwidth.

We will start with the high-frequency small-signal model as in Fig. 7.30.

Now, taking capacitance C_{gs} into consideration, we will disable the source and open-circuit the capacitance C_{gd} as in Fig. 7.31.

We will have to replace C_{gs} with a test source as in Fig. 7.32a.

Noting that v_{gs} is equal to v_{test}, we can redraw the circuit as in Fig. 7.32b.

Starting with a KVL:

$$- i_{test}R_{SRC} + v_{test} + (g_m v_{test} - i_{test})(R_{LD}//r_o) = 0 \tag{7.63}$$

Consequently, isolating v_{test} and i_{test}:

$$R_{gs} = \frac{v_{test}}{i_{test}} = \frac{R_{SRC} + (R_{LD}//r_o)}{1 + g_m(R_{LD}//r_o)} \tag{7.64}$$

Therefore:

$$\tau_{gs} = C_{gs}R_{gs} \tag{7.65}$$

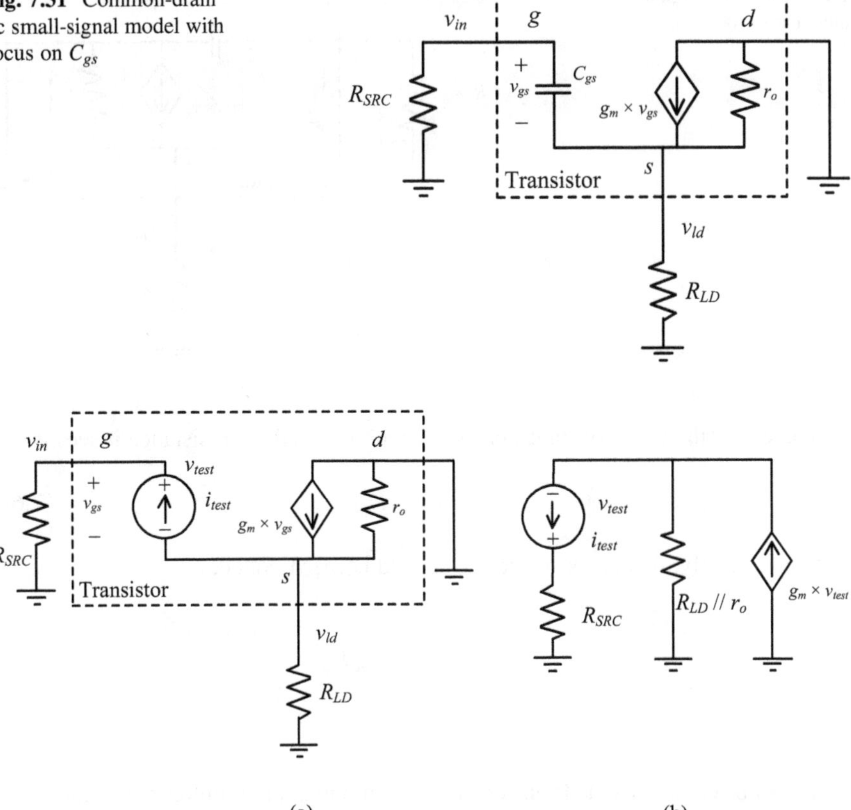

Fig. 7.31 Common-drain ac small-signal model with focus on C_{gs}

(a)   (b)

Fig. 7.32 Common-drain ac small-signal model with focus on C_{gs}—test source

Consequently, this gives rise to the first corner frequency

$$\omega_{gs} = \frac{1}{\tau_{gs}} = \frac{1}{C_{gs}R_{gs}}$$

$$f_{gs} = \frac{1}{2\pi \times C_{gs}R_{gs}} \tag{7.66}$$

Now, we will get the second time constant associated with C_{gd} and the resistance it sees. Redrawing the circuit again, now focusing on C_{gd}, we get the circuit as in Fig. 7.33.

By inspection, we can see clearly that:

$$R_{gd} = R_{SRC} \tag{7.67}$$

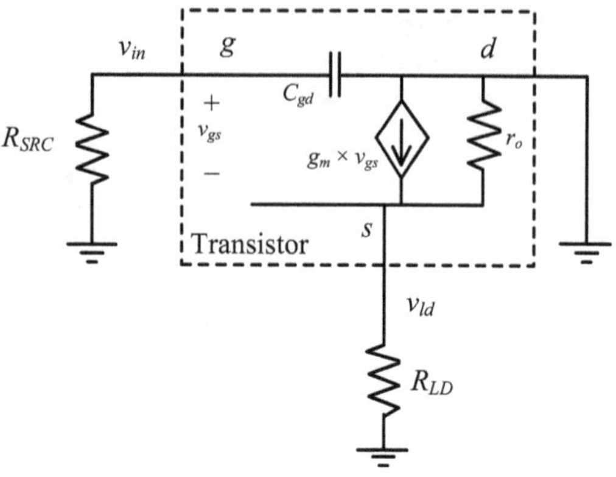

Fig. 7.33 Small-signal model with focus on C_{gd}

Therefore, the time constant associated with C_{gd} and the resistance it sees is

$$\tau_{gd} = C_{gd}R_{gd} \tag{7.68}$$

Consequently, this gives rise to the second corner frequency

$$\omega_{gd} = \frac{1}{\tau_{gd}} = \frac{1}{C_{gd}R_{gd}}$$
$$f_{gd} = \frac{1}{2\pi \times C_{gd}R_{gd}} \tag{7.69}$$

The same comments made above for the common-source configuration apply also here.

$$\frac{1}{f_{3db}} \approx \frac{1}{f_{gs}} + \frac{1}{f_{gd}} \tag{7.70}$$

7.5   Performance Summary

Tables 7.2, 7.3, and 7.4 show a summary of the different amplifier configurations' quantitative parameters presented above.

Table 7.5 shows a summary of some of the different amplifier configurations' qualitative properties presented in Table 7.3. Additionally, typical applications are also included.

Table 7.2 Quantitative summary of amplifier parameters 1

Configuration name	v_{LD_MAX}	v_{LD_MIN}
Common-Source	$V_{SUPPLY} - V_{CT\_REQ}$	V_{OV_REQ}
Common-Gate	$V_{SUPPLY} - V_{CT\_REQ}$	$V_{OV\_REQ} + v_{IN\_MAX}$
Common-Drain	$V_{SUPPLY} - V_{OV\_REQ}$	0

7.6   Cascaded Amplifiers and Transistor Pairing

In many applications, one amplification stage might not be enough. Therefore, we
end-up cascading several stages in order to meet the required overall specifications
as in an N-stage cascaded signal-processing chain shown in Fig. 7.34.

7.6.1   Example 1: 2-Stage Cascade

In this example, the requirements are:

- a given signal source,
- a large load (small R_{LD}),
- a high gain which is high.

We usually start with the simplest solution, so let us use a CS configuration since
we know it has the potential to give us a high gain. The result will be as in Fig. 7.35.
The gain will be as derived earlier:

$$\text{gain} = \frac{v_{ld}}{v_{src}} = -\frac{r_o//R_{CT}//R_{LD}}{\frac{1}{g_m}} \tag{7.71}$$

If the load is large (that is R_{LD} is small as in an audio speaker application for
example), then the total gain will be very small, rendering the setup useless. We say
in this case that the CS cannot drive this load.

In order to counter this, we will use a cascade of two stages. Figure 7.36 shows
such a signal-processing chain with a CS followed by a CD.

Here, the total gain is:

$$\text{gain} = \frac{v_{ld}}{v_{src}} = \left(\frac{v_{ld}}{v_x}\right)\left(\frac{v_x}{v_{src}}\right) \tag{7.72}$$

where.

v_{ld} / v_x is the gain of a CD with $R_{SRC} = 0$ so

Table 7.3 Quantitative summary of amplifier parameters 2

Configuration name	r_{in}	r_{out}	Voltage gain
Common-Source	∞	$r_o//R_{CT}$	$-\dfrac{r_o//R_{CT}//R_{LD}}{\frac{1}{g_m}} = -\dfrac{r_{out}//R_{LD}}{\frac{1}{g_m}}$

Common-Gate	$\dfrac{1 + \dfrac{R_{CT}//R_{LD}}{r_o}}{\dfrac{1}{r_o} + g_m}$
	
	 $\approx \dfrac{1}{g_m}$ if r_o is very large
		$\dfrac{\dfrac{R_{SRC}}{r_o} + 1 + g_m R_{SRC}}{\dfrac{R_{SRC}}{R_{CT}r_o} + \dfrac{(1 + g_m R_{SRC})}{R_{CT}} + \dfrac{1}{r_o}}$
		
		 $\approx r_o//R_{CT}$ if R_{SRC} is very small
		 OR
		 $\approx r_o + R_{SRC} + g_m r_o R_{SRC}$ if R_{CT} is very large
			$\left(\dfrac{g_m r_o + 1}{1 + \dfrac{r_o}{R_{LD}} + \dfrac{r_o}{R_{CT}}}\right)\left(\dfrac{r_{in}}{r_{in} + R_{SRC}}\right)$
			
			 $\approx \dfrac{r_o//R_{LD}//R_{CT}}{\dfrac{1}{g_m}}\left(\dfrac{r_{in}}{r_{in} + R_{SRC}}\right)$
			
			 $\approx \dfrac{r_{out}//R_{LD}}{\dfrac{1}{g_m}}$ if R_{SRC} is very small

Common-Drain	∞	$\dfrac{1}{g_m}//r_o$ $\approx \dfrac{1}{g_m}$ if r_o is very large	$\dfrac{r_o//R_{LD}}{\frac{1}{g_m} + (r_o//R_{LD})} \approx \dfrac{r_o//R_{LD}}{r_{out} + (r_o//R_{LD})}$

Table 7.4 Quantitative summary of amplifier parameters 3

Configuration name	R_{gs}	R_{gd}
Common-Source	R_{SRC}	$g_m R_{SRC}(r_o//R_{CT}//R_{LD})$ $+(r_o//R_{CT}//R_{LD})$ $+R_{SRC}$
Common-Gate	$\frac{R_{SRC}[r_o+(R_{CT}//R_{LD})]}{R_{SRC}+r_o+(R_{CT}//R_{LD})+g_m R_{SRC} r_o}$	$R_{CT}//R_{LD}//(r_o + R_{SRC} + g_m r_o R_{SRC})$
Common-Drain	$\frac{R_{SRC}+(r_o//R_{LD})}{1+g_m(r_o//R_{LD})}$	R_{SRC}

Table 7.5 Qualitative summary of amplifier configuration properties and typical applications

Configuration name	r_{in}	r_{out}	Voltage gain	Typical application
Common-Source	Large	Medium	InvertingHigh	Middle of the signal-processing chain providing high voltage gain
Common-Gate	Small	Medium	Non-inverting High	Beginning of the signal-processing chain such as in: Low-Noise Amplifiers (LNAs)
Common-Drain	Large	Small	Less than one	End of the signal-processing chain such as in: Output drivers for large loads (small R_{LD}) Power Amplifiers (PAs)

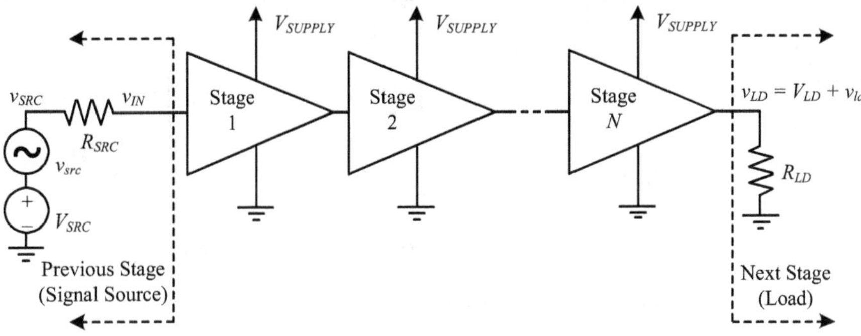

Fig. 7.34 An N-stage signal-processing chain

$$\frac{v_{ld}}{v_x} = \frac{r_o//R_{LD}}{\frac{1}{g_m} + (r_o//R_{LD})} \tag{7.73}$$

and

v_x / v_{src} is the gain of a CS with $R_{LD} = r_{in\_CD} = \infty$

$$\frac{v_x}{v_{src}} = -\frac{r_o//R_{CT}//\infty}{\frac{1}{g_m}} = -\frac{r_o//R_{CT}}{\frac{1}{g_m}} \tag{7.74}$$

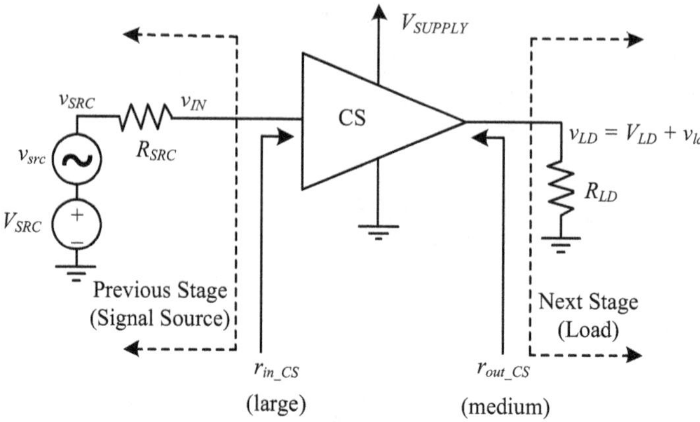

Fig. 7.35 A single-stage CS amplifier

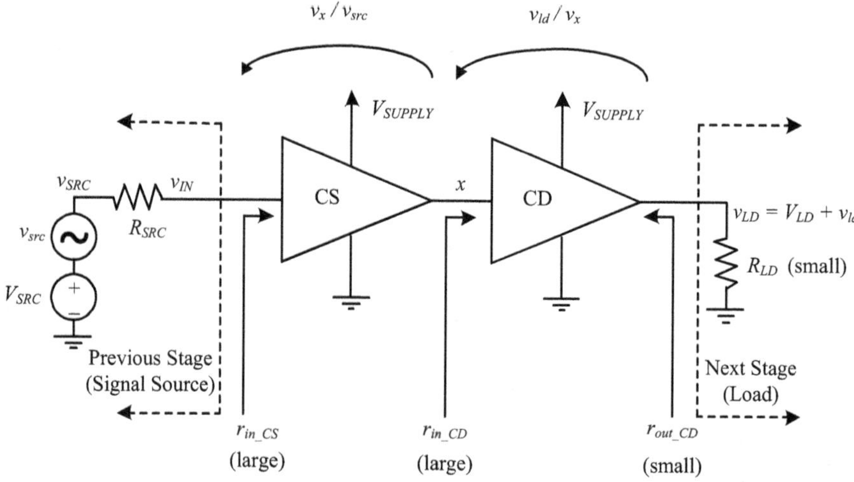

Fig. 7.36 A two-stage cascaded CS-CD amplifier chain

As a result, the total gain is:

$$\text{gain} = \frac{v_{ld}}{v_{src}} = \left(\frac{v_{ld}}{v_x}\right)\left(\frac{v_x}{v_{src}}\right) = \left(\frac{r_o//R_{LD}}{\frac{1}{g_m} + (r_o//R_{LD})}\right)\left(-\frac{r_o//R_{CT}}{\frac{1}{g_m}}\right) \tag{7.75}$$

In this case, even if the load is large (that is R_{LD} is small as in an audio speaker application for example), the total gain is still large. This might sound a bit counterintuitive at first since the gain of a stand-alone CD is slightly less than one. However, the CD acts as a buffer between the CS and R_{LD} so that the R_{LD} does not load the CS directly and the CS will be able to provide the gain it is capable of. As a result, the CD is very good as an output stage not because of its voltage gain but because of its high input resistance and low output resistance that make it capable of isolating the inner stages of the chain from the load.

7.6.2 Example 2: 3-Stage Cascade

In this example, the requirements are as in the previous example:

- a given signal source,
- a large load (small R_{LD}),
- a high gain which is high.

Additionally, we have the requirement that the input resistance of the amplifier should be matched to the small signal source resistance R_{SC}. This is a typical example when we are dealing with high frequencies.

In this case, the 2-stage solution given in Example 1 will not work since the input resistance of the CS is large, so it cannot match the small R_{SC}. Therefore, we need to precede the CS stage with another one with a small input resistance such as a CG. This results in a 3-stage chain as in Fig. 7.37.

Here, the total gain is:

$$\text{gain} = \frac{v_{ld}}{v_{src}} = \left(\frac{v_{ld}}{v_{x2}}\right)\left(\frac{v_{x2}}{v_{x1}}\right)\left(\frac{v_{x1}}{v_{src}}\right) \tag{7.76}$$

where.

v_{ld} / v_{x2} is the gain of a CD with $R_{SRC} = 0$ so

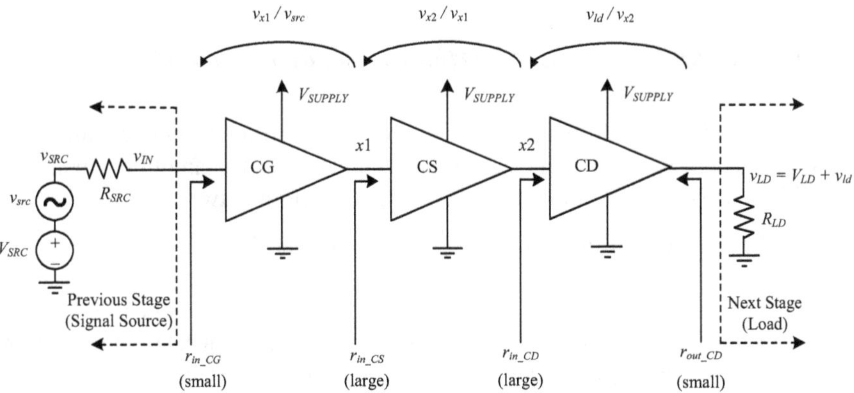

Fig. 7.37 A three-stage cascaded CG-CS-CD amplifier chain

$$\frac{v_{ld}}{v_{x2}} = \frac{r_o//R_{LD}}{\frac{1}{g_m} + (r_o//R_{LD})} \tag{7.77}$$

and

v_{x2} / v_{x1} is the gain of a CS with $R_{SRC} = 0$ $R_{LD} = r_{in\_CD} = \infty$

$$\frac{v_{x2}}{v_{x1}} = -\frac{r_o//R_{CT}//\infty}{\frac{1}{g_m}} = -\frac{r_o//R_{CT}}{\frac{1}{g_m}} \tag{7.78}$$

and

v_{x1} / v_{src} is the gain of a CG with $R_{LD} = r_{in\_CS} = \infty$

$$\frac{v_{x1}}{v_{src}} = \left(\frac{g_m r_o + 1}{1 + \frac{r_o}{\infty} + \frac{r_o}{R_{CT}}}\right)\left(\frac{r_{in}}{r_{in} + R_{SRC}}\right) \approx \frac{r_o//R_{CT}}{\frac{1}{g_m}}\left(\frac{\frac{1}{g_m}}{\frac{1}{g_m} + R_{SRC}}\right) \tag{7.79}$$

where we have used the approximation $g_m r_o >> 1$ and $r_o >> 1/g_m$ as in Table 7.3. As a result, the total gain is:

$$\begin{aligned} \text{gain} &= \frac{v_{ld}}{v_{src}} = \left(\frac{v_{ld}}{v_{x2}}\right)\left(\frac{v_{x2}}{v_{x1}}\right)\left(\frac{v_{x1}}{v_{src}}\right) \\[2mm] &= \left(\frac{r_o//R_{LD}}{\frac{1}{g_m} + (r_o//R_{LD})}\right)\left(-\frac{r_o//R_{CT}}{\frac{1}{g_m}}\right)\left[\frac{r_o//R_{CT}}{\frac{1}{g_m}}\left(\frac{\frac{1}{g_m}}{\frac{1}{g_m} + R_{SRC}}\right)\right] \end{aligned} \tag{7.80}$$

The first gain term is that of a CD, which is slightly less than one, the second is that of a CS, which is large, and the third is that of a CG, which is also large. As a result, this chain has the potential to meet the requirements.

7.6.3   Example 3: 2-Stage Vertical Cascade (Cascode)

In this example, we will present a popular chain called a Cascode. Figure 7.38a shows a single Cascode, which is essentially a 2-stage CS-CG amplifier chain as in Fig. 7.38b. This chain is unlike the ones presented before in two ways:

1. The chain is vertical, not horizontal. As a result, the CS and the CG share the same biasing current.
2. Although the information at the input and output is in the voltage mode, the intermediate information at the output of the CS is communicated in the current mode (i_X) to the CG, not in the voltage mode as in other chains.

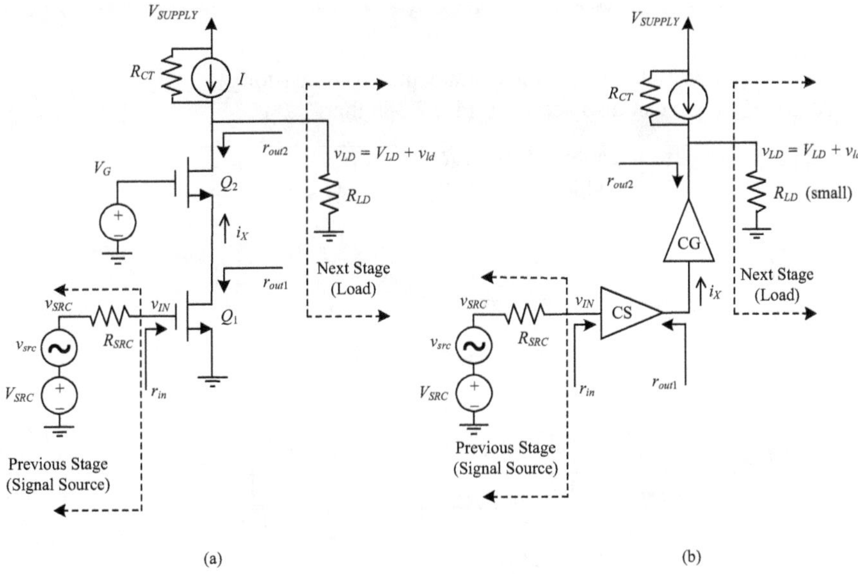

Fig. 7.38 A Cascode amplifier circuit (**a**) and conceptual (**b**)

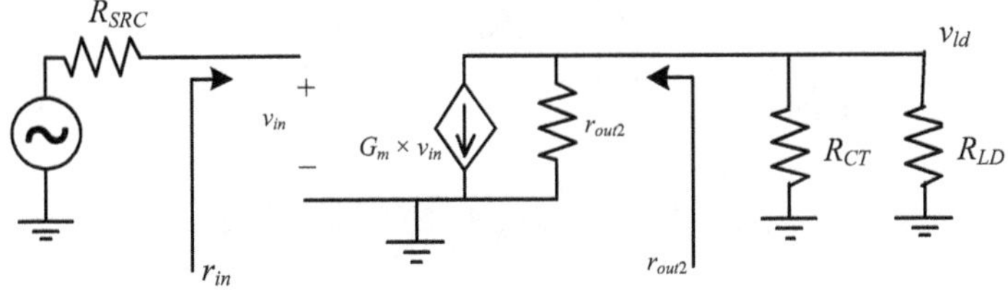

Fig. 7.39 AC small-signal model of the Cascode amplifier

In order to analyze for the total gain, we will draw the AC small-signal model of the circuit with the CS-CG combination treated as a single amplifier. This is shown in Fig. 7.39.

As can be seen, we know that r_{in} is that of a CS, so we have an open circuit at the input. As for the total gain, it is given as:

$$\text{gain} = \frac{v_{ld}}{v_{src}} = -G_m(r_{out2}//R_{CT}//R_{LD}) = -\frac{r_{out2}//R_{CT}//R_{LD}}{\frac{1}{G_m}} \tag{7.81}$$

Now, we need to get G_m and r_{out2}.

G_m is the transconductance, that is the factor by which v_{in} is transformed into a current. By looking at Fig. 7.38a, we can see that this is the job of transistor Q_1. As a result, we have:

$$G_m = g_{m1} \tag{7.82}$$

where g_{m1} is the small-signal transconductance of transistor Q_1.

As for r_{out2}, we can also see from Fig. 7.38a that:

- $r_{out1} = r_{out\_CS}$ from Table 7.3, with $R_{CT} = \infty$,
- $r_{out2} = r_{out\_CG}$ from Table 7.3, with $R_{SRC} = r_{out1}$ and $R_{CT} = \infty$.

Therefore:

$$r_{out1} = r_{o1} // R_{CT} = r_{o1} \tag{7.83}$$

and

$$r_{out2} = \frac{\dfrac{R_{SRC}}{r_{o2}} + 1 + g_{m2}R_{SRC}}{\dfrac{R_{SRC}}{R_{CT}r_{o2}} + \dfrac{(1 + g_{m2}R_{SRC})}{R_{CT}} + \dfrac{1}{r_{o2}}} = \frac{\dfrac{r_{o1}}{r_{o2}} + 1 + g_{m2}r_{o1}}{\dfrac{1}{r_{o2}}} \tag{7.84}$$

$$= r_{o1} + r_{o2} + g_{m2}r_{o1}r_{o2}$$

Usually, the third term dominates, so we have:

$$r_{out2} \approx g_{m2}r_{o1}r_{o2} \tag{7.85}$$

As a result, the total gain is:

$$\text{gain} = \frac{v_{ld}}{v_{src}} \approx -\frac{g_{m2}r_{o1}r_{o2} // R_{CT} // R_{LD}}{\dfrac{1}{g_{m1}}} \tag{7.86}$$

This result is much higher than that of a CS as in Table 7.3, not due to any change in the transconductance, but due to an increase in the output resistance by a factor of $g_{m2}r_{o2}$.

Therefore, Cascoding can also be considered as a method to increase the output resistance, a fact that will be very useful in applications other than amplifiers.

7.6.4 Example 4: BJT Transistor Pairing

From the previous examples, we have seen in several cases the benefits of having a high input resistance. BJTs have a current i_B that goes through their Base terminal. This makes their input resistance at the Base (r_π) finite. In order to increase this resistance, we can combine two transistors to form one "supertransistor." Two examples of such a combination are shown in Fig. 7.40.

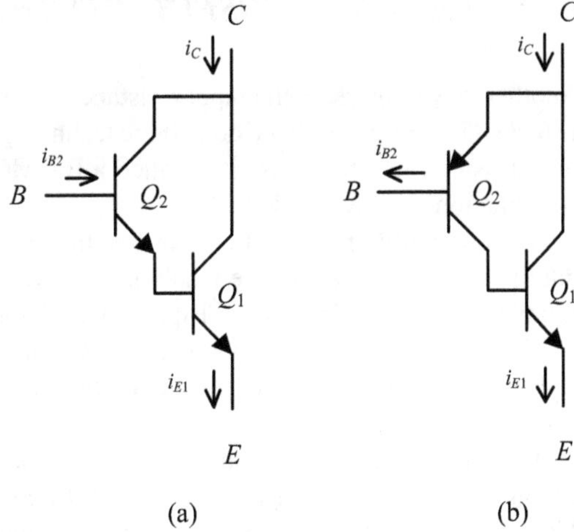

Fig. 7.40 BJT transistor pairing examples: two NPNs (Darlington Configuration) (**a**), and a PNP with an NPN (**b**)

Figure 7.40a makes use of two NPN transistors resulting in what is known as a Darlington Configuration. This configuration can also be viewed as two Common-Collector configurations in cascade.

Figure 7.40b makes use of a PNP transistor with an NPN transistor. This configuration can also be viewed as to Common-Emitter configuration followed by a Common-Collector configuration in cascade.

Focusing on Fig. 7.40a, the benefits can be seen by finding the relation between the i_E and i_B.

$$i_{E1} = (\beta_1 + 1)i_{B1}$$
$$i_{B1} = i_{E2} \tag{7.87}$$
$$i_{E2} = (\beta_2 + 1)i_{B2}$$

Therefore:

$$i_{E1} = (\beta_1 + 1)(\beta_2 + 1)i_{B2} \tag{7.88}$$

If the transistors are the same then:

$$i_{E1} \approx \beta^2 i_{B2} \tag{7.89}$$

Therefore, the combination acts as a "supertransistor" with a β equal to the square of that of the individual transistors, effectively increasing the input resistance at the Base terminal.

7.6.5　Example 5: MOSFET- BJT Transistor Pairing

Another way to increase the input resistance from Example 4 is by using a MOSFET with a BJT. This can only be done if the technology process used supports both types of transistors, in which case it is called a BiCMOS process. An example of such a combination is shown in Fig. 7.41.

In these configurations, it is clear that the input resistance at the Gate terminal is infinite since i_{G2} is zero. The resulting "supertransistor" has an insulated gate and behaves like an Insulated-Gate Bipolar Transistor (IGBT).

Figure 7.41a can also be viewed as a Common-Drain followed by a Common-Collector, while Fig. 7.41b can be viewed as a Common-Source followed by a Common-Collector.

The advantage of these configurations compared to using MOSFETs only is the exponential current–voltage relation in BJTs compared to the quadratic relation in MOSFETs. As a result, if these are used as switches, the switching process will be faster all while having an infinite input resistance at the Gate terminal.

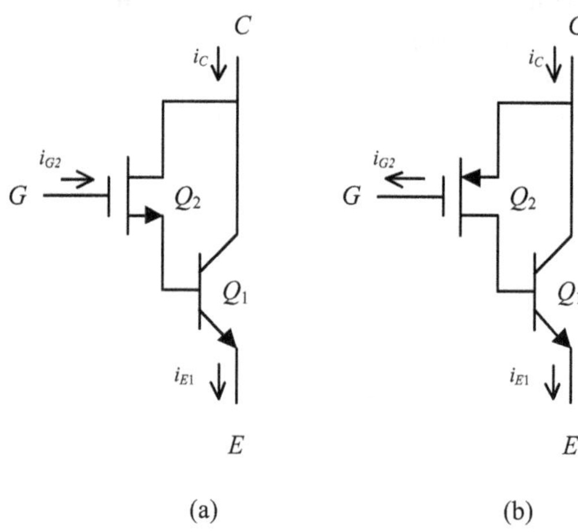

Fig. 7.41 MOSFET-BJT transistor pairing examples: an NMOS with an NPN (**a**), and a PMOS with an NPN (**b**)

Chapter 8
Logic Circuits

This chapter describes circuits for logic applications. These circuits are at the heart of modern information technology. The constant improvement required in their performance is key to the advancement of the digital age. The vast majority of today's logic circuits are implemented using Field-Effect Transistors (FETs). This is expected to be the case for several years to come. As a result, and without any loss of generality regarding other FET families, this chapter will concentrate particularly on MOSFET-based implementations. It starts by categorizing the different logic circuit families. This is followed by detailed discussions regarding each logic circuit family, starting with Static logic and finishing with Dynamic logic.

8.1 Introduction to Logic Circuit Families

There are two broad families of logic circuits. The first is called Static and the second is called Dynamic. Each one of them can be subdivided into other types as can be seen in Fig. 8.1.

PDN stands for Pull-Down Network and PUN stands for Pull-Up Network.

In Static logic families, the output changes as a direct result of the input, whereas in Dynamic logic families, the output changes based on the input as well as a clock signal. The general structure is shown in Fig. 8.2.

In integrated circuits, logic gates are cascaded in order to create a set of functionalities. As a result, and since at high frequencies the interconnections as well as the input to all these logic gates are capacitive, we will model the load as such and we will call it C_{LD}. Since a logic gate can drive several other gates, the number of gates at the load is called the fan-out. Additionally, since a logic gate can have several inputs, the number of inputs to the gate is called the fan-in.

J. G. Atallah, M. Ismail, *Integrated Electronic Circuits*,
https://doi.org/10.1007/978-3-031-62707-1_8

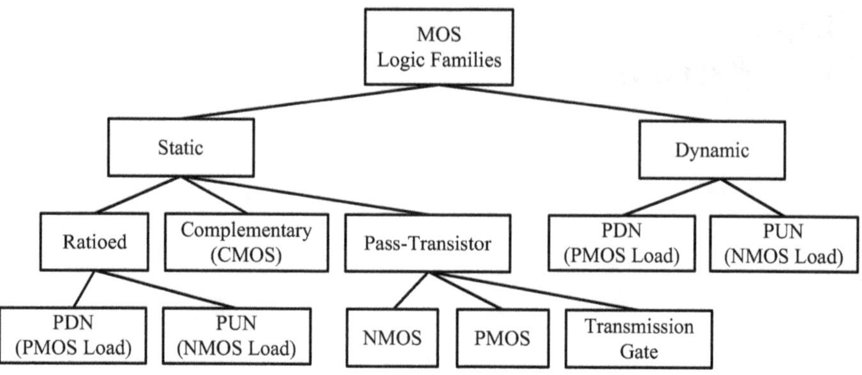

Fig. 8.1 Logic circuit families

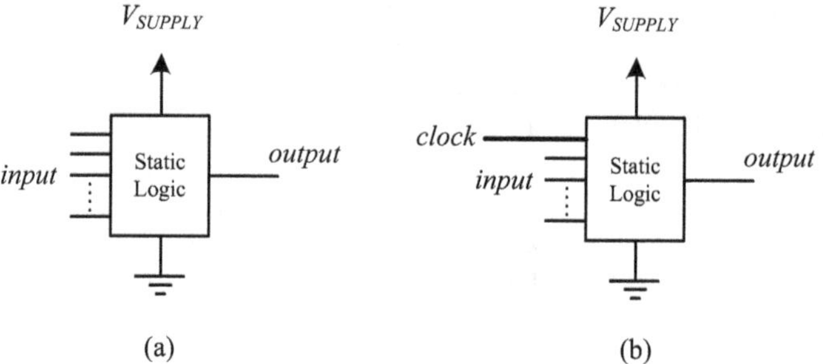

Fig. 8.2 Static logic (**a**) used to implement combinational relations and Dynamic logic (**b**) used to build sequential relations

8.2   General Properties

In this section, we will discuss some general logic circuits' properties and present the derivations for the ones that are common to all logic families.

8.2.1   Voltage-Transfer Characteristic

The simplest logic gate is an inverter, whose voltage-transfer characteristic is shown in Fig. 8.3.

As can be seen, the voltage-transfer characteristic is a plot of the output v_{LD} versus the input v_{SRC}.

V_{LD-H} is the maximum value at the output and V_{LD-L} is the minimum value at the output.

V_M is the point on the curve where v_{LD} is equal to v_{SRC}.

Fig. 8.3 Voltage-transfer characteristic of an inverter

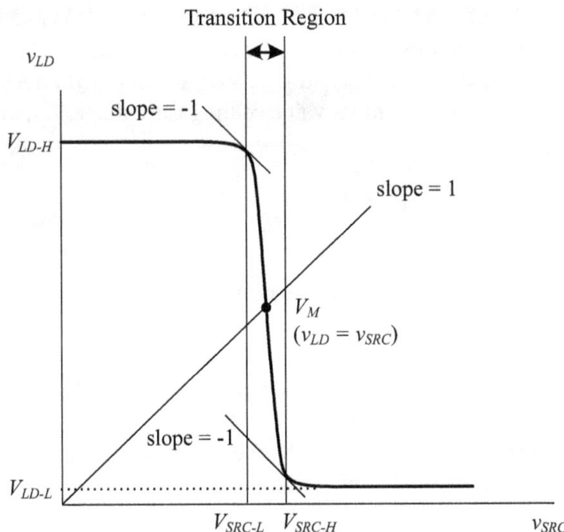

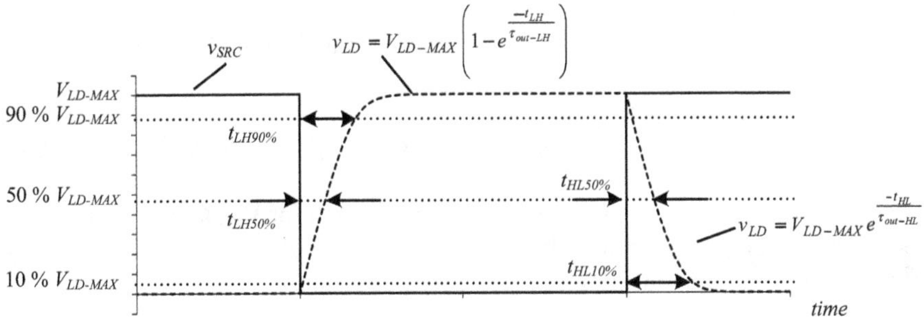

Fig. 8.4 Inverter gate transition time

$V_{SRC\text{-}L}$ is the value of the input before which the output is high, and $V_{SRC\text{-}H}$ is the value of the input beyond which the output is low. Between these two, the output is in a transition region that is neither low nor high.

8.2.2   Transition Time

The transition time is described by measuring the time it takes for the output to reach a certain percentage of its final value. This is done by taking as a reference an input signal that transitions instantaneously.

Given an inverter gate, the waveforms used to measure the transition time are shown in Fig. 8.4.

The time it takes for the output to change, also called the transition time, is related to the time constant due to the average resistance R_{out} and average capacitance C_{out} seen at that node. R_{out} and C_{out} are the combined effect of the gate itself and its load.

If we take the initial value of the load voltage V_{LD} to be 0 and the final value to be $V_{LD\text{-}MAX}$, then:

$$v_{LD} = V_{LD-MAX}\left(1 - e^{\frac{-t_{LH}}{\tau_{out-LH}}}\right) \tag{8.1}$$

where t_{LH} indicates that the output V_{LD} is transitioning from low to high.

Also, we have

$$\tau_{out-LH} = R_{out-LH}C_{out-LH} \tag{8.2}$$

On the other hand, if we take the initial value of the load voltage V_{LD} to be $V_{LD\text{-}MAX}$ and the final value to be 0, then:

$$v_{LD} = V_{LD-MAX}e^{\frac{-t_{HL}}{\tau_{out-HL}}} \tag{8.3}$$

where t_{HL} indicates that the output V_{LD} is transitioning from high to low.

Also, we have

$$\tau_{out-HL} = R_{out-HL}C_{out-HL} \tag{8.4}$$

For example, if we are interested in the time it takes for V_{LD} to start from 0 and reach half $V_{LD\text{-}MAX}$, also called the switching point for the load, which is usually another gate, then the transition time is derived as follows:

$$
\begin{aligned}
V_{LD} &= 0.5 \times V_{LD-MAX} \\
\Rightarrow 1 - e^{\frac{-t_{LH50\%}}{\tau_{out-LH}}} &= 0.5 \\
\Rightarrow t_{LH50\%} &= 0.69\tau_{out-LH}
\end{aligned}
\tag{8.5}
$$

If, however, we are interested in the time it takes V_{LD} to start from 0 and reach $0.9V_{LD\text{-}MAX}$, then the transition time is derived as follows:

$$
\begin{aligned}
V_{LD} &= 0.9 \times V_{LD-MAX} \\
\Rightarrow 1 - e^{\frac{-t_{LH90\%}}{\tau_{out-LH}}} &= 0.9 \\
\Rightarrow t_{LH90\%} &= 2.3\tau_{out-LH}
\end{aligned}
\tag{8.6}
$$

which is more than 3 times $t_{LH50\%}$.

Additionally, if we are interested in the time it takes V_{LD} to start from $0.1V_{LD\text{-}MAX}$ and reach $0.9V_{LD\text{-}MAX}$, then the transition time would be slightly shorter:

$$t_{LH10\%-90\%} = 2.2\tau_{out-LH} \tag{8.7}$$

Similarly, if we are interested in the time it takes for V_{LD} to start from $V_{LD\text{-}MAX}$ and reach $0.5V_{LD\text{-}MAX}$, also called the switching point for the load, which is usually another gate, then the transition time is derived as follows:

$$\begin{aligned} V_{LD} &= 0.5 \times V_{LD-MAX} \\ \Rightarrow e^{\frac{-t_{HL50\%}}{\tau_{out-HL}}} &= 0.5 \\ \Rightarrow t_{HL50\%} &= 0.69\tau_{out-HL} \end{aligned} \tag{8.8}$$

If, however, we are interested in the time it takes V_{LD} to start from $V_{LD\text{-}MAX}$ and reach $0.1V_{LD\text{-}MAX}$, then the transition time is derived as follows:

$$\begin{aligned} V_{LD} &= 0.1 \times V_{LD-MAX} \\ \Rightarrow e^{\frac{-t_{HL10\%}}{\tau_{out}}} &= 0.1 \\ \Rightarrow t_{HL10\%} &= 2.3\tau_{out-HL} \end{aligned} \tag{8.9}$$

which is more than 3 times $t_{HL50\%}$.

Additionally, if we are interested in the time it takes V_{LD} to start from $0.9V_{LD\text{-}MAX}$ and reach $0.1V_{LD\text{-}MAX}$, then the transition time would be slightly shorter:

$$t_{HL10\%-90\%} = 2.2\tau_{out-HL} \tag{8.10}$$

As a result, the transition time t_p is defined as:

$$t_p = \frac{t_{LH50\%} + t_{HL50\%}}{2} \tag{8.11}$$

8.2.3   Noise Margins

All electrical signals are subject to electrical noise from sources internal and external to the system. An important measure of how much a logic gate such as an inverter can cope with the noise is called the noise margin. In essence, the noise margin is the allowed range of an input signal such that the output signal stays at its desired value. Two of these margins exist for every gate, one for the high value and one for the low value.

From Fig. 8.3, the two noise margins can be calculated as follows:

$$NM_L = V_{SRC-L} - V_{LD-L}$$
$$NM_H = V_{LD-H} - V_{SRC-H} \tag{8.12}$$

NM_L and NM_H are the low and high noise margins respectively.

It is in our interest to have the highest noise margins possible.

The primary method to achieve that end is for the logic gate to have an output range from 0 to V_{SUPPLY}. As a result of that

$$V_{LD-L} = 0$$
$$V_{LD-H} = V_{SUPPLY}$$
$$NM_L = V_{SRC-L}$$
$$NM_H = V_{SUPPLY} - V_{SRC-H} \tag{8.13}$$

An additional method is for the transition region to be very small such that

$$V_{SRC-L} = V_{SRC-H} = V_M$$
$$NM_L = NM_H = \frac{V_{SUPPLY}}{2} \tag{8.14}$$

Both methods are targeted in modern logic gates to varying degrees of success depending on the logic family.

8.2.4 Power Consumption: Dynamic

The power consumed P_{SUPPLY} is the average power provided by the power supply. This power can be subdivided into two parts:

- $P_{DYNAMIC}$: dynamic power, which is the power consumed while the output is changing.
- P_{STATIC}: static power, which is power consumed while the gate is in standby mode, that is its output is not changing.

As a result, the following relation holds:

$$P_{SUPPLY} = P_{DYNAMIC} + P_{STATIC} \tag{8.15}$$

Note that these same relations that hold for power also hold for the energy.

We will now derive the expression for $P_{DYNAMIC}$ since it is common to all logic gates. As for P_{STATIC}, it turns out to be one of the main differences between logic families, and will be dealt with individually for every logic family.

When the logic gate turns ON, energy is transferred from the power supply to the load C_{LD} which gets charged up and its voltage changes from zero to the maximum value, V_{SUPPLY}. In order to do so, the amount of energy drawn from the power supply is

$$E_{SUPPLY} = V_{SUPPLY}Q \tag{8.16}$$

where Q is the charge delivered to the load. Knowing that

$$Q = C_{LD}V_{SUPPLY} \tag{8.17}$$

then

$$E_{SUPPLY} = C_{LD}V_{SUPPLY}^2 \tag{8.18}$$

At this point, and from electromagnetics, the energy stored in C_{LD} is

$$E_{LD} = \frac{1}{2}C_{LD}V_{SUPPLY}^2 \tag{8.19}$$

As a result, the energy dissipated is

$$E_{DYNAMIC\_ON} = E_{SUPPLY} - E_{LD} = \frac{1}{2}C_{LD}V_{SUPPLY}^2 \tag{8.20}$$

When the logic gate turns OFF, C_{LD} gets discharged and its voltage changes from V_{SUPPLY} to zero. As a result, the energy dissipated will be the energy that was stored in C_{LD}, which is

$$E_{DYNAMIC\_OFF} = \frac{1}{2}C_{LD}V_{SUPPLY}^2 \tag{8.21}$$

Combining (8.20) with (8.21), the total dissipated dynamic energy per cycle is

$$\begin{aligned} E_{DYNAMIC\_PERCYCLE} \\ = E_{DYNAMIC\_ON} + E_{DYNAMIC\_OFF} \\ = C_{LD}V_{SUPPLY}^2 \end{aligned} \tag{8.22}$$

If the gate switching frequency is f in Hz, and assuming that the probability of the gate turning ON is the same as turning OFF, the dynamic power dissipation can be written as

$$P_{DYNAMIC} = f \times E_{DYNAMIC\_PERCYCLE} = f C_{LD}V_{SUPPLY}^2 \tag{8.23}$$

Eventually, since the load is being charged, then discharged, all the dynamic power ends up as dissipated power.

8.3 Static Logic

Static logic does not require a clock signal to operate. It is used to create combinational logic circuits. We will start by describing ratioed implementations, followed by complementary implementations, and pass-transistor implementations.

8.3.1 Ratioed Logic

The discussion of ratioed logic starts from the common-source amplifier configuration. This configuration results in an inverting gain. This inversion property can be used in order to build an inverter out of a common-source amplifier by swinging the input between the extreme voltage values.

As for the current source used in this configuration, we need to implement it in the simplest way possible. In fact, any load that will allow the output to swing between extreme values can be used. Obviously, the simplest load would be a resistor, but we prefer to use transistors rather than resistors in integrated circuits. As a result, we always end up with two possibilities for the load that tries to mimic a current source: a PMOS transistor (PMOS load, thus a PDN implementation), or an NMOS transistor (NMOS load, thus a PUN implementation). As a result, all ratioed logic circuits can be depicted as in Fig. 8.5.

A ratioed logic inverter gate is essentially a CS amplifier driven to its extreme. The current source is implemented as a single transistor as shown in Fig. 8.6 followed by typical input and output waveforms in Fig. 8.7. In both cases, the input transistor is labeled Q_1 and the load is labeled Q_2. Notice that in the PDN case, the load is a PMOS transistor that needs to be kept ON, thus its gate is connected to ground. As for the PUN case, the load is an NMOS transistor that needs to be kept ON, thus its gate is connected to V_{SUPPLY}.

Fig. 8.5 Ratioed logic gates with a PDN (**a**) and a PUN (**b**)

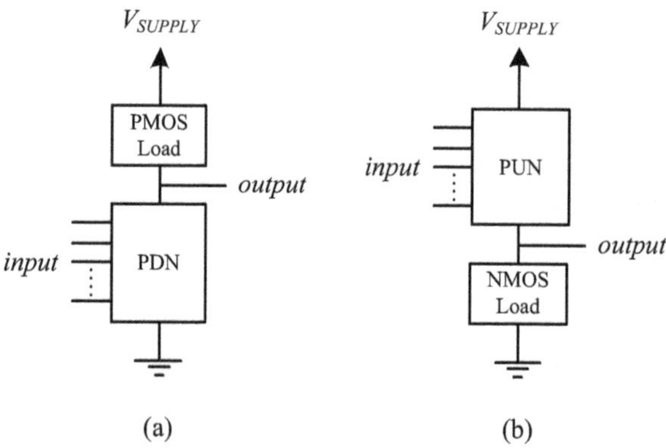

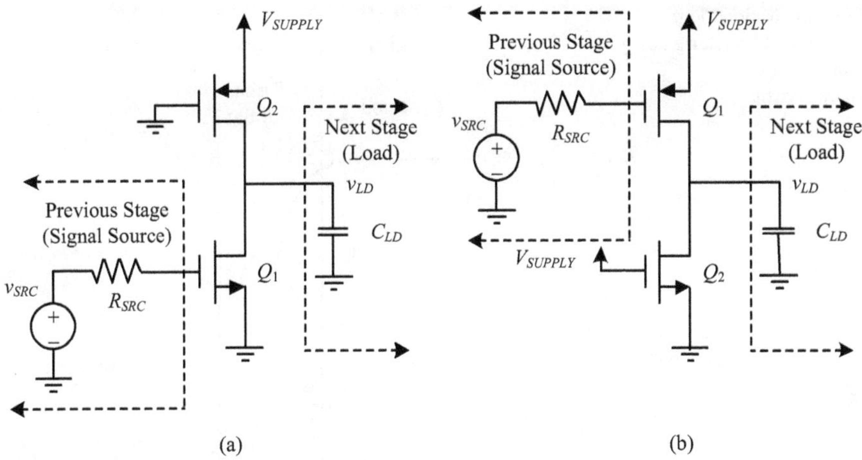

Fig. 8.6 Ratioed logic inverter gate PDN-based (**a**), PUN-based (**b**)

Fig. 8.7 Typical ratioed logic inverter gate input and output waveforms

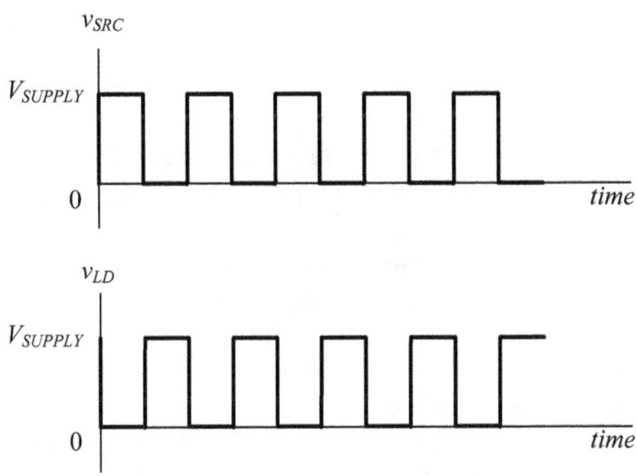

8.3.1.1 Ratioed Logic: Voltage-Transfer Characteristic

Without any loss of generalization, we will focus on the PDN-based ratioed logic inverter as in Fig. 8.6a.

In the following analysis, let

$$
\begin{aligned}
k_1 &= k'_1 \left(\frac{W}{L}\right)_1 \\
k_2 &= k'_2 \left(\frac{W}{L}\right)_2
\end{aligned}
\tag{8.24}
$$

Also, usually within the same technology, the NMOS and PMOS transistors have the same threshold voltage magnitudes:

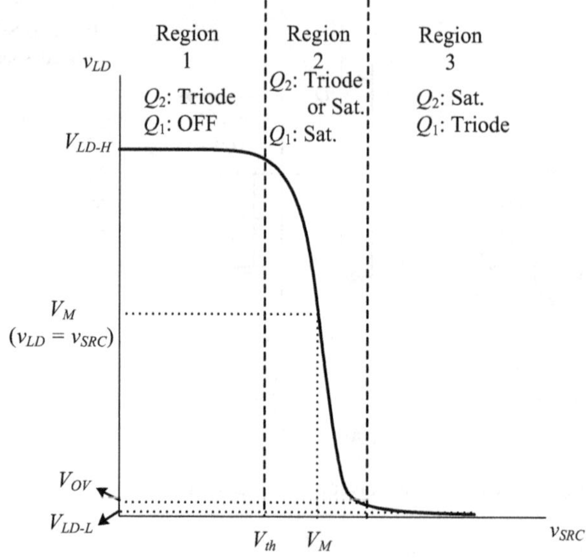

Fig. 8.8 Ratioed logic PDN-based inverter gate voltage-transfer characteristic

$$|V_{thN}| = |V_{thP}| = V_{th} \tag{8.25}$$

Additionally, we are going to assume that $V_M > V_{th}$, which is usually the case.

We will start with the input–output voltage-transfer characteristic as shown in Fig. 8.8.

The voltage-transfer characteristic can be divided into three regions. This division is based on the state of the input transistor Q_1.

In Region 1, v_{SRC} is less than V_{th}. As a result, Q_1 is OFF while Q_2 is in the triode region since v_{LD} is high, thus v_{DS2} is low. In this region, v_{LD} can take on its highest value $V_{LD\text{-}H}$, which is almost V_{SUPPLY}.

In Region 3, v_{LD} which is v_{DS1} is less than V_{OV}, so Q_1 is in the triode region and Q_2 is in the saturation region. In this region, v_{LD} can take on its lowest value $V_{LD\text{-}L}$. To calculate for $V_{LD\text{-}L}$, we equate the currents in Q_1 and Q_2 as follows:

$$i_{D1} = i_{D2}$$
$$k_1\left[(|v_{GS1}| - |V_{th}|)\,|v_{DS1}| - \frac{1}{2}|v_{DS1}|^2\right]$$
$$= \frac{1}{2}k_2(|v_{GS2}| - |V_{th}|)^2(1 + \lambda|v_{DS2}|) \tag{8.26}$$

Additionally, we have:

$$|v_{GS1}| = V_{SUPPLY}$$
$$|v_{DS1}| = V_{LD-L}$$
$$|v_{GS2}| = V_{SUPPLY} \qquad\qquad (8.27)$$
$$|v_{DS2}| = V_{SUPPLY} - V_{LD-L}$$

Replacing (8.27) in (8.26) we get:

$$k_1 \left[(V_{SUPPLY} - V_{th}) V_{LD-L} - \frac{1}{2} V_{LD-L}^2 \right]$$
$$= \frac{1}{2} k_2 (V_{SUPPLY} - V_{th})^2 (1 + \lambda(V_{SUPPLY} - V_{LD-L})) \qquad (8.28)$$

Solving for V_{LD-L} we get:

$$V_{LD-L} = (V_{SUPPLY} - V_{th})$$
$$\times \left(\frac{1 + \dfrac{k_2}{k_1} \lambda(V_{SUPPLY} - V_{th})^2 -}{\sqrt{1 - \dfrac{k_2}{k_1} - \dfrac{k_2}{k_1} \lambda \left(V_{th} - \dfrac{k_2}{4k_1} \lambda(V_{SUPPLY} - V_{th})^2 \right)}} \right) \qquad (8.29)$$

Note that if λ is very small, then:

$$V_{LD-L} \approx (V_{SUPPLY} - V_{th}) \left(1 + \sqrt{1 - \frac{k_2}{k_1}} \right) \qquad (8.30)$$

Either way, in order to have a low V_{LD-L}, k_2 / k_1 has to be kept small, thus:

$$\frac{k_2}{k_1} < < 1$$
$$\Rightarrow k_2 < < k_1$$
$$\Rightarrow k'_2 \left(\frac{W}{L} \right)_2 < < k'_1 \left(\frac{W}{L} \right)_1 \qquad (8.31)$$
$$\Rightarrow \frac{\left(\frac{W}{L} \right)_2}{\left(\frac{W}{L} \right)_1} < < \frac{k'_1}{k'_2}$$

As a result, the aspect ratio of the load transistor Q_2 must be kept much smaller than that of the input transistor Q_1 in order to reduce the voltage drop across Q_1. This decreases V_{LD-L} in PDN-based implementations and increases V_{LD-H} in PUN-based implementations.

Figure 8.9 shows how v_{LD} varies with the ratio k_2 / k_1 in the vicinity of V_{LD-L} for the case when λ is 0 and when λ is 1. As can be seen, in both cases, a small V_{LD-L} can be achieved if (8.31) is satisfied.

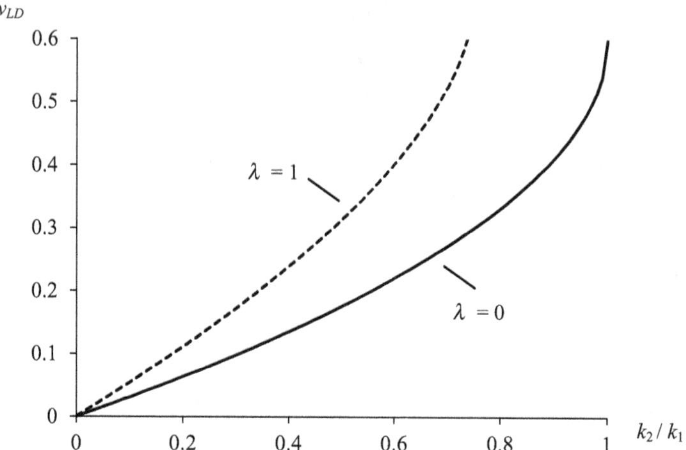

Fig. 8.9 Ratioed logic v_{LD} versus k_2/k_1 in the vicinity of $V_{LD\text{-}L}$

As a result, a PDN-based gate has a strong $V_{LD\text{-}H}$ and a weak $V_{LD\text{-}L}$, while a PUN-based gate has a weak $V_{LD\text{-}H}$ and a strong $V_{LD\text{-}L}$.

In Region 2, V_M is defined when the voltage v_{LD} is equal to v_{SRC}. Consequently, the voltages will be as follows:

$$\begin{aligned}
|v_{GS1}| &= V_M \\
|v_{DS1}| &= V_M \\
|v_{GS2}| &= V_{SUPPLY} \\
|v_{DS2}| &= V_{SUPPLY} - V_M
\end{aligned} \tag{8.32}$$

At this point, Q_1 is in the saturation region. The situation of Q_2 depends on λ. Therefore two cases arise:

Case 1: If λ is small, Q_2 would be in the triode region.
Case 2: Q_2 would be in the saturation region.

Taking Case 1 and equating the currents, we get:

$$\begin{aligned}
i_{D1} &= i_{D2} \\
\frac{1}{2}k_1(|v_{GS1}| &- |V_{th}|)^2(1 + \lambda|v_{DS1}|) \\
&= k_2\left[(|v_{GS2}| - |V_{th}|)\,|v_{DS2}| - \frac{1}{2}|v_{DS2}|^2\right]
\end{aligned} \tag{8.33}$$

Replacing (8.32) in (8.33) we get:

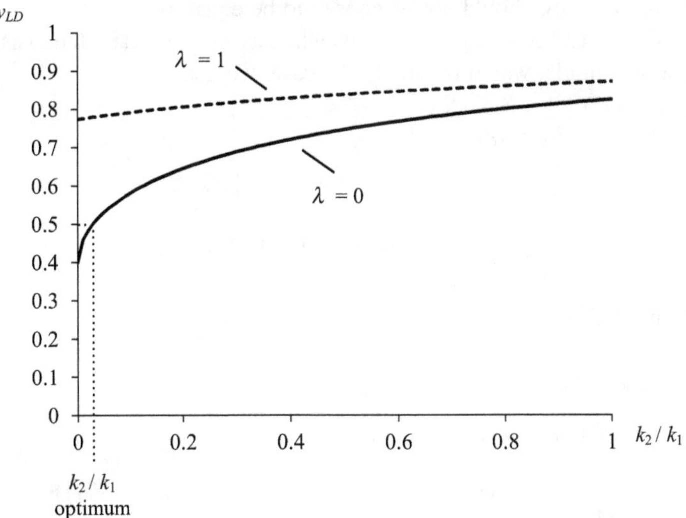

Fig. 8.10 Ratioed logic v_{LD} versus k_2 / k_1 in the vicinity of V_M - Case 1

$$\frac{1}{2}k_1(V_M - V_{th})^2(1 + \lambda V_M)$$

$$= k_2\left[(V_{SUPPLY} - V_{th})(V_{SUPPLY} - V_M) - \frac{1}{2}(V_{SUPPLY} - V_M)^2\right] \qquad (8.34)$$

$$\Rightarrow (V_M - V_{th})^2(1 + \lambda V_M)$$

$$= \frac{k_2}{2k_1}\left[(V_{SUPPLY} - V_{th})(V_{SUPPLY} - V_M) - \frac{1}{2}(V_{SUPPLY} - V_M)^2\right]$$

Plotting v_{LD} versus k_2 / k_1 in the vicinity of V_M with V_{SUPPLY} being 1 V, we get the graph shown in Fig. 8.10. As can be seen, if λ is small or zero for that matter, an optimum value for k_2 / k_1 exists so that v_{LD} is 0.5 V. However, if λ is high (equal to 1 in this case), no value exists and that is because Q_2 would be in the saturation region, which is one of the situations of Case 2.

As a result, in Case 1 and when λ is small (almost zero), V_M can be expressed as:

$$V_M \approx V_{th} + (V_{SUPPLY} - V_{th})\sqrt{\frac{k_2}{k_1 + k_2}} \qquad (8.35)$$

and k_2 / k_1 optimum would be:

$$\frac{k_2}{k_1}\text{ optimum} \approx \frac{\left(\dfrac{V_M - V_{th}}{V_{SUPPLY} - V_{th}}\right)^2}{1 - \left(\dfrac{V_M - V_{th}}{V_{SUPPLY} - V_{th}}\right)^2} \qquad (8.36)$$

As can be seen, V_M should not be chosen to be equal to V_{th}.

Taking Case 2 and equating the currents while assuming that both transistors have the same λ magnitude, which is usually the case, we get:

$$i_{D1} = i_{D2}$$
$$\frac{1}{2}k_1(|v_{GS1}| - |V_{th}|)^2(1 + \lambda|v_{DS1}|) \tag{8.37}$$
$$= \frac{1}{2}k_2(|v_{GS2}| - |V_{th}|)^2(1 + \lambda|v_{DS2}|)$$

Replacing (8.32) in (8.37) we get:

$$i_{D1} = i_{D2}$$
$$\frac{1}{2}k_1(V_M - V_{th})^2(1 + \lambda V_M) \tag{8.38}$$
$$= \frac{1}{2}k_2(V_{SUPPLY} - V_{th})^2(1 + \lambda(V_{SUPPLY} - V_M))$$

In Case 2, we have two situations: one of them if λ is small and the other one if λ is large.

If λ is small, almost zero, then V_M can be expressed as:

$$V_M \approx V_{th} + (V_{SUPPLY} - V_{th})\sqrt{\frac{k_2}{k_1}} \tag{8.39}$$

Note that this is slightly different from Eq. (8.35).

Also, in this situation, k_2 / k_1 optimum would be:

$$\frac{k_2}{k_1}\text{optimum} \approx \left(\frac{V_M - V_{th}}{V_{SUPPLY} - V_{th}}\right)^2 \tag{8.40}$$

On the other hand, if λ is large, then k_2 / k_1 optimum would be:

$$\frac{k_2}{k_1}\text{optimum} = \frac{(V_M - V_{th})^2(\lambda V_M + 1)}{(V_{SUPPLY} - V_{th})^2(\lambda(V_{SUPPLY} - V_M) + 1)} \tag{8.41}$$

Note that if V_M is chosen to be in the middle of V_{SUPPLY}, λ will have no effect in Case 2 and Eq. (8.41) degenerates into Eq. (8.40). This is shown in Fig. 8.11 where v_{LD} is plotted versus k_2 / k_1 in the vicinity of V_M with V_{SUPPLY} being 1 V.

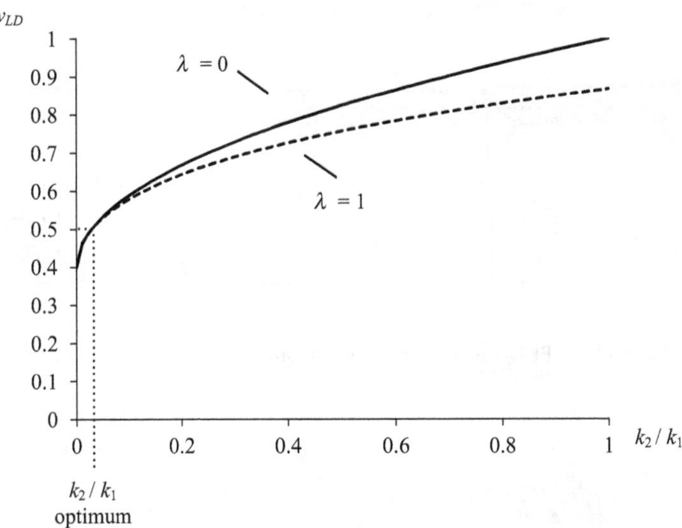

Fig. 8.11 Ratioed logic v_{LD} versus k_2 / k_1 in the vicinity of V_M - Case 2

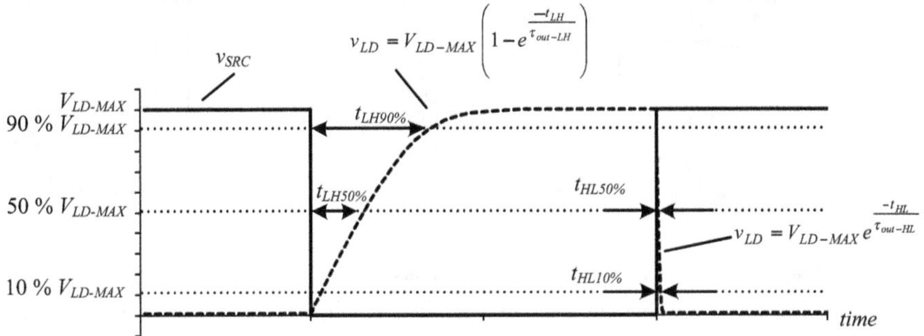

Fig. 8.12 Ratioed logic PDN-based inverter gate transition time

8.3.1.2   Ratioed Logic: Transition Time

Given the PDN-based inverter gate of Fig. 8.6a, the waveforms used to measure the transition time are shown in Fig. 8.12.

Also, given the PUN-based inverter gate of Fig. 8.6b, the waveforms used to measure the transition time are shown in Fig. 8.13.

The detailed derivation of the different time constants is presented in Sect. 8.2.2.

It is clear from Figs. 8.12 and 8.13 that τ_{out-LH} and τ_{out-HL} are different. Also, which one is larger depends on whether it is a PDN-based gate or a PUN-based gate.

Concentrating on a PDN-based gate, we can deduce that it has two stable extremes, namely Region 1 and Region 3. The equivalent circuits in both regions are shown in Fig. 8.14.

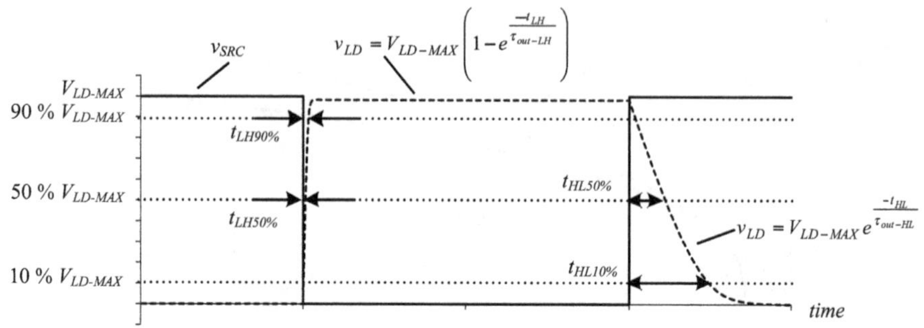

Fig. 8.13 Ratioed logic PUN-based inverter gate transition time

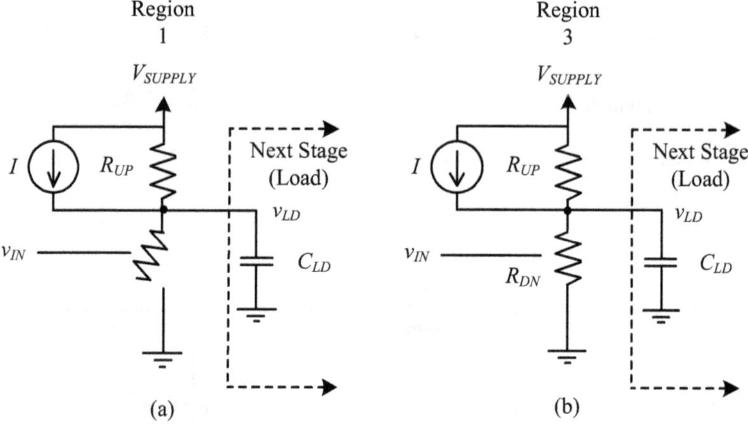

Fig. 8.14 Ratioed logic PDN-based gate time constants

Region 1 corresponds to $\tau_{out\text{-}LH}$ and Region 3 corresponds to $\tau_{out\text{-}HL}$. Deriving these time constants, we find that:

$$\tau_{out-LH} = R_{out-LH}C_{out-LH} = R_{UP}C_{LD} \tag{8.42}$$

$$\tau_{out-HL} = R_{out-HL}C_{out-HL} = (R_{UP}//R_{DN})C_{LD} \tag{8.43}$$

Note that we are assuming, and that is usually the case, that the load is capacitive with minimal resistance compared to that of the driving transistors. Also, the output capacitance is almost totally dominated by the load capacitance.

As can be clearly seen, for a PDN-based gate, $\tau_{out\text{-}HL}$ is always smaller than $\tau_{out\text{-}LH}$, which is why a PDN-based gate transitions to a low voltage faster than to a high voltage as can be seen in Fig. 8.12.

A similar analysis can be done for a PUN-based gate, where the time constants will be as follows:

$$\tau_{out-LH} = R_{out-LH}C_{out-LH} = (R_{UP}//R_{DN})C_{LD} \tag{8.44}$$

$$\tau_{out-HL} = R_{out-HL}C_{out-HL} = R_{DN}C_{LD} \tag{8.45}$$

In this case, for a PUN-based gate, τ_{out-HL} is always greater than τ_{out-LH}, which is why a PUN-based gate transitions to a high voltage faster than to a low voltage as can be seen in Fig. 8.13.

As a result of this analysis, the following can be deduced:

- PDN-based gates: strong high voltage, fast low voltage.
- PUN-based gate: strong low voltage, fast high voltage.

8.3.1.3   Ratioed Logic: Sample Gates

We have seen how an inverter can be implemented using ratioed logic. We will now showcase the implementation of other gates.

For example a 3-input ratioed NOR gate is shown in Fig. 8.15.

As can be seen, in order to keep the same functionality, the input transistors that are in parallel in the PDN version will become in series in the PUN version and vice versa.

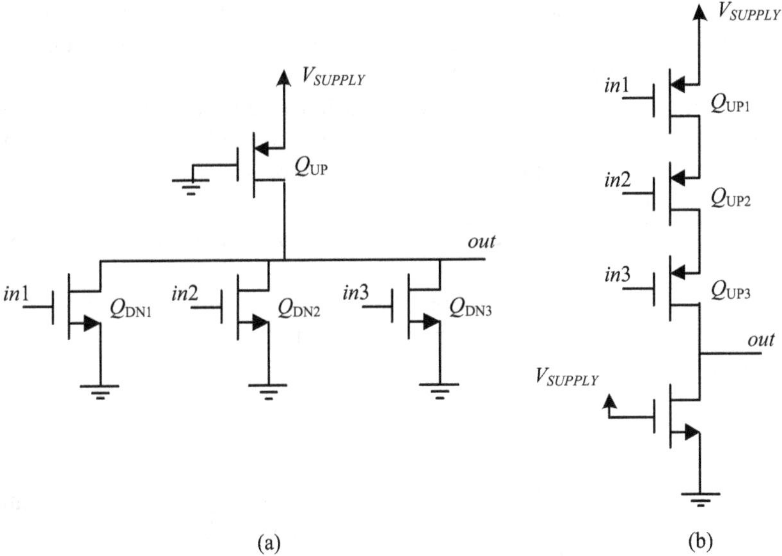

Fig. 8.15 Ratioed logic NOR gate PDN-based (**a**) and PUN-based (**b**)

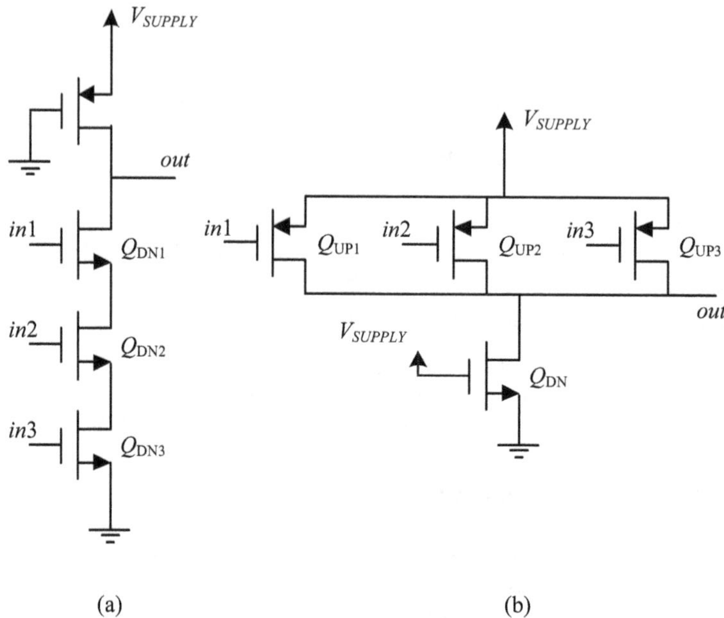

(a)   (b)

Fig. 8.16 Ratioed logic NAND gate PDN-based (**a**) and PUN-based (**b**)

The reason for this has to do with the fact that the PMOS transistors work in the opposite direction compared to NMOS transistors. When an input is high, a PMOS transistor is OFF whereas an NMOS transistor is ON and vice versa.

Also, in the PDN version, the NMOS transistors connect the output to ground when they are ON, while in the PUN version, the PMOS transistors connect the output to V_{SUPPLY} when they are ON.

Therefore, given the above two facts, and if we call the inputs to the NMOS transistors in the PDN version *inXn* and the inputs to the PMOS transistors in the PUN version *inXp*, then, following De Morgan's law, we have:

$$\overline{in1n + in2n + in3n} = \overline{in1n} \cdot \overline{in2n} \cdot \overline{in3n} = in1p \cdot in2p \cdot in3p \tag{8.46}$$

Another example is a 3-input ratioed NAND gate as shown in Fig. 8.16.

Yet another example of a more complex gate is shown in Fig. 8.17. This gate implements the function

$$out = \overline{(in1 \cdot in2) + in3} \tag{8.47}$$

The previously presented analysis for a ratioed inverter applies to all other ratioed gates. In this case, since there are several inputs that will affect which transistors are ON and which are OFF, the parameters that belong to a single input transistor in the analysis will represent the parameters of the input transistors on average in a complex logic gate.

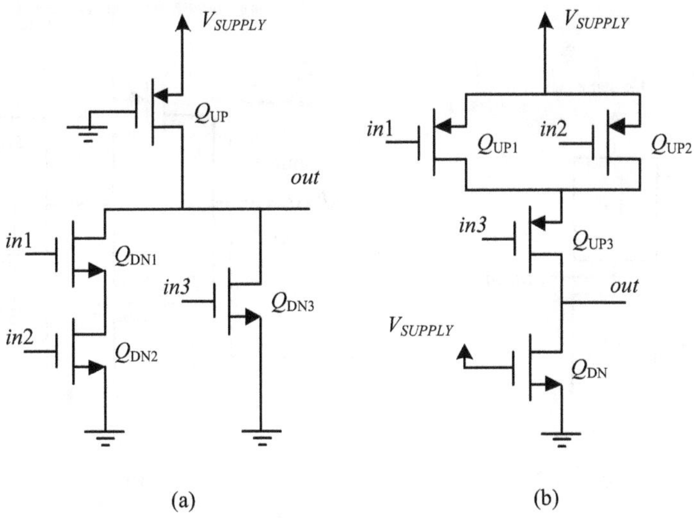

Fig. 8.17 Ratioed logic complex gate PDN-based (**a**) and PUN-based (**b**)

As a result, ratioed logic gates have several important drawbacks. The first is that the lack of symmetry above and below the output node as in Fig. 8.5 results in a lack of symmetry in the voltage-transfer characteristic as in Fig. 8.8 as well as in the transition time as in Figs. 8.12 and 8.13. Additionally, when the gate is ON, there is a direct path from V_{SUPPLY} to ground resulting in static power consumption.

8.3.2   Complementary (CMOS) Logic

Complementary logic aims at overcoming the drawbacks of ratioed logic by combining the PDN-based ratioed gates with the PUN-based ratioed gates with no current source. The PUN network is PMOS-based whereas the PDN network is NMOS-based, thus the name complementary or CMOS logic.

The general structure of a CMOS logic gate looks as in Fig. 8.18a. Note that in CMOS logic gates, the same input is replicated for the PUN and PDN. This is called a push–pull configuration since when the PUN is ON, the current is pushed from V_{SUPPLY} into the load, and when the PDN is ON, the current is pulled from the load into ground.

The circuit-level implementation of a CMOS inverter is shown in Fig. 8.18b. This will constitute the basis for our discussions in the coming sections.

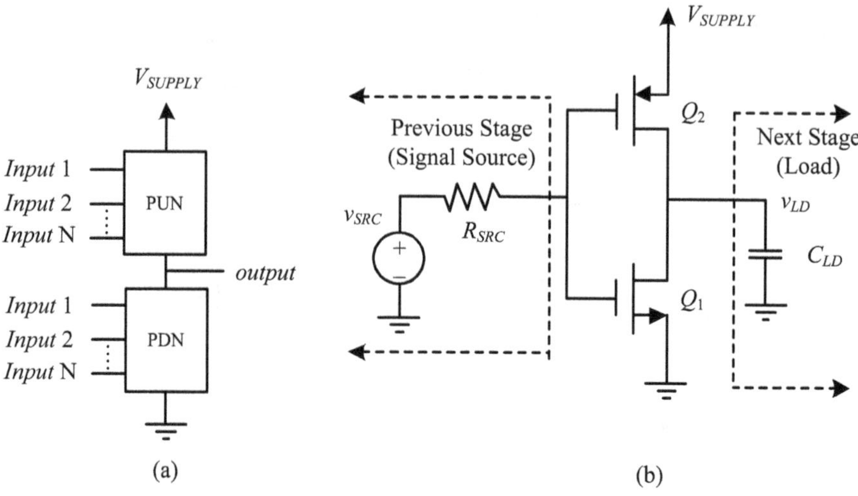

(a)   (b)

Fig. 8.18 General CMOS logic gate (**a**) and inverter (**b**)

8.3.2.1   CMOS Logic: Voltage-Transfer Characteristic

Without any loss of generalization, we will focus on the CMOS inverter as in Fig. 8.18b.

In the following analysis, let

$$
\begin{aligned}
k_1 &= k'_1 \left(\frac{W}{L}\right)_1 \\
k_2 &= k'_2 \left(\frac{W}{L}\right)_2
\end{aligned}
\tag{8.48}
$$

Also, usually within the same technology, the NMOS and PMOS transistors have the same threshold voltage magnitudes:

$$
|V_{thN}| = |V_{thP}| = V_{th}
\tag{8.49}
$$

Additionally, we are going to assume that $V_M > V_{th}$, which is usually the case.

We will start with the input–output voltage-transfer characteristic as shown in Fig. 8.19.

The voltage-transfer characteristic is symmetrical as opposed to that of ratioed logic and can be divided into three regions. This division is based on the state of the transistors.

In Region 1, v_{SRC} is less than V_{th}. As a result, Q_1 is OFF while Q_2 is in the triode region since v_{LD} is high, thus v_{DS2} is low. In this region, v_{LD} can take on its highest value V_{LD-H}, which is almost V_{SUPPLY}.

Fig. 8.19 CMOS inverter gate voltage-transfer characteristic

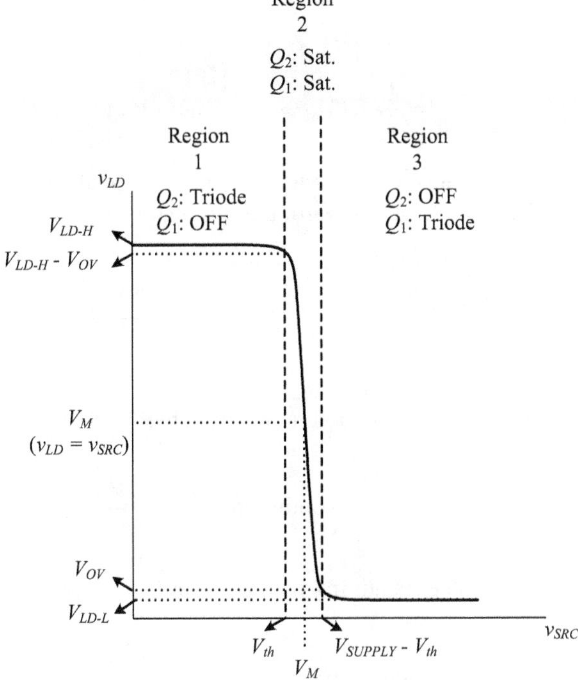

In Region 3, v_{SRC} is greater than $V_{SUPPLY} - V_{th}$. As a result, Q_2 is OFF while Q_1 is in the triode region since v_{LD} is low, thus v_{DS1} is low. In this region, v_{LD} can take on its lowest value $V_{LD\text{-}L}$, which is almost ground.

In Region 2, V_M is defined as the voltage v_{LD} is equal to v_{SRC}. Consequently, the voltages will be as follows:

$$\begin{aligned}
|v_{GS1}| &= V_M \\
|v_{DS1}| &= V_M \\
|v_{GS2}| &= V_{SUPPLY} - V_M \\
|v_{DS2}| &= V_{SUPPLY} - V_M
\end{aligned} \tag{8.50}$$

At this point, Q_1 and Q_2 are in the saturation region. Also, the same current that goes through Q_1 has to pass through Q_2. As a result:

$$\begin{aligned}
i_{D1} &= i_{D2} \\
\frac{1}{2}k_1(|v_{GS1}| &- |V_{th}|)^2(1 + \lambda|v_{DS1}|) \\
&= \frac{1}{2}k_2(|v_{GS2}| - |V_{th}|)^2(1 + \lambda|v_{DS2}|)
\end{aligned} \tag{8.51}$$

Replacing (8.50) in (8.51) we get:

$$i_{D1} = i_{D2}$$

$$\frac{1}{2}k_1(V_M - V_{th})^2(1 + \lambda V_M) \tag{8.52}$$

$$= \frac{1}{2}k_2((V_{SUPPLY} - V_M) - V_{th})^2(1 + \lambda(V_{SUPPLY} - V_M))$$

Two situations arise: one of them if λ is small and the other one if λ is large. If λ is small, almost zero, then V_M can be expressed as:

$$V_M \approx \frac{\sqrt{\frac{k_2}{k_1}}(V_{SUPPLY} - V_{th}) + V_{th}}{1 + \sqrt{\frac{k_2}{k_1}}} \tag{8.53}$$

As a result, k_2 / k_1 optimum would be:

$$\frac{k_2}{k_1}\text{optimum} \approx \left(\frac{V_M - V_{th}}{V_{SUPPLY} - V_M - V_{th}}\right)^2 \tag{8.54}$$

On the other hand, if λ is large, then k_2 / k_1 optimum would be:

$$\frac{k_2}{k_1}\text{optimum} = \frac{(V_M - V_{th})^2(\lambda V_M + 1)}{(V_{SUPPLY} - V_M - V_{th})^2(\lambda(V_{SUPPLY} - V_M) + 1)} \tag{8.55}$$

Note that if V_M is chosen to be in the middle of V_{SUPPLY}, λ will have no effect and Eq. (8.55) degenerates into Eq. (8.54), and they will both be equal to one.

As a result, in the absence of other constraints, we usually choose

$$k_2 = k_1 \tag{8.56}$$

in which case, we say that Q_1 and Q_2 are matched. This is what we will use in all subsequent discussions.

8.3.2.2 CMOS Logic: Transition Time

Given the CMOS inverter gate of Fig. 8.18b the waveforms used to measure the transition time are shown in Fig. 8.20.

The detailed derivation of the different time constants is presented in Sect. 8.2.2.

We can deduce that the voltage-transfer characteristic has two stable extremes, namely Region 1 and Region 3. The equivalent circuits in both regions are shown in Fig. 8.21.

Region 1 corresponds to $\tau_{out\text{-}LH}$ and Region 3 corresponds to $\tau_{out\text{-}HL}$. Deriving these time constants, we find that:

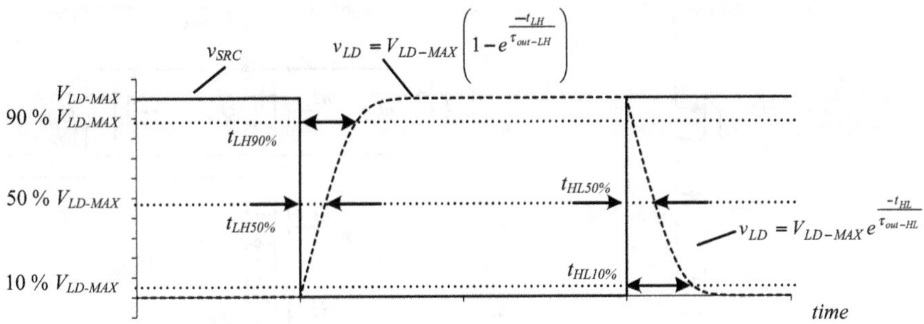

Fig. 8.20 CMOS logic inverter gate transition time

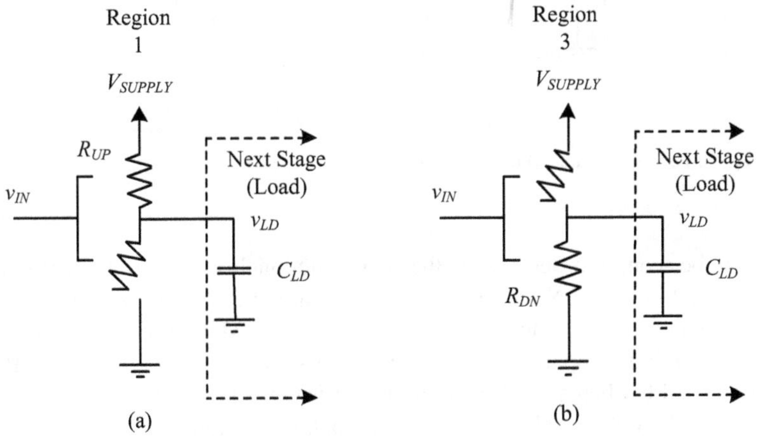

Fig. 8.21 CMOS logic gate time constants

$$\tau_{out-LH} = R_{out-LH}C_{out-LH} = R_{UP}C_{LD} \qquad (8.57)$$

$$\tau_{out-HL} = R_{out-HL}C_{out-HL} = R_{DN}C_{LD} \qquad (8.58)$$

Note that we are assuming, and that is usually the case, that the load is capacitive with minimal resistance compared to that of the driving transistors. Also, the output capacitance is almost totally dominated by the load capacitance.

8.3.2.3   CMOS Logic: Sample Gates

We have seen how an inverter can be implemented using CMOS logic. We will now showcase the implementation of other gates.

For example 3-input CMOS NOR and NAND gates are shown in Fig. 8.22a, b respectively.

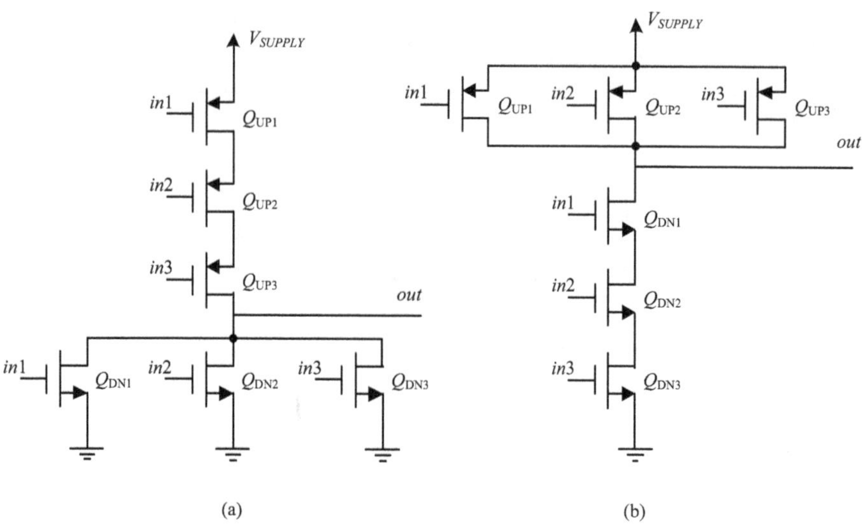

(a) (b)

Fig. 8.22 CMOS NOR gate (**a**) and NAND gate (**b**)

As can be seen, in order to keep the same functionality, the input transistors that are in parallel in the PDN will become in series in the PUN and vice versa.

The reason for this has to do with the fact that the PMOS transistors work in the opposite direction compared to NMOS transistors. When an input is high, a PMOS transistor is OFF whereas an NMOS transistor is ON and vice versa.

Also, the NMOS transistors connect the output to ground when they are ON, while the PMOS transistors connect the output to V_{SUPPLY} when they are ON.

Therefore, given the above two facts, and for the NOR gate, if we call the inputs to inX and the inputs to the PMOS transistors $inXp$, then, following De Morgan's law, we have:

$$\overline{in1n + in2n + in3n} = \overline{in1n} \cdot \overline{in2n} \cdot \overline{in3n} = in1p \cdot in2p \cdot in3p \tag{8.59}$$

Yet another example of a more complex gate is shown in Fig. 8.23. This gate implements the function

$$out = \overline{(in1 \cdot in2) + in3} \tag{8.60}$$

The previously presented analysis for a CMOS inverter applies to all other CMOS gates. In this case, since there are several inputs that will affect which transistors are ON and which are OFF, the parameters that belong to a single input transistor in the analysis will represent the parameters of the input transistors on average in a complex logic gate.

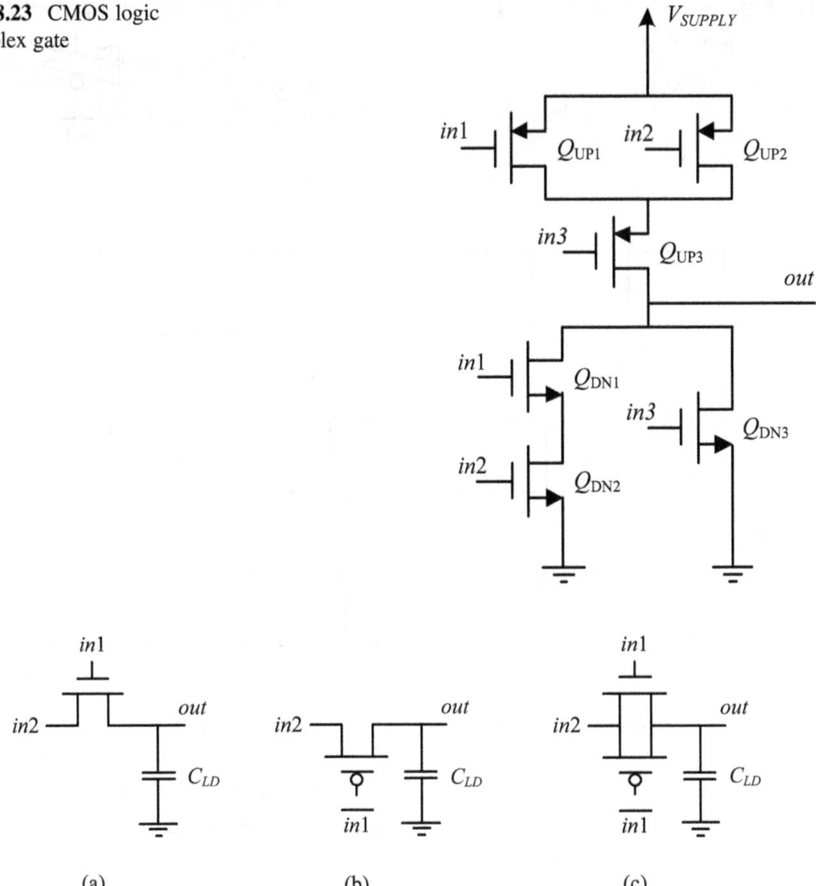

Fig. 8.23 CMOS logic complex gate

Fig. 8.24 Pass-transistor logic NMOS-based (**a**), PMOS-based (**b**), and transmission gate (**c**)

Another perspective of ratioed and CMOS logic is that the input is merely an enable signal that allows either V_{SUPPLY} or ground to show up at the output. With this in mind, we can substitute V_{SUPPLY} and ground with an actual signal. This leads to pass-transistor logic.

8.3.3 Pass-Transistor Logic

Pass-transistor logic relies on switches that are in-line with the signal. Additionally, pass-transistor gates draw their power from the inputs. As a result, they do not have a separate connection to the power supply.

Fig. 8.25 Pass-transistor logic transmission gate symbol

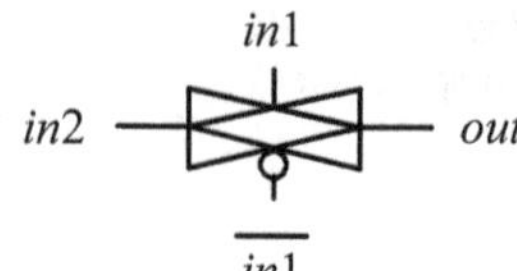

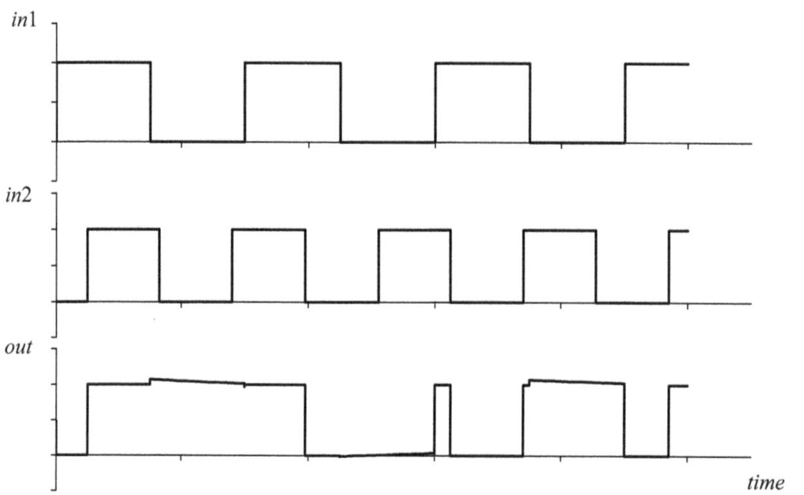

Fig. 8.26 Waveforms around a pass-transistor logic transmission gate

The simplest pass-transistor gate is shown in Fig. 8.24. NMOS-based implementation is shown in Fig. 8.24a, PMOS-based implementation in Fig. 8.24b, and transmission gate in Fig. 8.24c.

In all of the cases above, *in*1 acts as an enable signal. When *in*1 is high, *in*2 propagates to the output, otherwise, the output retains its previous value.

Following the discussion above, an NMOS can pass a strong low signal, while a PMOS can pass a strong high signal. As such, the transmission gate which uses both types has the advantage of both enabling to pass a strong high signal and a strong low signal. Owing to its usefulness, the transmission gate in Fig. 8.24c has its own symbol as in Fig. 8.25.

A sample waveform of the transmission gate in Fig. 8.24c is shown in Fig. 8.26.

When *in*1 is high, *in*2 propagates to the output, and when *in*1 is low, *out* keeps its previous value irrespective of *in*2.

Pass-transistor logic is not functionally complete. For example, it is not capable of implementing inversion. As a result, if a functionality more elaborate than a simple selection needs to be implemented, pass-transistor logic has to be implemented in conjunction with another logic family, most notably CMOS logic.

The use of CMOS logic in conjunction with pass-transistor logic has the added benefit of cleaning-up the signal, effectively to take more precise output levels. For example, if the transmission gate in Fig. 8.24c is followed by an inverter whose output is called "out'" as in Fig. 8.27, the waveforms will be as in Fig. 8.28.

Fig. 8.27 Pass-transistor logic transmission gate followed by a CMOS inverter

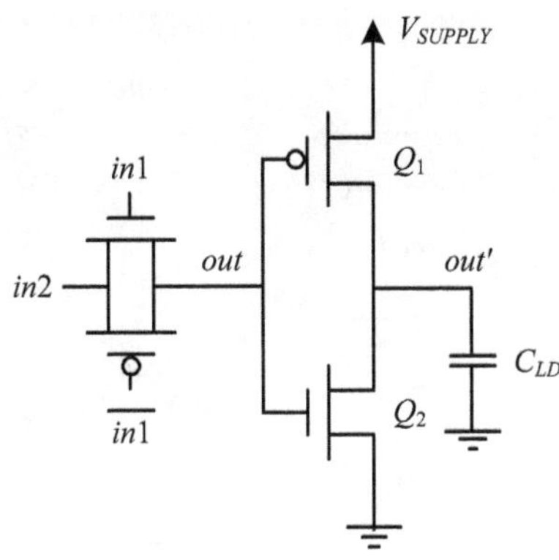

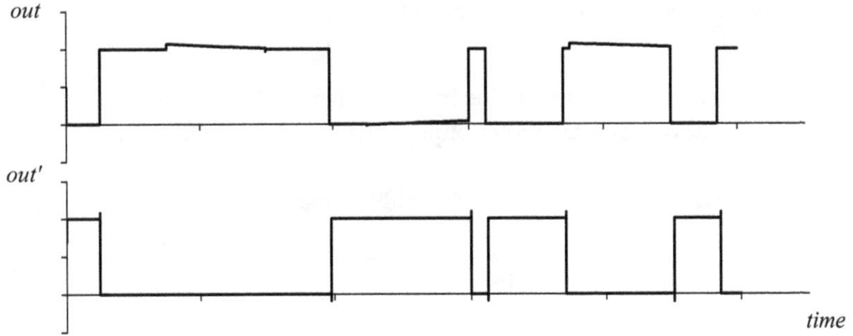

Fig. 8.28 Waveforms at the output of a transmission gate followed by a CMOS inverter

Fig. 8.29 Pass-transistor logic with CMOS logic

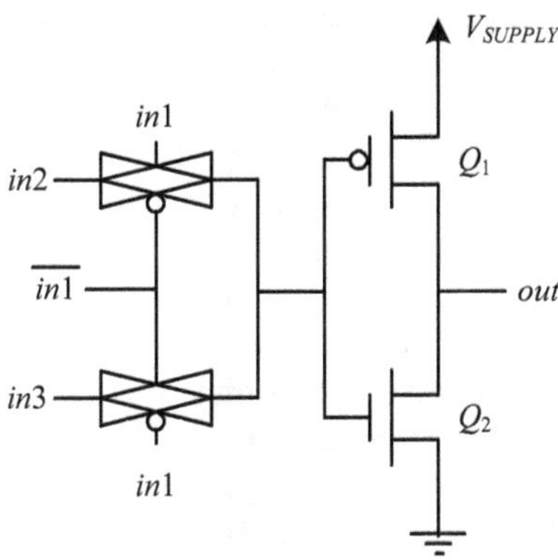

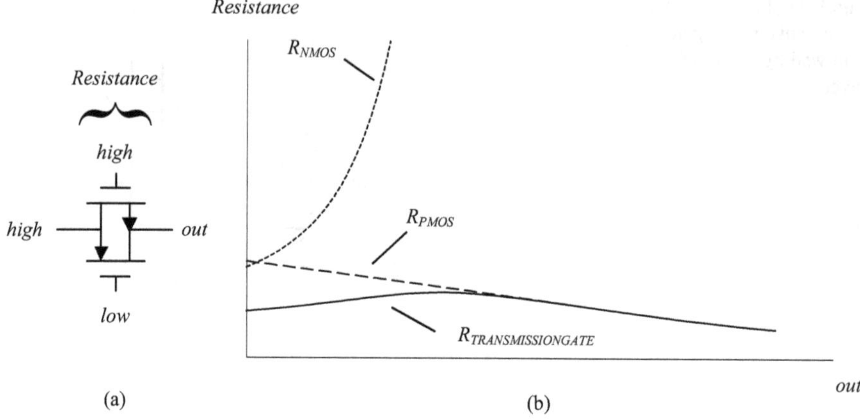

Fig. 8.30 Pass-transistor logic transmission gate ON resistance

This combination can be used in the implementation of sequential logic where $in1$ is the clock signal which either allows or prevents $in2$ from reaching the CMOS logic gates.

As another example, the circuit in Fig. 8.29 implements the function

$$out = \overline{(in2 \cdot in1) + (in3 \cdot \overline{in1})} \tag{8.61}$$

When a transmission gate is ON, it has some resistance ($R_{TRANSMISSIONGATE}$). This resistance is the combination of the NMOS resistance (R_{NMOS}) in parallel with the PMOS resistance (R_{PMOS}).

A sample setup to measure this resistance is shown in Fig. 8.30a. The individual resistance along with the combined resistance is shown in Fig. 8.30b.

This setup mimics the situation when we turn ON the transmission gate, allowing the high input to charge the initially low output.

Directly after enabling the gate, out is low and both transistors are in the saturation region.

As out increases, both transistors start to go into the triode region.

When out reaches the value of $high - V_{th}$, the NMOS transistor turns OFF, whereas the PMOS transistor stays in the triode region.

As can be seen, $R_{TRANSMISSIONGATE}$ does not vary a lot compared to R_{NMOS} and R_{PMOS}. Additionally, the wider the transistors, the smaller is the resulting $R_{TRANSMISSIONGATE}$. However, this $R_{TRANSMISSIONGATE}$ is going to interact with the surrounding parasitic capacitances, which increase with the transistors' widths. As a result, a compromise needs to be made in order to keep the time constant, that is the product between $R_{TRANSMISSIONGATE}$ and the parasitic capacitances, within range.

Fig. 8.31 Dynamic logic using a PDN (**a**) and PUN (**b**)

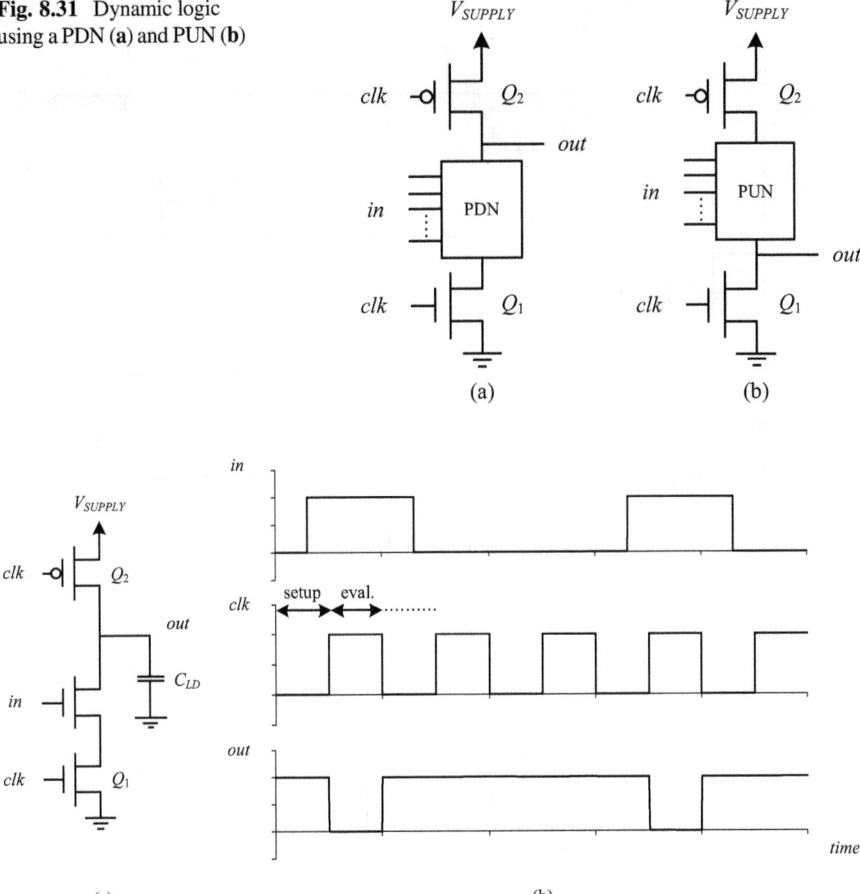

Fig. 8.32 Dynamic logic PDN-based inverter (**a**) and associated waveforms (**b**)

8.4   Dynamic Logic

Enabling and disabling a logic gate using a clock signal is an important requirement in the implementation of sequential logic. This can be achieved either by using a cascade of pass-transistor logic followed by CMOS logic, where the clock signal goes into the transmission gate as in Fig. 8.27, or by using dynamic logic, which results in a more compact circuit.

Dynamic logic is based on ratioed logic and its general implementation can be PDN-based as in Fig. 8.31a or PDN-based as in Fig. 8.31b.

The PDN-based dynamic logic circuit operates based on two phases:

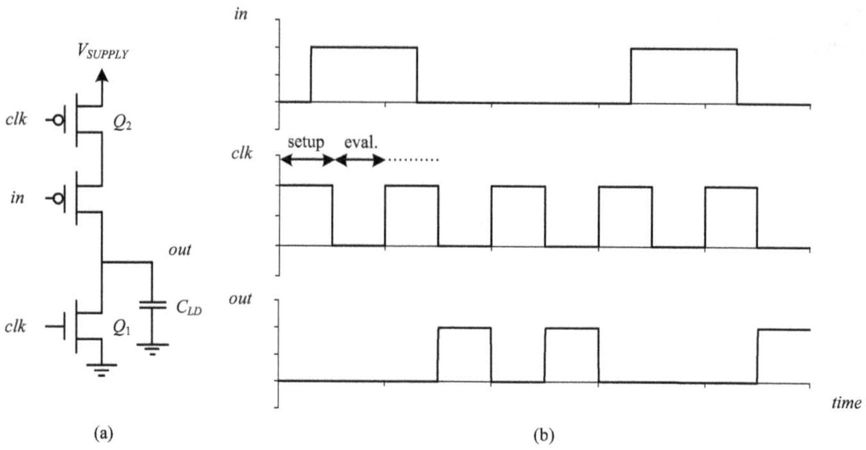

Fig. 8.33 Dynamic logic PUN-based inverter (**a**) and associated waveforms (**b**)

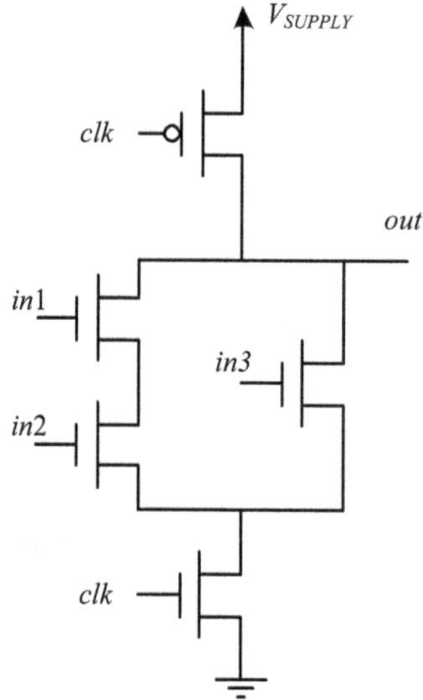

Fig. 8.34 Dynamic logic PDN-based complex logic gate

- Setup phase: *clk* is low, Q_1 is OFF and Q_2 is ON. As a result, *out* is charged to *VSUPPLY*. In this phase, *in* is allowed to change.
- Evaluation phase: *clk* is high, Q_1 is ON and Q_2 is OFF. *in* is not allowed to change. As a result, *out* either keeps its value of *VSUPPLY* or goes to *GROUND* depending on the PDN network.

An example of such a circuit and the associated waveforms are shown in the PDN-based dynamic inverter shown in Fig. 8.32.

The PUN-based dynamic logic circuit also operates based on two phases:

- Setup phase: *clk* is high, Q_1 is ON and Q_2 is OFF. As a result, *out* is discharged to *GROUND*. In this phase, *in* is allowed to change.
- Evaluation phase: *clk* is low, Q_1 is OFF and Q_2 is ON. *in* is not allowed to change. As a result, *out* either keeps its value of *GROUND* or goes to *VSUPPLY* depending on the PUN network.

An example of such a circuit and the associated waveforms are shown in the PDN-based dynamic inverter shown in Fig. 8.33.

More complex dynamic gates can be implemented. For example, the gate in Fig. 8.34 implements the function:

$$out = \overline{(in1 \cdot in2) + in3} \tag{8.62}$$

Chapter 9
Differential Structures

This chapter describes differential structures for analog and digital applications. Differential circuits provide several advantages compared to their single-ended counterparts such as insensitivity to reference ground mismatch between the transmitter and the receiver, better noise resilience, as well as higher-speed performance owing to better interference suppression. This chapter starts by introducing the concept of electrical noise. Afterwards, the main idea behind differential circuits is presented before detailing analog followed by digital differential structures.

9.1 Electrical Noise

All electronic signals are subject to electrical noise. Noise is any unwanted, random, and usually tiny signal that couples to our own signal. All electronic components, internal and external to our system, generate noise. Therefore, at any instance of time, the sources of noise are numerous, unlike interference where the sources are numbered and well-known. As a result, we cannot disable the sources of noise, but we can reduce their effect on our signal by taking precautions such as using differential structures.

Figure 9.1 shows a typical noise signal versus time. Note the randomness of noise and its small root mean square value, typically in the range of nanovolts to microvolts. Also, typically, noise has an average value of zero, so it has no DC shift.

When noise couples to our signal, typically in an additive manner, then the result will be as shown in Fig. 9.2. v_{noise} is the noise voltage and v_{signal} is the signal voltage.

An important measure of the quality of a signal in the presence of noise is the signal-to-noise ratio (SNR). SNR is typically expressed in dB and is defined as follows:

J. G. Atallah, M. Ismail, *Integrated Electronic Circuits*,
https://doi.org/10.1007/978-3-031-62707-1_9

Fig. 9.1 Sample noise signal

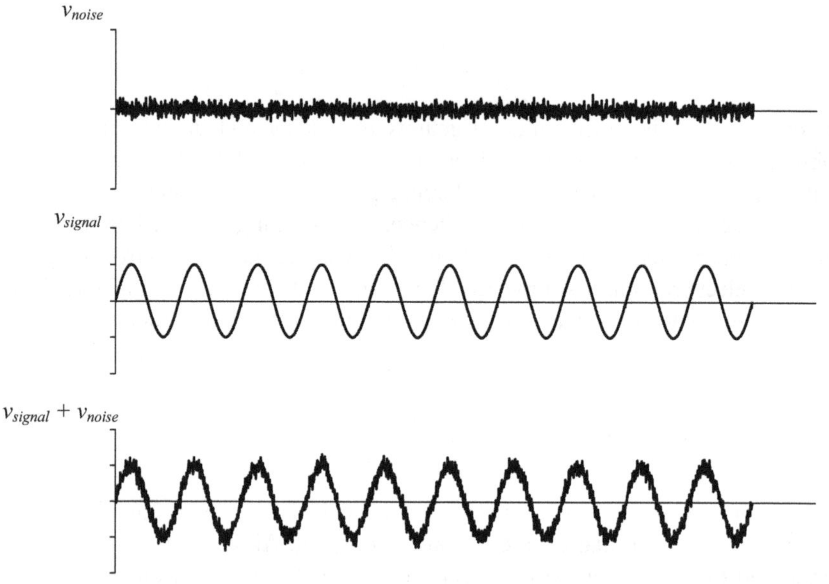

Fig. 9.2 Noise coupling to our signal

$$SNR = 10\log\left(\frac{v_{signal\_rms}^2}{v_{noise\_rms}^2}\right) = 20\log\left(\frac{v_{signal\_rms}}{v_{noise\_rms}}\right) \qquad (9.1)$$

Consider an ideal noiseless amplifier, which is an amplifier that does not add any noise to the signals at its input. Realistically this is not the case, but this assumption will serve its purpose in explaining the situation. If a signal with noise is amplified using this noiseless amplifier, the signal and the noise at the input will be amplified with the same gain resulting in an *SNR* at the output of the amplifier equal to the *SNR* at the input. This does not look bad until we consider a cascade of such amplifiers.

Consider now a signal passing through a cascade of noiseless amplifiers. At every step of the way, the amplifiers amplify the signal and the noise. However, along the

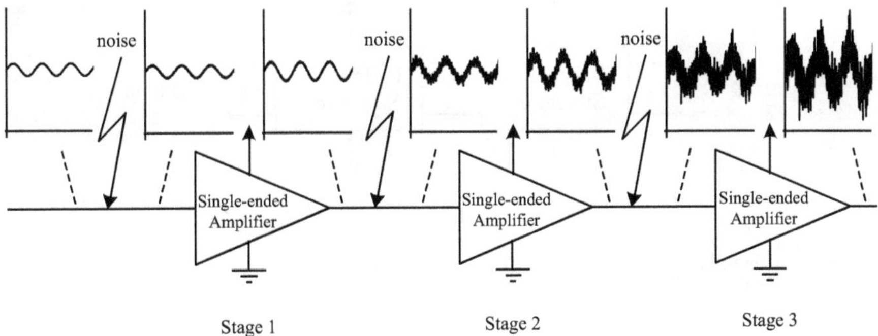

Fig. 9.3 Cascade of noiseless single-ended amplifiers with injected noise

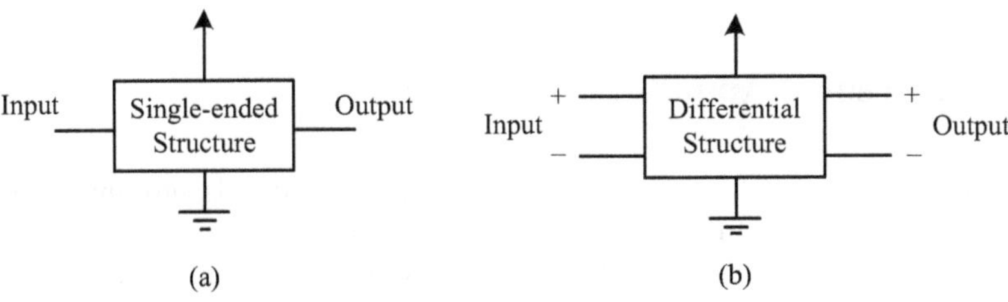

Fig. 9.4 Single-ended structure (**a**) and differential structure (**b**)

way, noise from external noise sources is injected into the signal. This is depicted in Fig. 9.3.

As we proceed down the cascaded stages, more noise will be injected along the path, which will then be amplified along with the signal. The end result is a signal quality significantly inferior, that is with much lower *SNR*, than the input signal we started with.

To remedy this situation, we will resort to differential structures as opposed to single-ended structures.

9.2 Introduction to Differential Structures

Compared to single-ended structures as seen in Fig. 9.4a, differential structures have two signal input lines and two signal output lines in addition to the power supply lines as seen in Fig. 9.4b.

The input and output signals of a single-ended structure are referenced to ground, whereas those of the differential structure are not. This allows the reference grounds of the differential transmitter and the differential receiver to be at different potential levels, which greatly eases the implementation (Fig. 9.5).

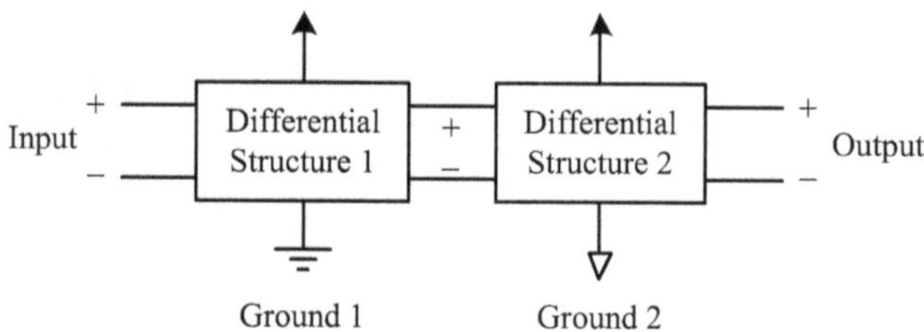

Fig. 9.5 Differential structures with different reference grounds

In the next section, we will detail how differential amplifiers in particular can achieve noise resilience, thus improving on the situation depicted in Fig. 9.3.

9.3 Differential Amplifiers

The general structure of a differential amplifier and the signals surrounding it are shown in Fig. 9.6. The differential amplifier is made of two single-ended amplifiers with a special link between them. Without this link, the differential amplifier will still work, but with lower performance as will be discussed later. Ideally, the two amplifiers have to be perfectly matched.

The main aim of the differential amplifier is to distinguish between the data part of the signal and the noise part, while treating them differently. Ideally, we need this amplifier to amplify the data part and block the noise part. Notice that how, even in the presence of noise at the input, ideally the final output signal V_{OUT_DIFF} is noiseless.

The use of capital and small letters for the signals is indicative of their nature. Therefore, a capital letter followed by a capital subscript indicates a DC signal. A small letter followed by a small subscript indicates an AC (usually small) signal, and a small letter followed by a capital subscript indicates a combined signal.

The inputs and outputs of a differential amplifier are indicated as either '+' or '−'. This is indicative of the sequence in which the subtraction is taken in case a single-ended signal such as v_{OUT_DIFF} needs to be outputted.

The signals can be categorized in two ways. The first way, which also applies to single-ended structures is to determine whether the signal is:

- DC.
- AC (usually small-signal).
- combined,

The second way, which is exclusive to differential structures, is to determine whether the signal is

- common,
- differential.

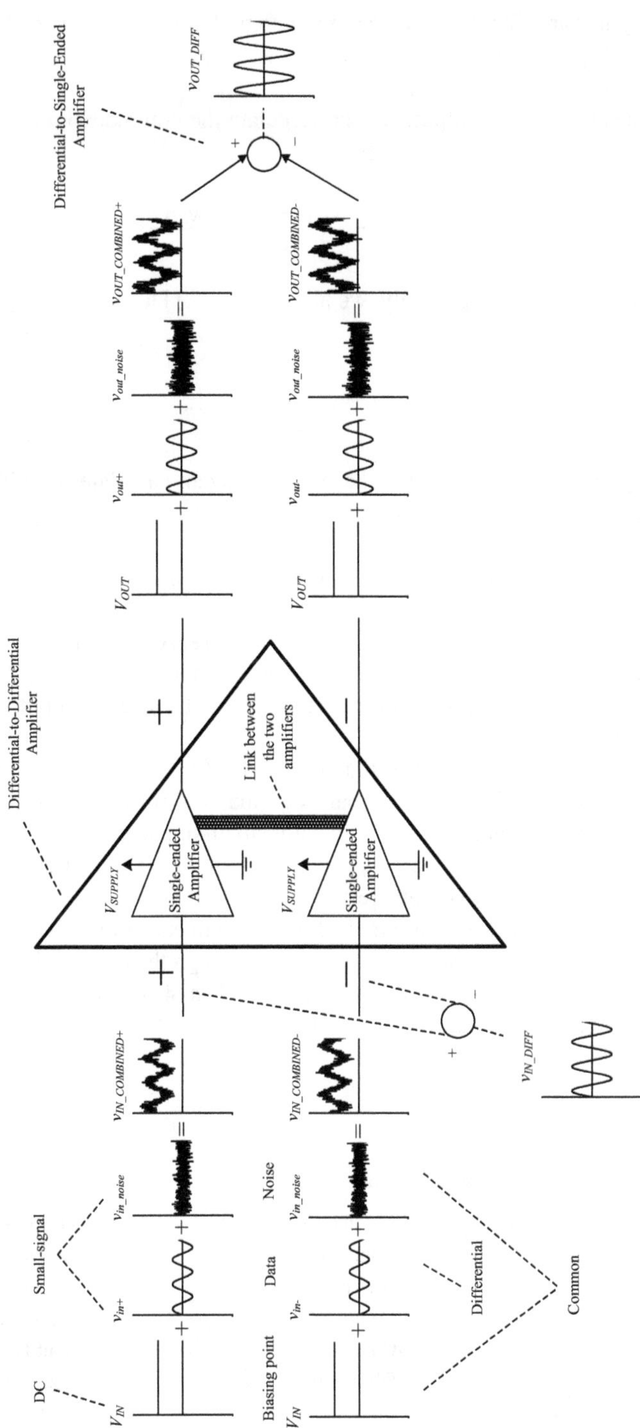

Fig. 9.6 Differential amplifier signals overview

Common signals are signals that are common to both input lines, '+' and '−'. For example, V_{IN} and v_{in_noise} are common. On the other hand, differential signals are the ones that differ between the input lines.

At the input, just like the output, we can calculate the common signals from the combined signal as follows:

$$v_{COMMON} = \frac{v_{COMBINED+} + v_{COMBINED-}}{2} \tag{9.2}$$

If we need the small signal part only, we just remove the DC part from $v_{COMBINED}$ in Eq. (9.2). As a result:

$$v_{common} = \frac{v_{combined+} + v_{combined-}}{2} \tag{9.3}$$

Additionally, at the input, just like the output, we can calculate the differential signal from the combined signal as follows:

$$v_{differential} = v_{COMBINED+} - v_{COMBINED-} \tag{9.4}$$

Both amplifiers have a single power supply in our case. As a result, they can only deal with a signal that exists between V_{SUPPLY} and ground. Therefore, the signal must have a DC shift, thus the role of the first component V_{IN} at the input. This DC shift is common to both.

The second component of the input signal consists of v_{in+} and v_{in-}. These signals model our data. They are inputted in a balanced manner, meaning that one is the inverse of the other. They are of the AC (usually small-signal) type.

The third component is the noise v_{in_noise}. Noise is of the AC type (usually small-signal). By implementation, we try our best to make sure that this noise is common to both input lines, for example, by routing the lines close to each other so that they are subject to the same external noise. With this, we can design the differential amplifier to have a different behavior towards the AC differential small signal (data) compared to the AC common small signal (noise), which sets it apart from the single-ended amplifier (Table 9.1).

Table 9.1 Differential amplifier signal components

Signal	DC/AC	Common/Differential	What does it model
VIN	DC	Common	Input DC shift
$vin+$, $vin-$	AC	Differential	Input data
vin_noise	AC	Common	Input noise
$VOUT$	DC	Common	Output DC shift
$vout+$, $vout-$	AC	Differential	Output data
$vout\_noise$	AC	Common	Output noise

As a result we can categorize our signal components and what they model as in Table 4.1.

9.3.1   *Differential-to-Differential Amplifier*

The most common implementation of a differential-to-differential amplifier is the differential pair. It is based on two single-ended common-source amplifiers connected together in order to ensure a differential behavior between them. An NMOS implementation is shown in Fig. 9.7.

The link between the two single-ended amplifiers is the current source at the bottom, otherwise known as the tail current source. This current source should be able to provide double the amount of current provided by each one of the current sources at the top, owing to the behavior of the current at node X. KCL at this node ensures that the sum of the currents coming from Q_1 and Q_2 is always equal to $2I_D$. As a result, for any decrease in the current happening in one branch, an equal increase in the current is happening in the other. It is this link that makes the amplifier fully differential as opposed to having two separate amplifiers sitting side by side.

Notice that the left branch of the differential amplifier is indicated as '+', while its output is indicated as '−'. A similar inversion in the notation exists in the right branch. This is due to the fact that these single-ended amplifiers are of the common-source type, which is inverting by nature.

Now, we are going to analyze the behavior of this amplifier given each one of its input signal constituents separately.

We can figure out the maximum and minimum value of $v_{IN\_COMBINED}$ and $v_{OUT\_COMBINED}$, whether on the positive side or on the negative side.

The maximum value for $v_{IN\_COMBINED}$ can be derived by noting that Q_1 and Q_2 have to stay in saturation. In order to do so, and knowing that each of the current sources at the drains has to have a minimum of v_{CT_MIN}, we have

$$v_{DS} > v_{GS} - V_{th}$$
$$v_D - v_S > v_G - v_S - V_{th}$$
$$v_D > v_G - V_{th}$$
$$v_{OUT\_COMBINED} > v_{IN\_COMBINED} - V_{th}$$
$$v_{IN\_COMBINED} < v_{OUT\_COMBINED} + V_{th}$$
$$\Rightarrow v_{IN\_COMBINED\_MAX} = V_{SUPPLY} - v_{CT\_MIN} + V_{th}$$

$$(9.5)$$

The minimum value for $v_{IN\_COMBINED}$ can be derived by noting that as $v_{IN\_COMBINED}$ decreases, v_X decreases, and so does the voltage v_{CTD} across the tail current source. No matter how that current source is implemented, there is a

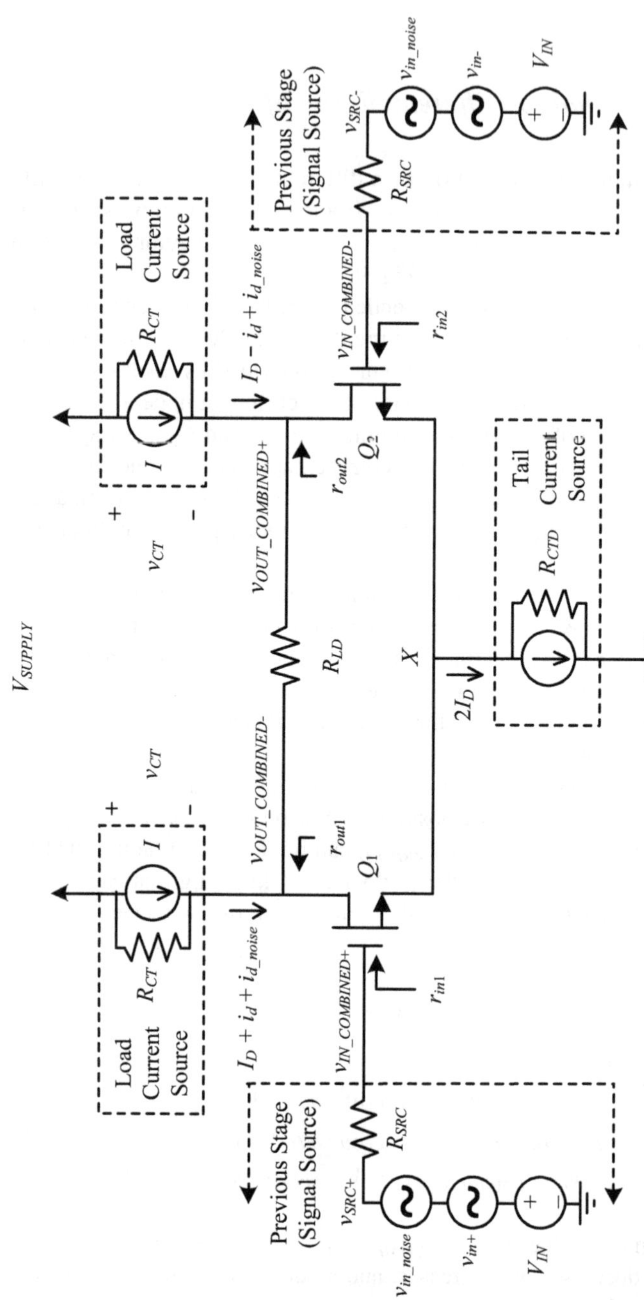

Fig. 9.7 The differeential pair

minimum voltage V_{CTD_MIN} that needs to be across it. As a result, and given that the minimum value of v_{GS} corresponds to i_{D_MIN}:

$$v_{IN\_COMBINED\_MIN} = v_{GS\_MIN} + v_{CTD\_MIN} \qquad (9.6)$$

Similarly, we can figure out the maximum and minimum value of $v_{OUT\_COMBINED}$.

The maximum value of $v_{OUT\_COMBINED}$ is dictated by the minimum value of v_{CT}. Therefore, we have

$$v_{OUT\_COMBINED} = V_{SUPPLY} - v_{CT}$$
$$\Rightarrow v_{OUT\_COMBINED\_MAX} = V_{SUPPLY} - v_{CT\_MIN} = v_{IN\_COMBINED\_MAX} - V_{th} \qquad (9.7)$$

The minimum value of $v_{OUT\_COMBINED}$ is dictated by the minimum value of v_{DS} and the minimum value of v_{CTD}. Therefore, we have

$$v_{OUT\_COMBINED} = v_{DS} + v_{CTD}$$
$$v_{OUT\_COMBINED\_MIN} = v_{DS\_MIN} + v_{CTD\_MIN}$$
$$\Rightarrow v_{OUT\_COMBINED\_MIN} = v_{GS\_MIN} - V_{th} + v_{CTD\_MIN} \qquad (9.8)$$
$$= v_{IN\_COMBINED\_MIN} - V_{th}$$

Obviously, to get the maximum swing at the output, we would like to place the DC value V_{OUT} in the middle between $v_{OUT\_COMBINED\_MIN}$ and $v_{OUT\_COMBINED\_MAX}$.

9.3.1.1 Common-Mode DC Behavior (Biasing Point)

Consider the differential pair with only V_{IN} at its input. The corresponding output is V_{OUT} (Fig. 9.8).

For Q_1 and Q_2, V_{GS} is the same since their V_G is V_{IN} and their V_S is V_X. Assuming they are in saturation, then they are conducting the same current. This current is I_D since KCL at node X dictates that the sum of these currents is $2I_D$.

Since the two current sources at the drains of Q_1 and Q_2 are the same, then the voltage drop across them is the same too. As a result, the voltages at the drains of Q_1 and Q_2 are both equal to V_{OUT} where

$$V_{OUT} = V_{SUPPLY} - V_{CT} \qquad (9.9)$$

Consequently, there is no DC current passing through R_{LD}.

As V_{IN} is increased or decreased, and as long as the transistors are in saturation, the current passing through them will always be equal to I_D owing to KCL at node X, and the voltages at their drains will always be fixed to V_{OUT}. Their current being constant means that their V_{GS} is constant too. However, since V_G is V_{IN}, and as V_{IN} is

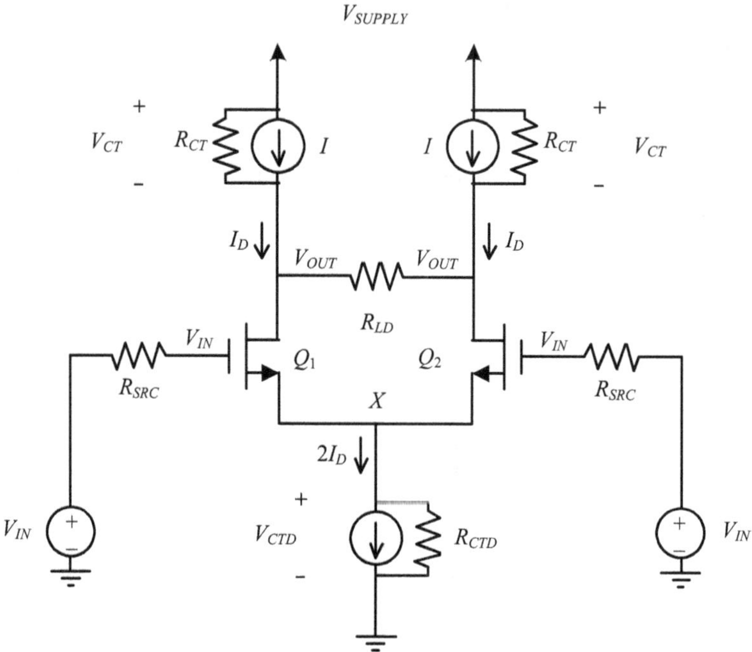

Fig. 9.8 Differential pair: common-mode DC

changing, V_X is changing with it to keep V_{GS} constant. As a result, the changes in V_{IN} are seen in V_X in the same direction while V_{OUT} stays constant.

9.3.1.2 Common-Mode AC Small-Signal Behavior (Noise)

Now we will discuss the behavior of the differential pair when a common-mode AC small signal is applied to its inputs. This signal mimics that of noise, which the designer is supposed to ensure that it is common between the two input terminals, for example, by routing the connections to the two input terminals close to each other.

The small-signal model in this case is shown in Fig. 9.9. Note that as a first-hand approximation, we have neglected r_o.

We need to get the ratio $v_{out\_noise}/v_{in\_noise}$. This ratio is called the common-mode gain when the output is taken single-endedly. This will be indicated as A_{cm_single}.

The way to analyze this circuit is by cutting it in half and analyzing only one side, either the left or the right one. Both of these sides give the same results.

Therefore, by cutting it in half, we get the common-mode small-signal half circuit. Taking the left side, we get the circuit in Fig. 9.10.

Note that owing to the equal voltages on both sides of R_{LD}, no current will flow into this resistor, so it acts as an open-circuit.

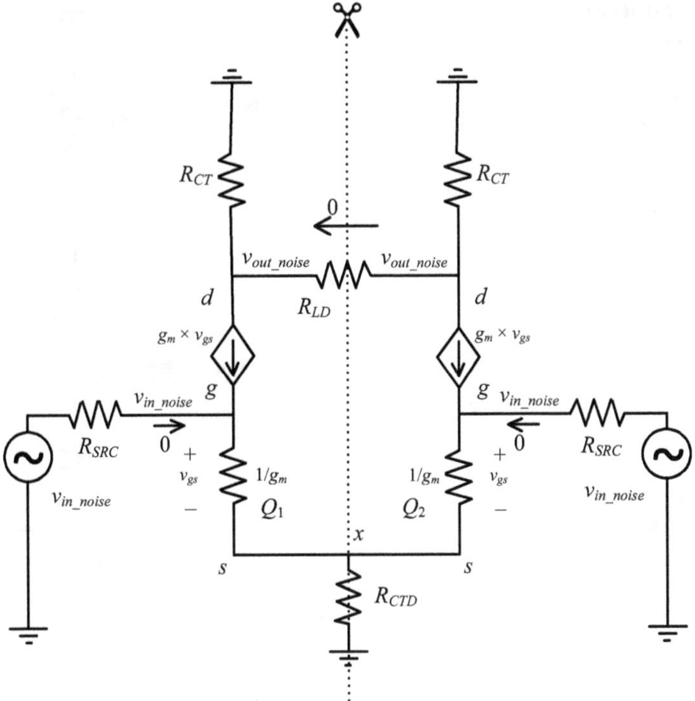

Fig. 9.9 Differential pair: common-mode small-signal AC

However, R_{CTD} is split into two pieces, that, when combined in parallel, result in R_{CTD}. Therefore, each piece is $2R_{CTD}$.

Calculating the common-mode single-ended gain A_{cm_single}, while realizing that this is a common-source amplifier whose gain is the total resistance at the drain divided by the total resistance at the source, we get

$$\left|A_{cm\_single}\right| = \left|\frac{v_{out\_noise}}{v_{in\_noise}}\right| \approx \frac{R_{CT}}{(1/g_m) + 2R_{CTD}} \tag{9.10}$$

Note that we are only interested in the magnitude of A_{cm}, since inverted noise is still noise.

Since $2R_{CTD}$, which is double the output resistance of the lower current source, is much larger than $1/g_m$, then we get

$$\left|A_{cm\_single}\right| \approx \frac{R_{CT}}{2R_{CTD}} \tag{9.11}$$

As can be seen, the common-mode gain is less than one, owing to the typically large value of R_{CTD}. Therefore, the noise seen at every output of the differential amplifier is an attenuated version of the noise at the input. Another conclusion is that

Fig. 9.10 Differential pair: common-mode small-signal half-circuit

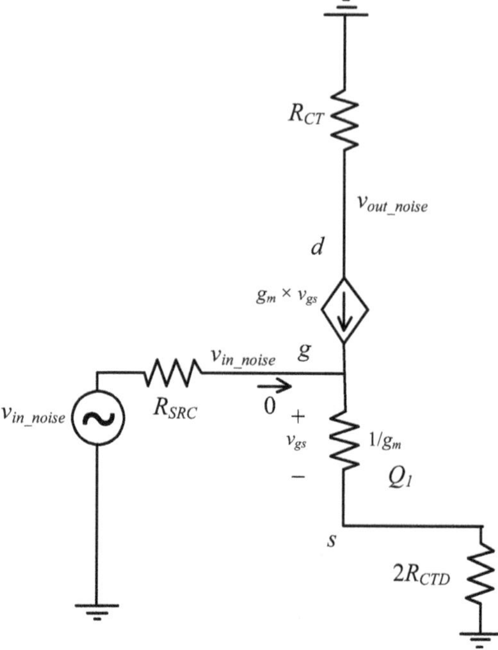

it pays off to have a lower current source with a high output resistance, since this will decrease the amount of noise seen at each output terminal.

Additionally, if we take the difference between the two outputs, and since both of them carry the same noise, then the difference will not contain any noise. As a result, we can say that the common-mode gain when the output is taken differentially, A_{cm_diff}, is

$$\left| A_{cm\_diff} \right| = \left| \frac{v_{out\_noise} - v_{out\_noise}}{v_{in\_noise}} \right| = 0 \tag{9.12}$$

As a conclusion, we can say that under perfectly matched conditions, and when the output is taken differentially, the common-mode gain is zero, meaning that the differential amplifier rejects common-mode small-signals at the input.

9.3.1.3   Differential-Mode Small-Signal Behavior (Data)

We will apply a differential AC small signal to the amplifier that models our data. The signal is applied in a balanced manner. The small-signal model of the amplifier in this case is shown in Fig. 9.11. As a first-hand approximation, we have neglected r_o.

In this situation, what is happening in the right branch is the opposite of what is happening in the left branch in terms of both voltage and current. Therefore, if the

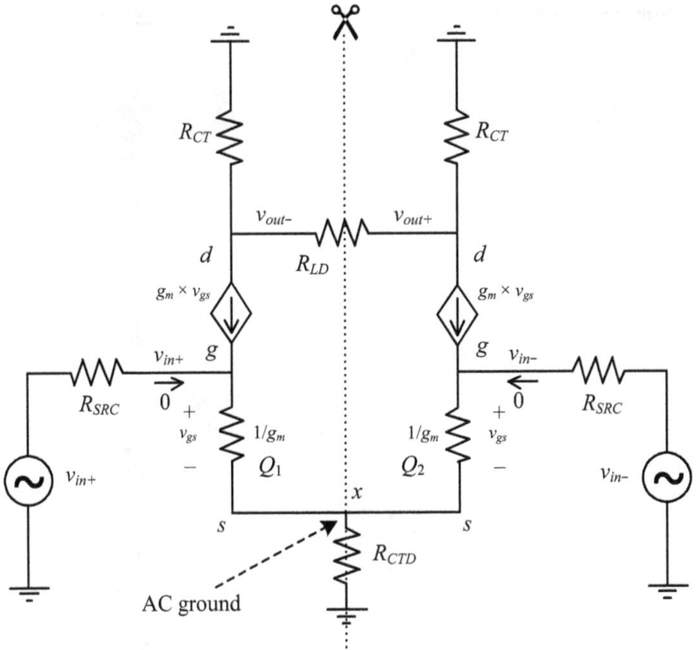

Fig. 9.11 Differential pair: differential mode small-signal circuit

small-signal current is going down in the left branch, then the current in the right branch is equal and going up owing to the equal magnitudes and opposite polarities of v_{gs} on both sides. As a result, the AC voltage at node x is zero. Therefore, node x is called an AC ground.

We need to get the ratio v_{out}/v_{in} on the right or left branch. This ratio is called the differential gain when the input and output are taken single-endedly. This will be indicated as A_{d_single}.

Additionally, we need to get the ratio $(v_{out+} - v_{out-}) / (v_{in+} - v_{in-})$. This ratio is called the differential gain when the input and output are taken differentially. This will be indicated as A_{d_diff}.

The way to analyze this circuit is by cutting it in half and analyzing only one side, either the left or the right one. Both of these sides give the same results.

Therefore, by cutting it in half, we get the differential small-signal half circuit. Note that the main difference between this and the common-mode small-signal half circuit of Fig. 9.10 is the voltage at node x. Taking the left side, we get the circuit in Fig. 9.12.

Note that when the cut is made, R_{LD} is split into two pieces that, when combined in series, result in R_{LD}. Therefore, each piece is $R_{LD}/2$.

Calculating the differential gain with single-ended input and output A_{d_single}, while realizing that this is a common-source amplifier whose gain is the total resistance at the drain divided by the total resistance at the source, we get

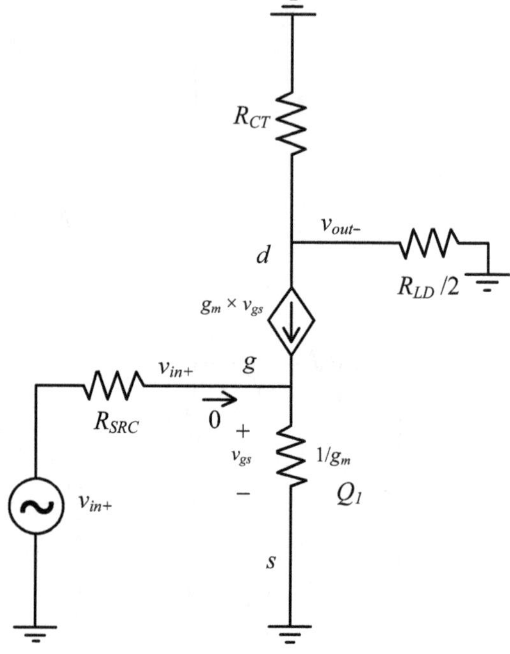

Fig. 9.12 Differential pair: differential-mode small-signal half-circuit

$$\left|A_{d\_single}\right| = \left|\frac{v_{out-}}{v_{in+}}\right| = \left|\frac{v_{out+}}{v_{in-}}\right| = \frac{R_{CT}//(R_{LD}/2)}{(1/g_m)} \tag{9.13}$$

From Fig. 9.6, we observe that

$$v_{out+} = -v_{out-} = \frac{v_{out\_diff}}{2}$$
$$v_{in+} = -v_{in-} = \frac{v_{in\_diff}}{2} \tag{9.14}$$

As a result, the differential gain A_{d_diff} with differential input and output is the same as A_{d_single}

$$\left|A_{d\_diff}\right| = \left|\frac{v_{out+} - v_{out-}}{v_{in+} - v_{in-}}\right| = \left|\frac{v_{out\_diff}}{v_{in\_diff}}\right| = \frac{R_{CT}//(R_{LD}/2)}{(1/g_m)} \tag{9.15}$$

Since node x is a virtual ground, taking r_o of Q_1 and Q_2 into account does not complicate things and the result will be

$$\left|A_{d\_diff}\right| = \left|\frac{v_{out+} - v_{out-}}{v_{in+} - v_{in-}}\right| = \left|\frac{v_{out\_diff}}{v_{in\_diff}}\right| = \frac{R_{CT}//(R_{LD}/2)//r_o}{(1/g_m)} \tag{9.16}$$

Therefore we can say that this differential amplifier amplifies the differential signal with the same gain as that of a common-source amplifier.

If $(R_{CT}//r_o) << (R_{LD}/2)$, then

$$\left| A_{d\_diff} \right| = \left| \frac{v_{out+} - v_{out-}}{v_{in+} - v_{in-}} \right| = \left| \frac{v_{out\_diff}}{v_{in\_diff}} \right| \approx g_m (R_{CT}//r_o) \tag{9.17}$$

An important figure of merit for differential amplifiers is their Common-Mode Rejection Ratio (*CMRR*). This is defined as the ratio of the differential gain to the common-mode gain. The higher the *CMRR*, the better is the performance of the amplifier. *CMRR* is typically expressed in decibels as in

$$CMRR = 20 \log \left| \frac{A_{d\_diff}}{A_{cm\_diff}} \right| \tag{9.18}$$

For the ideal case with no mismatches between the two branches, we have a *CMRR* which is infinite since A_{cm_diff} is zero.

However, when mismatches are present, the *CMRR* becomes finite as will be discussed next.

9.3.1.4   Behavior in the Presence of Mismatches

In spite of the designers doing their best in order to have matched left and right branches, usually using proper layout techniques, some mismatches will still occur. We will now analyze how some of the major mismatches affect the *CMRR* and draw some conclusions regarding how to reduce these effects by design.

The first mismatch to take into account is that between g_{m1} of transistor Q_1 and g_{m2} of transistor Q_2. We need to calculate the $CMRR_{mis\_trans}$ defined as follows

$$CMRR_{mis\_trans} = 20 \log \left| \frac{A_{d\_diff\_mis\_trans}}{A_{cm\_diff\_mis\_trans}} \right| \tag{9.19}$$

Let g_{m1} be the transconductance of Q_1 and g_{m2} be the transconductance of Q_2 in such a way that their average is called g_m and

$$\begin{aligned} g_{m1} &= g_m + \frac{\Delta g_m}{2} \\ g_{m2} &= g_m - \frac{\Delta g_m}{2} \end{aligned} \tag{9.20}$$

Note that Δg_m can be either positive or negative.

The relation in (9.20) is depicted visually in Fig. 9.13.

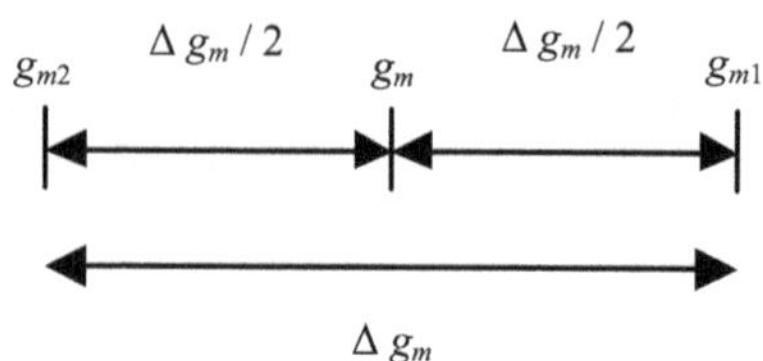

Fig. 9.13 Relation between g_{m1} and g_{m2}

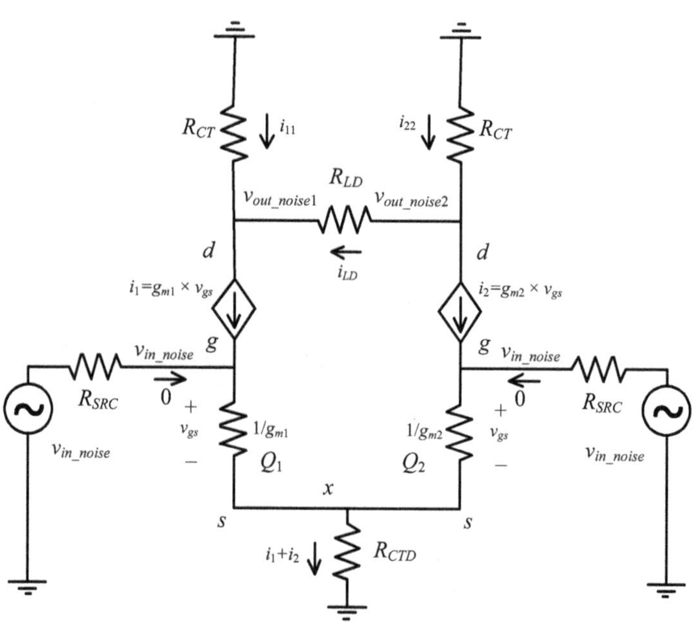

Fig. 9.14 Differential pair: common-mode small-signal AC model with g_m mismatch between Q_1 and Q_2

From the common-mode AC small-signal model as shown in Fig. 9.14 where we have neglected r_o as a first-hand approximation, $A_{cm\_diff\_mis\_trans}$ is defined as

$$\left| A_{cm\_diff\_mis\_trans} \right| = \left| \frac{v_{out\_noise2} - v_{out\_noise1}}{v_{in\_noise}} \right| \tag{9.21}$$

From Fig. 9.14, we have

$$i_1 + i_2 = \frac{v_x}{R_{CTD}} \tag{9.22}$$

and

$$v_{in\_noise} = v_{gs} + v_x \tag{9.23}$$

Combining (9.22) and (9.23), we get

$$i_1 + i_2 = \frac{v_{in\_noise} - v_{gs}}{R_{CTD}} \tag{9.24}$$

Also, since we have the same v_{gs} on both sides, we can see that

$$\frac{i_1}{i_2} = \frac{g_{m1}}{g_{m2}} \tag{9.25}$$

Combining (9.24) and (9.25) and, we get

$$
\begin{aligned}
i_1 &= \frac{v_{in\_noise} - v_{gs}}{R_{CTD}} \left(\frac{g_{m1}}{g_{m1} + g_{m2}} \right) = \frac{v_{in\_noise} - v_{gs}}{2R_{CTD}} \left(\frac{g_{m1}}{g_m} \right) \\
i_2 &= \frac{v_{in\_noise} - v_{gs}}{R_{CTD}} \left(\frac{g_{m2}}{g_{m1} + g_{m2}} \right) = \frac{v_{in\_noise} - v_{gs}}{2R_{CTD}} \left(\frac{g_{m2}}{g_m} \right)
\end{aligned}
\tag{9.26}
$$

A KVL in the upper portion of the circuit gives us

$$
\begin{aligned}
&i_{22}R_{CT} + v_{out\_noise2} - v_{out\_noise1} - i_{11}R_{CT} = 0 \\
&v_{out\_noise2} - v_{out\_noise1} = R_{CT}(i_{11} - i_{22}) = i_{LD}R_{LD}
\end{aligned}
\tag{9.27}
$$

Expressing i_{11} and i_{22} in terms of i_1, i_2, and i_{LD} starting with a KCL at the output nodes, we get

$$
\begin{aligned}
i_{11} &= i_1 - i_{LD} \\
i_{22} &= i_2 + i_{LD} \\
\Rightarrow i_{11} - i_{22} &= i_1 - i_2 - 2i_{LD}
\end{aligned}
\tag{9.28}
$$

Combining (9.27) with (9.28) we get

$$
\begin{aligned}
v_{out\_noise2} - v_{out\_noise1} &= R_{CT}(i_1 - i_2) - 2i_{LD}R_{CT} \\
&= R_{CT}(i_1 - i_2) - 2\left(\frac{v_{out\_noise2} - v_{out\_noise1}}{R_{LD}} \right) R_{CT}
\end{aligned}
\tag{9.29}
$$

Combining (9.20) with (9.26) we get

$$i_1 - i_2 = \frac{v_{in\_noise} - v_{gs}}{2R_{CTD}} \left(\frac{g_{m1} - g_{m2}}{g_m} \right) \tag{9.30}$$

Substituting for $i_1 - i_2$ from (9.30) into (9.29) we get

$$v_{out\_noise2} - v_{out\_noise1} =$$

$$R_{CT}\left[\frac{v_{in\_noise} - v_{gs}}{2R_{CTD}}\left(\frac{g_{m1} - g_{m2}}{g_m}\right)\right] - 2\left(\frac{v_{out\_noise2} - v_{out\_noise1}}{R_{LD}}\right)R_{CT} \tag{9.31}$$

Rearranging (9.31) knowing that $g_{m1} - g_{m2}$ is Δg_m, we get

$$v_{out\_noise2} - v_{out\_noise1} =$$

$$\frac{1}{2R_{CTD}}\left(\frac{R_{CT}}{1 + \frac{2R_{CT}}{R_{LD}}}\right)\left(\frac{\Delta g_m}{g_m}\right)v_{in\_noise} - \frac{1}{2R_{CTD}}\left(\frac{R_{CT}}{1 + \frac{2R_{CT}}{R_{LD}}}\right)\left(\frac{\Delta g_m}{g_m}\right)v_{gs} \tag{9.32}$$

We can see that the noise at the output is a linear function of the noise at the input with a sensitivity factor of

$$\frac{\partial\left(v_{out\_noise2} - v_{out\_noise1}\right)}{\partial v_{in\_noise}} = \frac{1}{2R_{CTD}}\left(\frac{R_{CT}}{1 + \frac{2R_{CT}}{R_{LD}}}\right)\left(\frac{\Delta g_m}{g_m}\right) \tag{9.33}$$

This sensitivity factor is less than one, which means that the noise at the output is an attenuated version of the noise at the input. This factor can be reduced in two ways by:

1. reducing the mismatch factor ($\Delta g_m/g_m$), for example, by doing a proper layout
2. having a large R_{CTD} for example by implementing the lower current source with a very high output resistance.

If v_{gs} is much smaller than v_{in_noise}, that is, from Fig. 9.14, $1/g_{m1} << R_{CTD}$, then

$$v_{out\_noise2} - v_{out\_noise1} \approx \frac{1}{2R_{CTD}}\left(\frac{R_{CT}}{1 + \frac{2R_{CT}}{R_{LD}}}\right)\left(\frac{\Delta g_m}{g_m}\right)v_{in\_noise} \tag{9.34}$$

and, as a result

$$\left|A_{cm\_diff\_mis\_trans}\right| = \left|\frac{v_{out\_noise2} - v_{out\_noise1}}{v_{in\_noise}}\right| \approx \frac{1}{2R_{CTD}}\left(\frac{R_{CT}}{1 + \frac{2R_{CT}}{R_{LD}}}\right)\left(\frac{\Delta g_m}{g_m}\right) \tag{9.35}$$

Additionally, if $R_{LD} >> 2R_{CT}$, then

$$\left| A_{cm\_diff\_mis\_trans} \right| = \left| \frac{v_{out\_noise2} - v_{out\_noise1}}{v_{in\_noise}} \right| \approx \frac{R_{CT}}{2R_{CTD}} \left(\frac{\Delta g_m}{g_m} \right) \qquad (9.36)$$

As for $A_{d\_diff\_mis\_trans}$, its value stays the same as in (9.16) since the difference between the two branches is very small. Therefore, making use of (9.16) and (9.35), the $CMRR_{mis\_trans}$ becomes

$$CMRR_{mis\_trans} = 20 \log \left| \frac{A_{d\_diff\_mis\_trans}}{A_{cm\_diff\_mis\_trans}} \right|$$

$$\approx 20 \log \left| \frac{\dfrac{R_{CT}//(R_{LD}/2)//r_o}{(1/g_m)}}{\dfrac{1}{2R_{CTD}} \left(\dfrac{R_{CT}}{1 + \dfrac{2R_{CT}}{R_{LD}}} \right) \left(\dfrac{\Delta g_m}{g_m} \right)} \right| \qquad (9.37)$$

With the approximations above including $R_{CT} << r_o$, if we make use of (9.17) and (9.36), the $CMRR_{mis\_trans}$ becomes

$$CMRR_{mis\_trans} = 20 \log \left| \frac{A_{d\_diff\_mis\_trans}}{A_{cm\_diff\_mis\_trans}} \right|$$

$$\approx 20 \log \left| \frac{g_m(R_{CT}//r_o)}{\dfrac{R_{CT}}{2R_{CTD}} \left(\dfrac{\Delta g_m}{g_m} \right)} \right| \approx 20 \log \left| \frac{g_m R_{CT}}{\dfrac{R_{CT}}{2R_{CTD}} \left(\dfrac{\Delta g_m}{g_m} \right)} \right| \qquad (9.38)$$

$$= 20 \log \left| \frac{2 g_m R_{CTD}}{\left(\dfrac{\Delta g_m}{g_m} \right)} \right|$$

We can see that the qualitative result for improving the $CMRR_{mis\_trans}$ is the same as that for reducing the sensitivity of the output noise to the input noise, which consists of reducing $(\Delta g_m/g_m)$ and having a large R_{CTD}.

The second mismatch to take into account is that between the two current sources at the load. This can be modeled by using two current sources with different output resistances R_{CT1} and R_{CT2}.

We need to calculate the $CMRR_{mis\_ct}$ defined as follows

Fig. 9.15 Relation between R_{CT1} and R_{CT2}

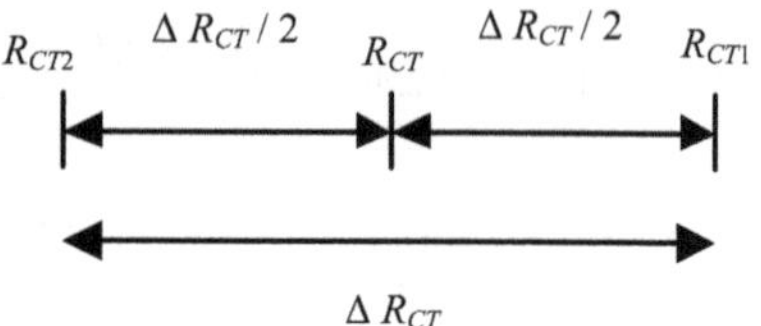

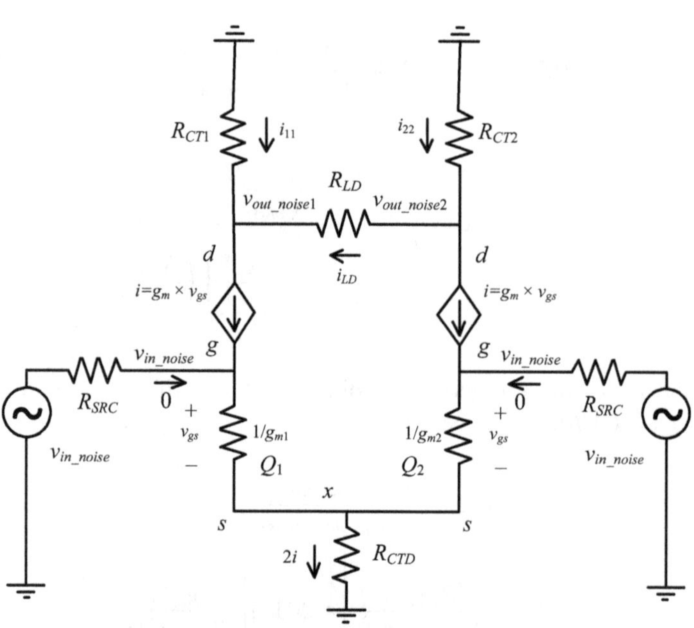

Fig. 9.16 Differential pair: common-mode small-signal AC model with R_{CT} mismatch

$$CMRR_{mis\_ct} = 20\log\left|\frac{A_{d\_diff\_mis\_ct}}{A_{cm\_diff\_mis\_ct}}\right| \qquad (9.39)$$

Let R_{CT1} be the output resistance of the current source in the left branch and R_{CT2} the output resistance of the current source in the right branch in such a way that their average is called R_{CT} and

$$R_{CT1} = R_{CT} + \frac{\Delta R_{CT}}{2}$$
$$R_{CT2} = R_{CT} - \frac{\Delta R_{CT}}{2} \qquad (9.40)$$

Note that ΔR_{CT} can be either positive or negative.

The relation in (9.40) is depicted visually in Fig. 9.15.

From the common-mode AC small-signal model as shown in Fig. 9.16, where we have neglected r_o as a first-hand approximation, $A_{cm\_diff\_mis\_ct}$ is defined as

$$\left| A_{cm\_diff\_mis\_ct} \right| = \left| \frac{v_{out\_noise2} - v_{out\_noise1}}{v_{in\_noise}} \right| \tag{9.41}$$

From Fig. 9.16, we have

$$v_{in\_noise} = i \left(\frac{1}{g_m} + 2R_{CTD} \right) \tag{9.42}$$

Expressing i in terms of i_{11} and i_{22} starting with a KCL at the output nodes, we get

$$\begin{aligned} i_{11} &= i - i_{LD} \\ i_{22} &= i + i_{LD} \\ \Rightarrow i &= \frac{i_{11} + i_{22}}{2} \end{aligned} \tag{9.43}$$

Combining (9.42) and (9.43), we get

$$v_{in\_noise} = \left(\frac{i_{11} + i_{22}}{2} \right) \left(\frac{1}{g_m} + 2R_{CTD} \right) \tag{9.44}$$

A KVL in the upper portion of the circuit gives us

$$\begin{aligned} i_{22}R_{CT2} + v_{out\_noise2} - v_{out\_noise1} - i_{11}R_{CT1} &= 0 \\ v_{out\_noise2} - v_{out\_noise1} &= i_{11}R_{CT1} - i_{22}R_{CT2} \end{aligned} \tag{9.45}$$

As can be seen from (9.44) and (9.45), in order to relate v_{in_noise} to $v_{out\_noise2} - v_{out\_noise1}$ and infer some qualitative attributes, we will assume as a first-hand approximation that R_{LD} is very large. From (9.43), this would mean that

$$\begin{aligned} i_{LD} &\approx 0 \\ i_{11} &\approx i_{22} \approx i \end{aligned} \tag{9.46}$$

Using (9.43) to replace i_{11} and i_{22} with i in (9.44) and (9.45) we get

$$\begin{aligned} v_{in\_noise} &\approx i \left(\frac{1}{g_m} + 2R_{CTD} \right) \\ \Rightarrow i &\approx \frac{v_{in\_noise}}{\dfrac{1}{g_m} + 2R_{CTD}} \end{aligned} \tag{9.47}$$

and

$$v_{out\_noise2} - v_{out\_noise1} \approx i(R_{CT1} - R_{CT2})$$

$$\Rightarrow i \approx \frac{v_{out\_noise2} - v_{out\_noise1}}{R_{CT1} - R_{CT2}} \tag{9.48}$$

Combining (9.47) and (9.48), we get

$$\frac{v_{in\_noise}}{\frac{1}{g_m} + 2R_{CTD}} \approx \frac{v_{out\_noise2} - v_{out\_noise1}}{R_{CT1} - R_{CT2}}$$

$$\Rightarrow \frac{v_{out\_noise2} - v_{out\_noise1}}{v_{in\_noise}} \approx \frac{R_{CT1} - R_{CT2}}{\frac{1}{g_m} + 2R_{CTD}} = \frac{\Delta R_{CT}}{\frac{1}{g_m} + 2R_{CTD}} \tag{9.49}$$

As a result, we get

$$\left| A_{cm\_diff\_mis\_ct} \right| = \left| \frac{v_{out\_noise2} - v_{out\_noise1}}{v_{in\_noise}} \right| \approx \frac{R_{CT}}{\frac{1}{g_m} + 2R_{CTD}} \left(\frac{\Delta R_{CT}}{R_{CT}} \right) \tag{9.50}$$

If v_{gs} is much smaller than v_{in_noise}, that is, from Fig. 9.16, $1/g_m << R_{CTD}$, then

$$\left| A_{cm\_diff\_mis\_ct} \right| = \left| \frac{v_{out\_noise2} - v_{out\_noise1}}{v_{in\_noise}} \right| \approx \frac{R_{CT}}{2R_{CTD}} \left(\frac{\Delta R_{CT}}{R_{CT}} \right) \tag{9.51}$$

Note how similar this result is to $A_{cm\_diff\_mis\_trans}$ in (9.36). This goes to show that mismatches in the current sources at the load have a similar effect as mismatches in the transistors Q_1 and Q_2 as a first-hand approximation.

As a result, similar comments can be made as before, namely that the noise at the output is an attenuated version of the noise at the input and that $A_{cm\_diff\_mis\_ct}$ can be reduced in two ways by:

1. reducing the mismatch factor ($\Delta R_{CT}/R_{CT}$), for example, by doing a proper layout,
2. having a large R_{CTD} for example by implementing the lower current source with a very high output resistance.

As for $A_{d\_diff\_mis\_ct}$, its value stays the same as in (9.16) since the difference between the two branches is very small. Therefore, making use of (9.17) and (9.50), the $CMRR_{mis\_ct}$ becomes

$$CMRR_{mis\_ct} = 20 \log \left| \frac{A_{d\_diff\_mis\_ct}}{A_{cm\_diff\_mis\_ct}} \right|$$

$$\approx 20 \log \left| \frac{g_m(R_{CT}//r_o)}{\frac{R_{CT}}{2R_{CTD}} \left(\frac{\Delta R_{CT}}{R_{CT}} \right)} \right| \tag{9.52}$$

With the approximations above including $R_{CT} << r_o$, if we make use of (9.17) and (9.52), the $CMRR_{mis\_ct}$ becomes

$$CMRR_{mis\_tct} = 20 \log \left| \frac{A_{d\_diff\_mis\_ct}}{A_{cm\_diff\_mis\_ct}} \right|$$

$$\approx 20 \log \left| \frac{g_m(R_{CT}//r_o)}{\dfrac{R_{CT}}{2R_{CTD}}\left(\dfrac{\Delta R_{CT}}{R_{CT}}\right)} \right| \approx 20 \log \left| \frac{g_m R_{CT}}{\dfrac{R_{CT}}{2R_{CTD}}\left(\dfrac{\Delta R_{CT}}{R_{CT}}\right)} \right| \qquad (9.53)$$

$$= 20 \log \left| \frac{2g_m R_{CTD}}{\left(\dfrac{\Delta R_{CT}}{R_{CT}}\right)} \right|$$

We can see that the qualitative result for improving the $CMRR_{mis\_ct}$ is the same as before, which consists of reducing $(\Delta R_{CT}/R_{CT})$ and increasing R_{CTD}.

9.3.1.5 Output Common-Mode Stabilization

As described in the sections above, the common-mode output DC voltage V_{OUT} is supposed to be well-known and stable so that the load sees a stable DC voltage at its input.

As was shown in (9.9), V_{OUT} depends on V_{CT}, which itself depends on I_D. This I_D is part of the tail current source output. As a result, we need to make sure that the tail current source is providing us with the appropriate current $2I_D$ that results in the needed V_{OUT}.

This is done by employing a technique called the Common-Mode Feedback (CMFB) as seen in Fig. 9.17. With this technique, the average output DC voltage V_{OUT} is measured and compared to a reference voltage V_{REF}. Based on the result of the comparison, a control voltage V_{CTRL} is issued to the tail current to reduce the gap between V_{OUT} and V_{REF}.

A sample implementation of the CMFB circuit is shown in Fig. 9.18. Here, two resistors, named R_F, are used to measure V_{OUT}, a comparator compares V_{OUT} to V_{REF}, and issues V_{CTRL} that controls the tail current source.

9.3.2 Differential-to-Single-Ended Amplifier

We have seen the benefits of differential signaling in terms of resilience in the presence of noise. However, in some cases, we might need to convert a differential input into a single-ended one, which is what a differential-to-single-ended amplifier does.

A general implementation of a differential-to-single-ended amplifier is shown in Fig. 9.19.

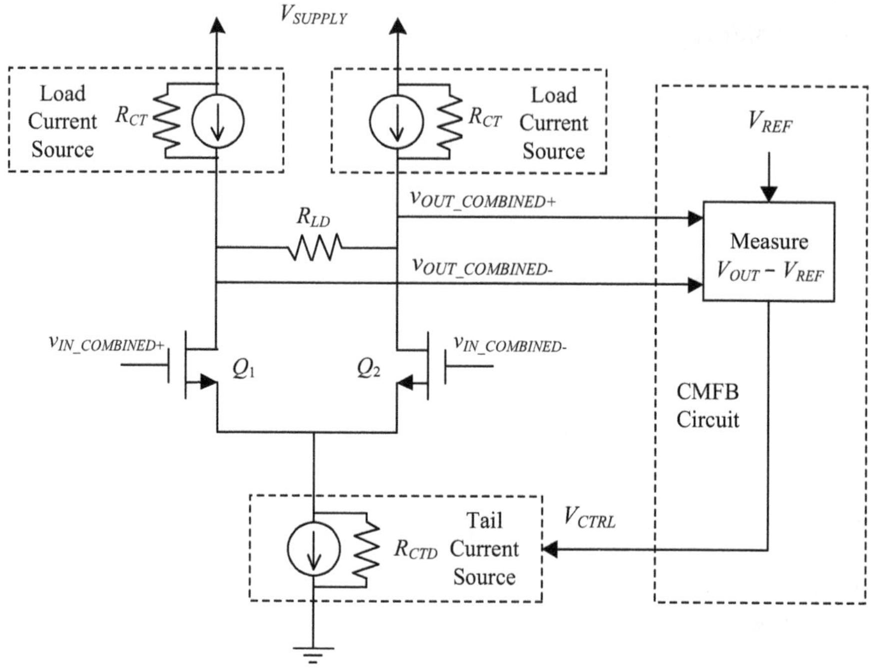

Fig. 9.17 Differential pair: CMFB circuit general implementation

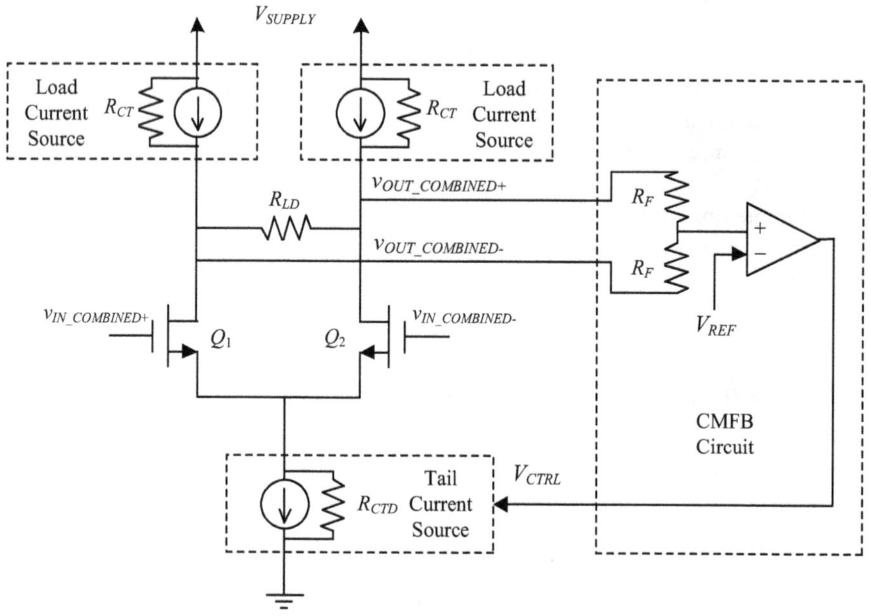

Fig. 9.18 Differential pair: CMFB circuit sample implementation

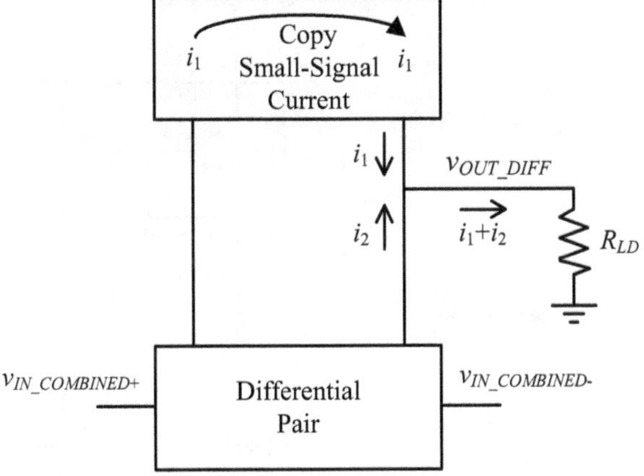

Fig. 9.19 Differential-to-single-ended amplifier general implementation

The main idea consists of the following. The single-ended output has to be a function of what is happening in both branches. As such, ideally, we would like to copy the AC small-signal current from the left branch, add it to the AC small-signal current in the right branch, and output the resulting current into the load, thus creating the output voltage.

In order to achieve this, we employ a current mirror on top of a differential pair. This current mirror will copy i_1 from the left branch into the right branch. Since i_1 and i_2 have opposite senses, they will add up at the output node creating v_{OUT_DIFF}. Note that since the current mirror copies the AC small-signal current ideally as it is, then we have:

$$i_1 = i_2 = i$$
$$i_1 + i_2 = 2i$$

$$(9.54)$$

A sample example of such an implementation is shown in Fig. 9.20.

Now, we will get the *CMRR* of this structure. In order to do so, we will start by getting the common-mode gain followed by the differential gain.

9.3.2.1   Common-Mode AC Small-Signal Behavior (Noise)

When noise is applied at the inputs, the small-signal model will be as in Fig. 9.21. Note that as a first-hand approximation, we have neglected r_o for the transistors which we have modeled using the T-model. Also, note also that diode-connected transistor Q_3 acts as a resistance whose value is $1/g_{m3}$.

The structure is not perfectly symmetrical owing to Q_3, unlike Q_4, being diode-connected. However, since the lower part is the same, with the same input v_{in_noise}, then v_{d1} is equal to v_{d2} and

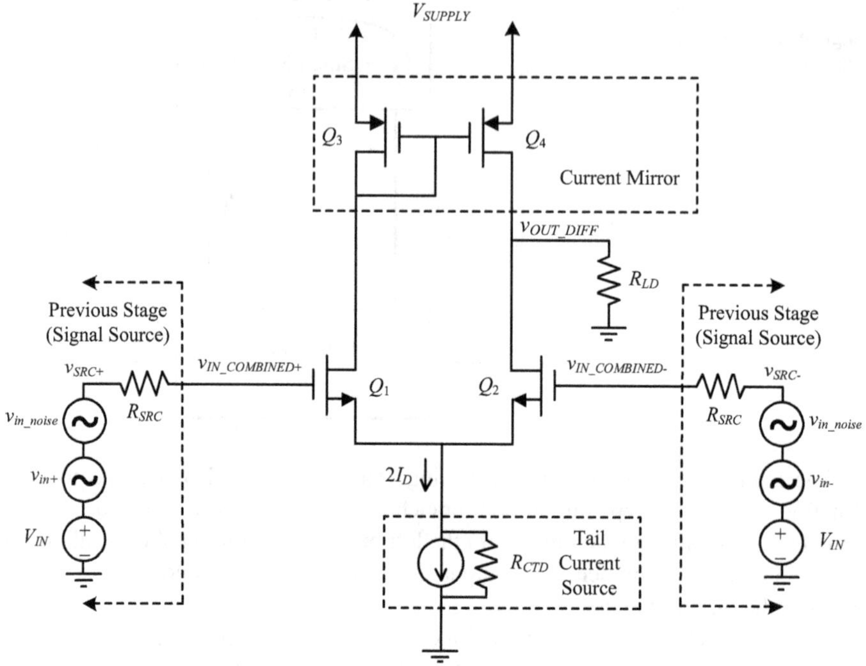

Fig. 9.20 Differential-to-single-ended amplifier sample implementation

$$v_{d1} = v_{gs3}$$

$$v_{gs3} = v_{gs4} \tag{9.55}$$

$$v_{out\_noise} = v_{d2} = v_{d1}$$

Additionally, if R_{LD} is large, then each branch will be biased around almost the same current I_D. In this case, the output resistance of the tail current can be split into two, each having a value of $2R_{CTD}$ so that when combined in parallel the result is R_{CTD}.

Taking into account the above considerations and calculating the common-mode gain A_{cm}, while realizing that this is a common-source amplifier whose gain is the total resistance at the drain divided by the total resistance at the source, we get

$$|A_{cm}| = \left| \frac{v_{out\_noise}}{v_{in\_noise}} \right| \approx \frac{v_{gs4}}{v_{in\_noise}} = \frac{v_{gs3}}{v_{in\_noise}} = \frac{\frac{1}{g_{m3}}}{\frac{1}{g_{m1}} + 2R_{CTD}} \tag{9.56}$$

Typically $2R_{CTD} >> 1/g_{m1}$ leading to

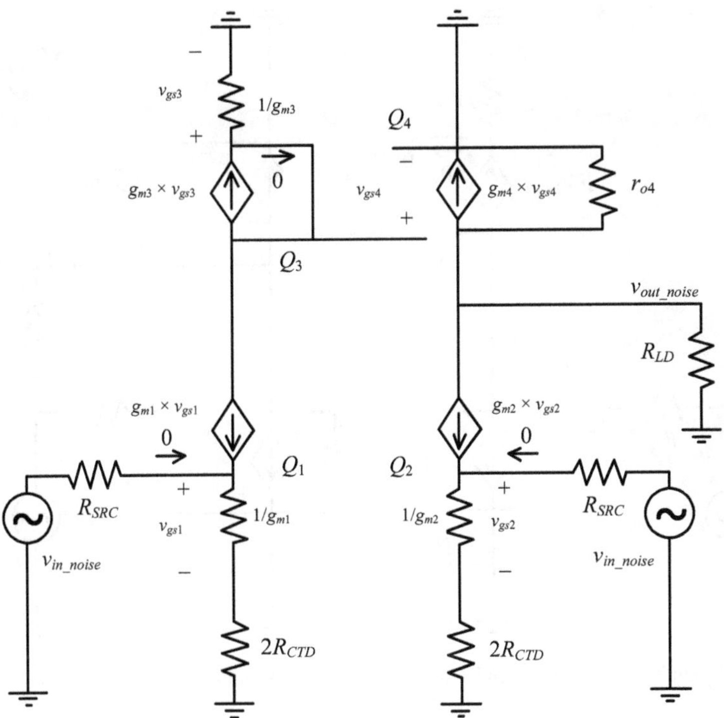

Fig. 9.21 Differential-to-single-ended amplifier: common-mode small-signal AC

$$|A_{cm}| = \left| \frac{v_{out\_noise}}{v_{in\_noise}} \right| \approx \frac{1}{2g_{m3}R_{CTD}} \tag{9.57}$$

which obviously is much less than one, but not zero either. This means that this differential-to-single-ended amplifier passes an attenuated version of the noise at its input, unlike the ideal one in Fig. 9.6.

9.3.2.2 Differential-Mode Small-Signal Behavior

The small-signal model of the amplifier in this case is shown in Fig. 9.21. As a first-hand approximation, we have neglected r_{o3} (Fig. 9.22).

Similarly to the common-mode case, if R_{LD} is large, then each branch will be biased around almost the same current I_D.

We need to get the ratio

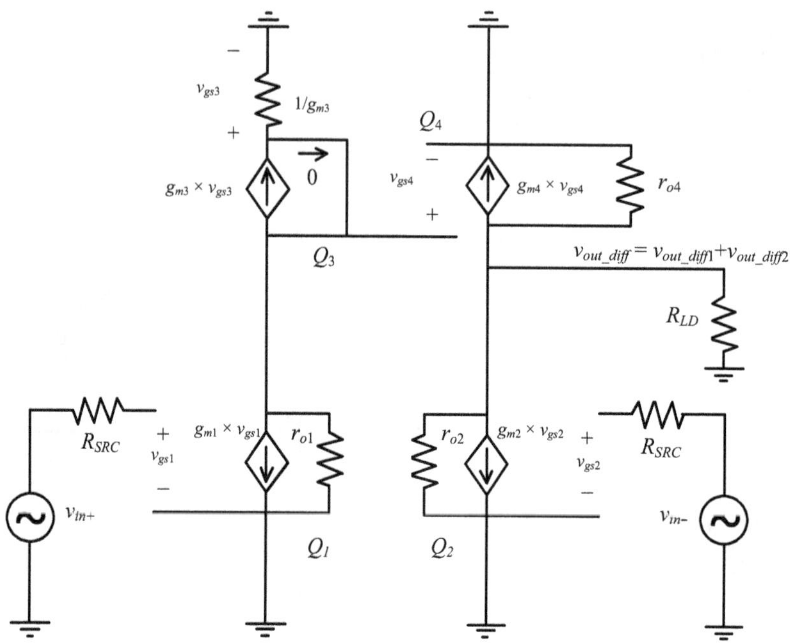

Fig. 9.22 Differential-to-single-ended amplifier: differential mode small-signal circuit

$$|A_d| = \left| \frac{v_{out\_diff}}{v_{in+} - v_{in-}} \right| = \left| \frac{v_{out\_diff1} + v_{out\_diff2}}{v_{in+} - v_{in-}} \right| \tag{9.58}$$

We will use the superposition approach, which consists of the following steps:

- disable v_{in-} and get v_{out_diff1} as a function of v_{in+},
- disable v_{in+} and get v_{out_diff2} as a function of v_{in-},
- express v_{out_diff} which is equal to $(v_{out\_diff1} + v_{out\_diff2})$ as a function of $(v_{in+} - v_{in-})$.

By disabling v_{in-}, and from Fig. 9.22, we see that the left branch is a common-source amplifier with the input being v_{in+} and the output being v_{gs3}. Since usually $1/g_{m3} << r_{o1}$, and since the $g_{m1} \approx g_{m3}$, then

$$v_{gs3} = -\frac{r_{o1} // \frac{1}{g_{m3}}}{\frac{1}{g_{m1}}} v_{in+} \approx -g_{m1}\left(\frac{1}{g_{m3}}\right) v_{in+} \approx -v_{in+} \tag{9.59}$$

Continuing with the right branch, this is also a common-source amplifier with the input being v_{gs4} and the output being v_{out_diff1}.

Noting that

$$v_{gs4} = v_{gs3} \approx -v_{in+} \tag{9.60}$$

we get

$$v_{out\_diff1} = -\frac{r_{o2}//r_{o4}//R_{LD}}{\frac{1}{g_{m4}}} v_{gs4} \approx \frac{r_{o2}//r_{o4}//R_{LD}}{\frac{1}{g_{m4}}} v_{in+} \tag{9.61}$$

As a result,

$$v_{out\_diff1} \approx g_{m4}(r_{o2}//r_{o4}//R_{LD})v_{in+} \tag{9.62}$$

By disabling v_{in+}, and from Fig. 9.22, we see that from the left branch that

$$v_{gs3} = v_{gs4} = 0 \tag{9.63}$$

As for the right branch, it is a common-source amplifier with the input being v_{in-} and the output being v_{out_diff2}. Therefore,

$$v_{out\_diff2} \approx -g_{m2}(r_{o2}//r_{o4}//R_{LD})v_{in-} \tag{9.64}$$

Combining (9.62) with (9.64) knowing that $g_{m2} \approx g_{m4} \approx g_m$ and that $r_{o2} \approx r_{o4} \approx r_o$ then

$$|A_d| = \left|\frac{v_{out\_diff1} + v_{out\_diff2}}{v_{in+} - v_{in-}}\right| \approx g_m(r_{o2}//r_{o4}//R_{LD}) \approx \frac{1}{2}g_m(r_o//R_{LD}) \tag{9.65}$$

Therefore, the differential-to-single-ended amplifier has the same gain as that of a common-source amplifier.

From another perspective, we could have gotten the result in (9.65) differently.

We see from Fig. 9.22 that the left branch is a common-source amplifier with a diode-connected transistor Q_3 as a load. This means that its output v_{gs4} is an inverted version of the input v_{in+}, Thus

$$v_{gs4} \approx -v_{in+} \tag{9.66}$$

The right branch consists of two common-source amplifiers on top of each other. The one on the top has an input transistor Q_4 and an input signal $-v_{in+}$ and the one at the bottom has an input transistor Q_2 and an input signal v_{in-}.

From (9.14), we know that

$$v_{in+} = -v_{in-} = \frac{v_{in\_diff}}{2} \tag{9.67}$$

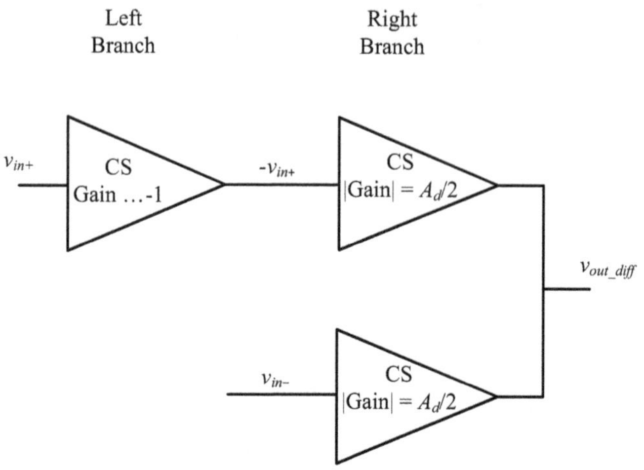

Fig. 9.23 Differential-to-single-ended amplifier block-level behavior

We can now envision the block-level model of the circuit in Fig. 9.22 as in Fig. 9.23.

The small-signal AC model of the right branch amplifier is depicted in Fig. 9.24. As a result, we can see that

$$|A_d| = \left| \frac{v_{out\_diff1} + v_{out\_diff2}}{v_{in+} - v_{in-}} \right| = \left| \frac{v_{out\_diff1} + v_{out\_diff2}}{v_{vin\_diff}} \right| \approx g_m(r_{o2}//r_{o4}//R_{LD}) \quad (9.68)$$

which is the same result as in (9.65).

9.4 Differential Logic Circuits

Differential logic circuits are based on ratioed logic and have the general structure as in Fig. 9.25. This structure is also known as Differential Cascode Voltage Switch Logic (DCVSL).

Differential logic circuits provide the logic function as well as its inverse. As a result, an inverter is not needed. Additionally, since the output of one branch enhances the input to the other branch, differential logic circuits tend to have shorter transition times than single-ended logic circuits. Of course, on the downside, they are larger and more power hungry than their single-ended counterparts.

Another advantage is that when using DCVSL, only one gate structure is needed to implement AND/NAND as well as OR/NOR. A PDN-based version of this gate is shown in Fig. 9.26.

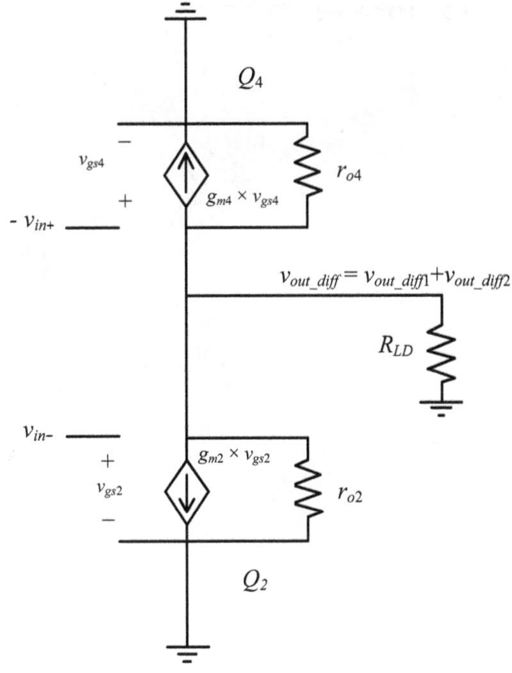

Fig. 9.24 Differential-to-single-ended amplifier: differential mode small-signal circuit, right branch

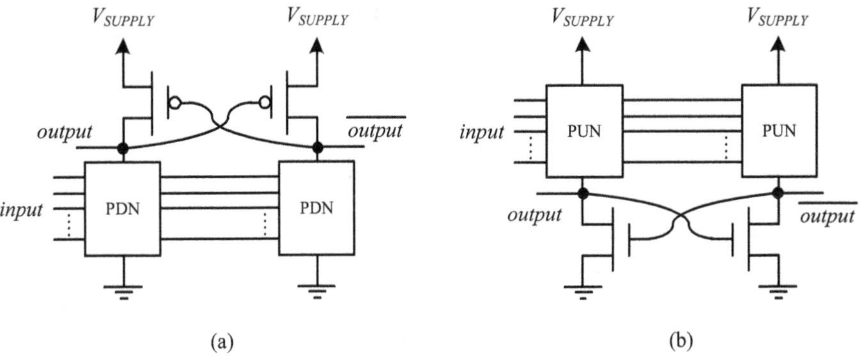

Fig. 9.25 DCVSL gates with a PDN (**a**) and a PUN (**b**)

If our inputs are *in*1 and *in*2, this gate implements the AND/NAND function as follows

$$\overline{out} = \overline{in1} + \overline{in2} = \overline{in1 \cdot in2}$$
$$out = in1 \cdot in2$$

(9.69)

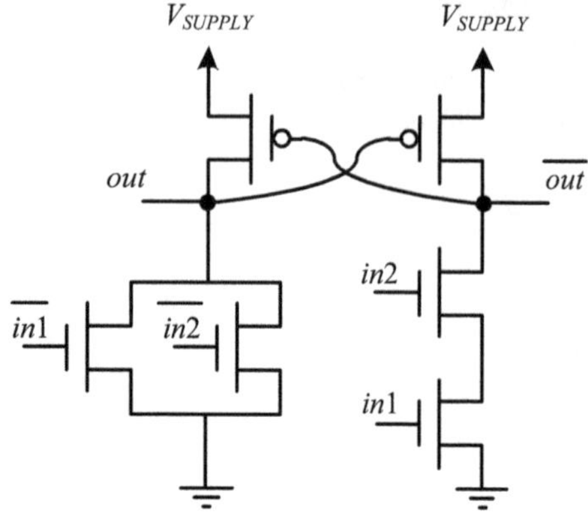

Fig. 9.26 PDN-based OR/NOR and AND/NAND DCVSL gate

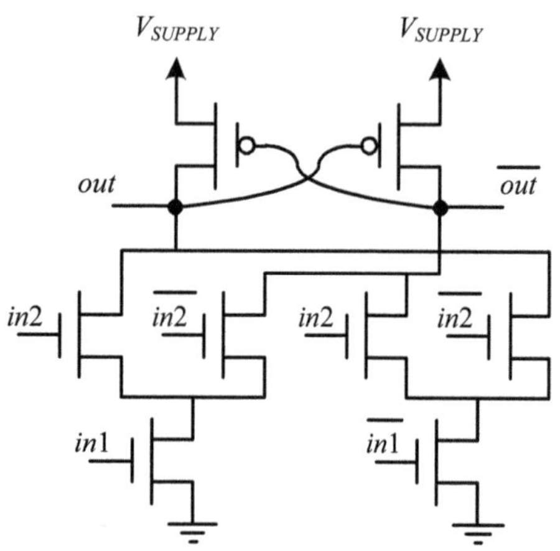

Fig. 9.27 PDN-based XOR/NXOR DCVSL gate

However, if our inputs are flipped as follows

$$in3 = \overline{in1}$$
$$in4 = \overline{in2}$$

(9.70)

then this gate implements the OR/NOR function as follows

$$\overline{out} = in3 + in4$$
$$out = \overline{in3} \cdot \overline{in4} = \overline{in3 + in4}$$

(9.71)

Another example is the PDN-based XOR/NXOR DCVSL gate shown in Fig. 9.27.

This gate implements the XOR/NXOR function as follows

$$\overline{out} = (in1 \cdot in2) + \left(\overline{in1} \cdot \overline{in2}\right) = \overline{in1 \oplus in2}$$
$$out = \left(in1 \cdot \overline{in2}\right) + \left(\overline{in1} \cdot in2\right) = in1 \oplus in2$$

(9.72)

Chapter 10
Voltage Regulators and Converters, Reference and Bias Circuits

Electronic circuits require a source that provides them with power as well as reference and bias voltages and currents. Special circuits are interposed between the power source and the core circuits in order to make this happen. This chapter provides an overview of the general power distribution scheme in integrated circuits, followed by a discussion about regulators and converters, reference, and bias circuits.

10.1 Power Distribution Overview

Electronic circuits require power to operate. This power might come from a DC source such as a battery or from an AC source such as the mains electricity that gets converted into DC using a rectifier. Either way, in this chapter, we will assume that a raw DC power source of some kind is already present.

This DC power is generally unstable and often quite noisy. Additionally, the DC value might not be exactly the one that is needed by the circuit. As a result, a voltage regulator is used in order to decrease the variations in this DC value, including those due to noise, and to stabilize it around the desired value. According to Fig. 10.1, and following the voltage regulator, sometimes a reference circuit is needed in order to create a fixed voltage or current. This, in turn, is sometimes followed by a bias circuit required by some core circuitry.

Based on Fig. 10.1, a Type 1 circuit such as an amplifier needs biasing, whereas a Type 2 circuit such as a data converter only needs a reference. As for a Type 3 circuit such as a logic gate, it only requires a stable supply voltage.

In the following sections, we will go over the details of voltage regulators and converters, reference circuits, and bias circuits.

© The Editor(s) (if applicable) and The Author(s), under exclusive license to Springer Nature Switzerland AG 2024
J. G. Atallah, M. Ismail, *Integrated Electronic Circuits*,
https://doi.org/10.1007/978-3-031-62707-1_10

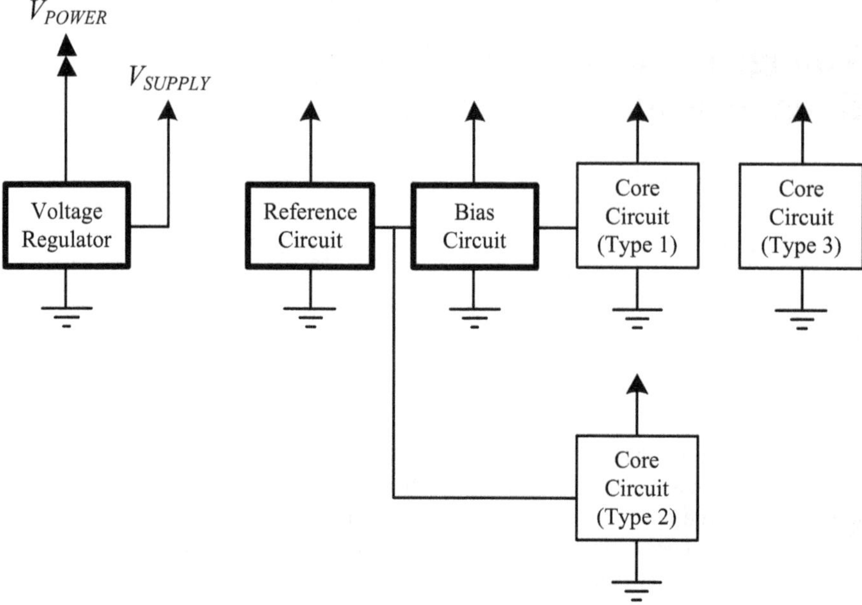

Fig. 10.1 General role of the voltage regulator, reference circuit, and bias circuit

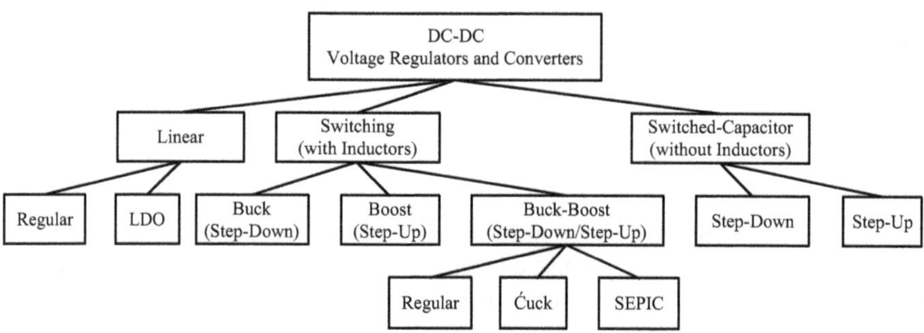

Fig. 10.2 Overview of voltage regulator and converter families

10.2 Voltage Regulators and Converters

A voltage regulator can exist either built into a given IC or as a separate IC shared by many others. Its primary purpose is to provide a given IC a suitable DC supply voltage with minimum variations including those due to noise. Since we are assuming, as mentioned earlier, that a raw DC power source of some kind is already present, then the voltage regulators we will be discussing are of the DC-DC type.

There are several ways to implement voltage regulators and converters. An overview map is shown in Fig. 10.2. Since the input DC value is different from the output DC value, these voltage regulators are also referred to as converters.

However, in practice, the linear ones are usually referred to as regulators whereas the switching ones are usually referred to as converters.

Linear regulators do not require a clock whereas switching converters do. Additionally, a buck converter is one whose output voltage is lower than its input voltage whereas the situation is opposite in a boost converter.

Table 10.1 Voltage regulator performance parameters

Parameter	Symbol	Explanation	Preference
Input voltage	V_I	Range of acceptable voltages at the input	Design-dependent
Output voltage	V_O	Designated output voltage	Design-dependent
Output current	I_O	Typical output current	High
Line regulation (Input voltage regulation)	$Reg_{line} = \Delta V_O / \Delta V_I$	Output voltage variations following the variations of the input voltage	Low
Load regulation (output voltage regulation)	$Reg_{load} = \Delta V_O / \Delta I_O$	Output voltage variations following the variations of the current drawn from the regulator	Low
Quiescent (bias) current	I_Q	The current drawn by the regulator under a no load condition	Low
Quiescent (bias) current change	ΔI_q	The amount of change in the bias current depending on the change in V_I and I_O.	Low
Supply voltage rejection (ripple rejection) or (power supply rejection ratio)	$PSRR = v_i / v_o$	The amount of noise attenuation between the input voltage and the output voltage, usually expressed in dB	High
Dropout voltage	$V_{drop} = V_I - V_O$	The difference between the input voltage and the output voltage	Low
Output resistance	R_O	The output resistance of the regulator	Low
Short circuit current	I_{SC}	The amount of current that the regulator will output if the load is short-circuited	Low
Short circuit peak current	I_{SCP}	The amount of peak current that the regulator will output if the load is short-circuited	Low
Output voltage drift (temperature coefficient of output voltage)	$\Delta V_O / \Delta T$	The amount of voltage change per degree Celsius away from the typical temperature	Low

10.2.1 Linear Voltage Regulators

Table 10.1 lists the electrical characteristics parameters of a voltage regulator along with a short explanation for each and whether the parameter value is preferred to be high or low. V_I consists of the input voltage, which is V_{POWER} in Fig. 10.1, and V_O consists of the output voltage, which is V_{SUPPLY} in Fig. 10.1.

Linear voltage regulators come in two flavors: regular and Low Dropout (LDO). A regular linear regulator has a relatively high V_{drop}, whereas an LDO has a relatively low V_{drop}. Typical implementations of such regulators are shown in Fig. 10.3. Both implementations require a reference voltage. The op-amp will ensure that V_O is as close as possible to the reference voltage V_{REF}, while providing enough current to the load.

A regular linear voltage regulator makes use of an n-type transistor such as an NMOS as in Fig. 10.3a. This NMOS is dealing with two signals: one coming from the op-amp and the other from V_I.

Starting with the input coming from the op-amp, it is connected to the gate of the NMOS and the output is at its source. Therefore, the NMOS is in the common-drain configuration, which is non-inverting. In order for the feedback to be negative, the source of the NMOS should be connected to the negative input of the op-amp.

As for the input coming from V_I, since it is connected to the drain while the output is at the source, the NMOS will have a gain less than one, meaning that noise on V_I will be attenuated before reaching V_O, leading to a high *PSRR*, which is an advantage.

However, for the NMOS to stay ON, its V_{GS} has to be greater than V_{th}. If V_G takes the highest value of V_I, then V_O will have to be V_{th} less than that. This limits the maximum value of V_O to $V_I - V_{th}$, preventing V_{drop} from being small, which is a drawback.

On the other hand, an LDO voltage regulator makes use of a p-type transistor such as a PMOS as in Fig. 10.3b. This PMOS is also dealing with two signals: one coming from the op-amp and the other from V_I.

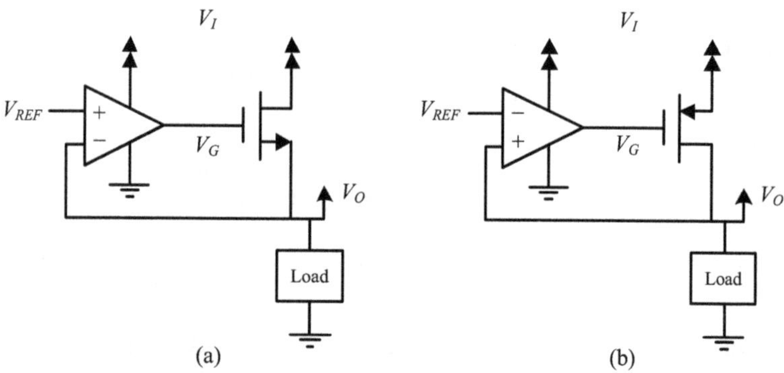

Fig. 10.3 Linear voltage regulator: regular (**a**) and LDO (**b**)

Starting with the input coming from the op-amp, it is connected to the gate of the PMOS and the output is at its drain. Therefore, the PMOS is in the common-source configuration, which is inverting. In order for the feedback to be negative, then the drain of the PMOS should be connected to the positive input of the op-amp.

As for the input coming from V_I, it is connected to the source of the PMOS, whereas the output is at the drain. Therefore, the PMOS is in the common-gate configuration, whose gain is quite high. Therefore, noise on V_I will be amplified before reaching V_O leading to a low *PSRR*, which is a drawback.

However, for the PMOS to stay ON, its $|V_{GS}|$ has to be greater than $|V_{th}|$. In spite of this, V_O can be greater than V_G as long as it is smaller than V_I by more than $|V_{OV}|$ to keep the transistor in the saturation region. Therefore, the maximum value of V_O is $V_I - |V_{OV}|$, which is higher than in the previous case since, in the absence of a small $|V_{th}|$, V_{OV} can be made smaller than $|V_{th}|$. As a result, V_{drop} can be made small, thus the name LDO, which is an advantage.

10.2.2 Switching Converters (With Inductors)

We will now look into the most popular DC converters, namely Buck, Boost, and Buck-Boost converters. All these converters make use of inductors.

10.2.2.1 Buck Converters

A Buck converter gives a DC output lower than its DC input, resulting in a step-down behavior. A typical Buck converter is shown in Fig. 10.4.

The signals in the Buck converter are shown in Fig. 10.5.

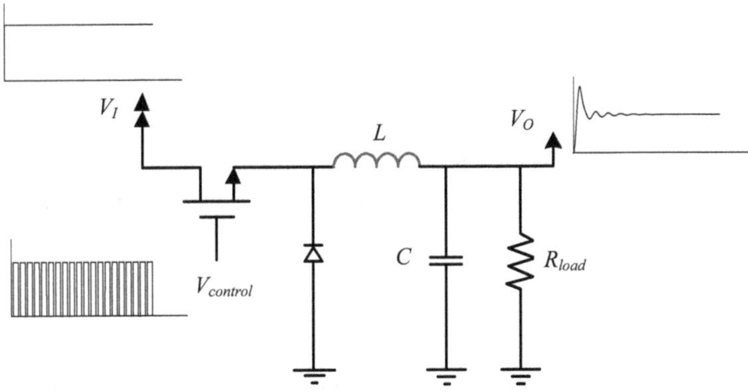

Fig. 10.4 A Buck converter

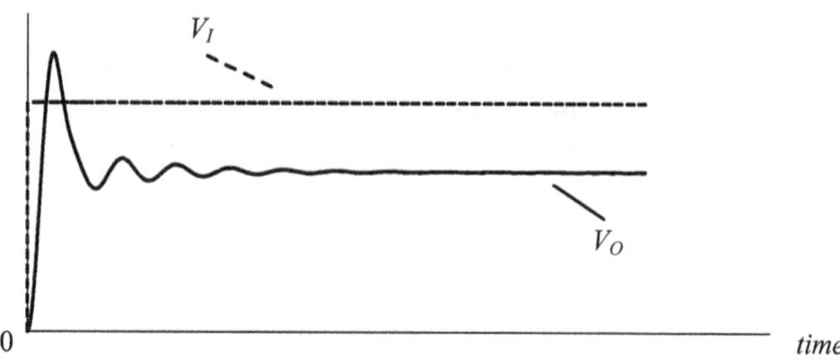

Fig. 10.5 Signals in a Buck converter

Let D be the duty cycle of the control voltage $V_{control}$, that is, if one period of the control signal is called T, then $V_{control}$ is ON $D \times T$ amount of time and OFF $(1 - D) \times T$ amount of time within a period T.

In this case, the relation between V_I and the average value of V_O in steady state is as follows:

$$V_O = D \times V_I \qquad (10.1)$$

This assumes that the current through the inductor always goes in one direction. In order to achieve that, the minimum value of the inductor is

$$L_{min} = \frac{(1-D)R_{load}}{2f} \qquad (10.2)$$

where f is the switching frequency of $V_{control}$.

The capacitor C and the load R_{load} form a low-pass filter. Therefore, in order to reduce the output ripple especially in the presence of a large load (small R_{load}), a large capacitor is needed.

10.2.2.2 Boost Converters

A Boost converter gives a DC output higher than its DC input, resulting in a step-up behavior. A typical Boost converter is shown in Fig. 10.6.

The signals in the Boost converter are shown in Fig. 10.7.

If D is the duty cycle of the control voltage $V_{control}$, then the relation between V_I and V_O is as follows:

$$V_O = \frac{1}{1-D} \times V_I \qquad (10.3)$$

This assumes that the current through the inductor always goes in one direction. In order to achieve that, the minimum value of the inductor is

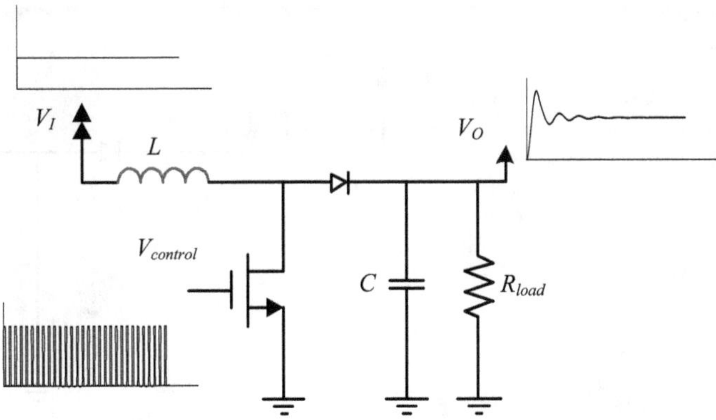

Fig. 10.6 A boost converter

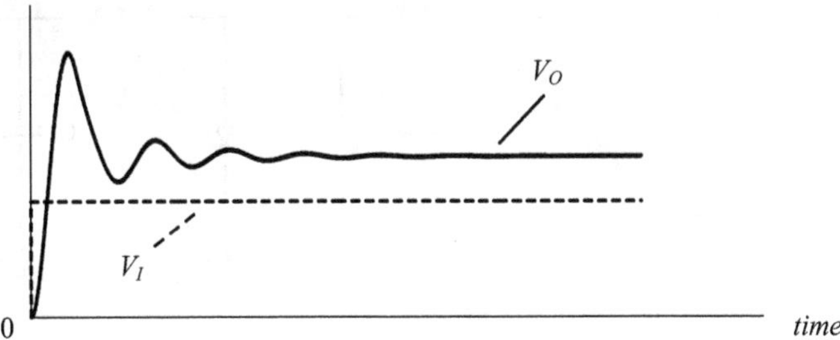

Fig. 10.7 Signals in a buck converter

$$L_{\min} = \frac{D(1-D)^2 R_{load}}{2f} \tag{10.4}$$

where f is the switching frequency of $V_{control}$.

As before, the capacitor C and the load R_{load} form a low-pass filter. Therefore, in order to reduce the output ripple especially in the presence of a large load (small R_{load}), a large capacitor is needed.

10.2.2.3   Buck-Boost Converters

A Buck-Boost converter gives a DC output that can be either higher or lower than its DC input, depending on the duty cycle D. As in Fig. 10.2, a Buck-Boost converter has three flavors:

- Regular.
- Ćuck.
- SEPIC.

Fig. 10.8 A regular buck-boost converter

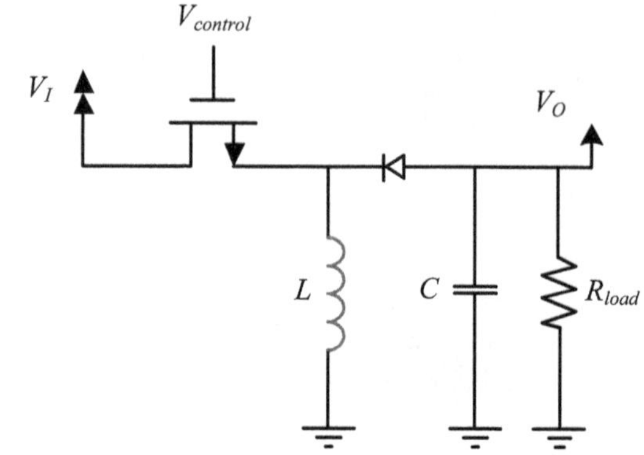

Fig. 10.9 A Ćuck converter

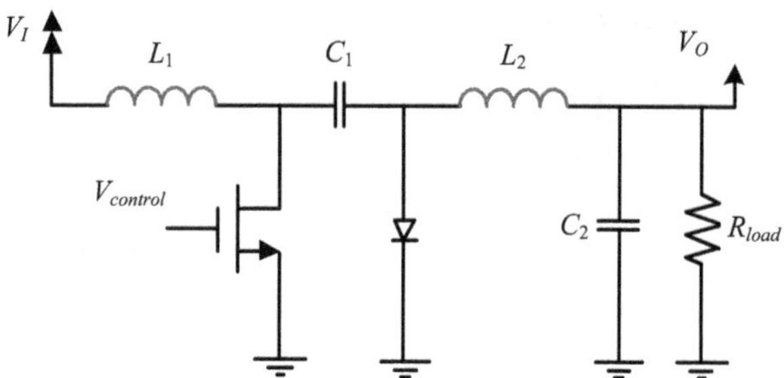

A typical Regular Buck-Boost converter is shown in Fig. 10.8.

If D is the duty cycle of the control voltage $V_{control}$, then the relation between V_I and V_O is as follows:

$$V_O = -\frac{D}{1-D} \times V_I \qquad (10.5)$$

Note that depending on the duty cycle D, the Buck-Boost converter can act either as a step-up or a step-down converter. Also note that the output has the opposite polarity of the input.

The current through the inductor always goes in one direction. In order to achieve that, the minimum value of the inductor is

$$L_{min} = \frac{(1-D)^2 R_{load}}{2f} \qquad (10.6)$$

where f is the switching frequency of $V_{control}$.

As before, the capacitor C and the load R_{load} form a low-pass filter. Therefore, in order to reduce the output ripple especially in the presence of a large load (small R_{load}), a large capacitor is needed.

A typical Ćuck converter is shown in Fig. 10.9.

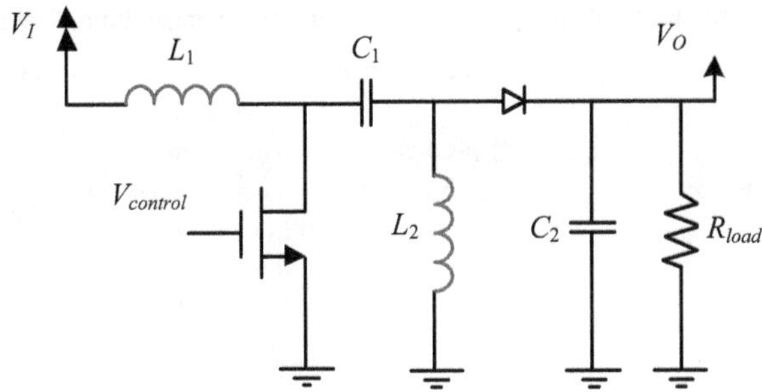

Fig. 10.10 A SEPIC converter

If D is the duty cycle of the control voltage $V_{control}$, then the relation between V_I and V_O is as follows:

$$V_O = -\frac{D}{1-D} \times V_I \qquad (10.7)$$

Note that depending on the duty cycle D, the Ćuck converter can act either as a step-up or a step-down converter. Also note that the output has the opposite polarity of the input.

These results are the same as those for the Regular Buck-Boost converter. The major difference is that a Ćuck converter makes use of two inductors and two capacitors. Inductor L_1 acts as a filter that attenuates unwanted frequencies on the input V_I.

The current through both inductors always goes in one direction. In order to achieve that, the minimum values of the inductors are

$$L_{1\,min} = \frac{(1-D)^2 R_{load}}{2Df}$$
$$L_{2\,min} = \frac{(1-D)R_{load}}{2f} \qquad (10.8)$$

where f is the switching frequency of $V_{control}$.

As before, the capacitor C_2 and the load R_{load} form a low-pass filter. Therefore, in order to reduce the output ripple, especially in the presence of a large load (small R_{load}), a large capacitor is needed.

A Single-Ended Primary Inductance (SEPIC) converter is shown in Fig. 10.10.

If D is the duty cycle of the control voltage $V_{control}$, then the relation between V_I and V_O is as follows:

$$V_O = \frac{D}{1-D} \times V_I \qquad (10.9)$$

Table 10.2 Characteristics of switching converters with inductors

Converter	Step-down step-up	Output polarity	Number of inductors	Number of capacitors
Buck	Step-down	Same as input	1	1
Boost	Step-up	Same as input	1	1
Regular buck–boost	Both	Inverse of input	1	1
Ćuk	Both	Inverse of input	2	2
SEPIC	Both	Same as input	2	2

Note that depending on the duty cycle D, the SEPIC converter can act either as a step-up or a step-down converter. Also note that, unlike in a Ćuk converter, the output has the same polarity as the input.

The SEPIC converter, similarly to a Ćuk converter, makes use of two inductors and two capacitors. Inductor L_1 acts as a filter that attenuates unwanted frequencies on the input V_I.

As before, the capacitor C_2 and the load R_{load} form a low-pass filter. Therefore, in order to reduce the output ripple, especially in the presence of a large load (small R_{load}), a large capacitor is needed.

A summary of the characteristics of switching converters with inductors is presented in Table 10.2.

10.2.3   Switched-Capacitor Converters (Without Inductors)

Switched-capacitor converters do not make use of inductors. The general way switched-capacitor converters work is as follows: Based on the control voltage $V_{control}$, the capacitors in the circuit are first charged, then reconnected in a different manner resulting in an output that is either lower or higher than the input. In the following, we will discuss Buck Switched-Capacitor Converters followed by Boost Switched-Capacitor Converters.

10.2.3.1   Buck Switched-Capacitor Converters

A Buck switched-capacitor converter gives a DC output lower than its DC input, resulting in a step-down behavior. Conceptually, this can be done by taking two capacitors, charging them in series, then reconnecting them in parallel as can be seen in Fig. 10.11.

Switch sets 1 and 2 are never closed at the same time. When switch set 1 is closed and switch set 2 is open, the capacitors are connected in series and charged to V_I. When switch set 1 is open and switch set 2 is closed, the capacitors will be connected

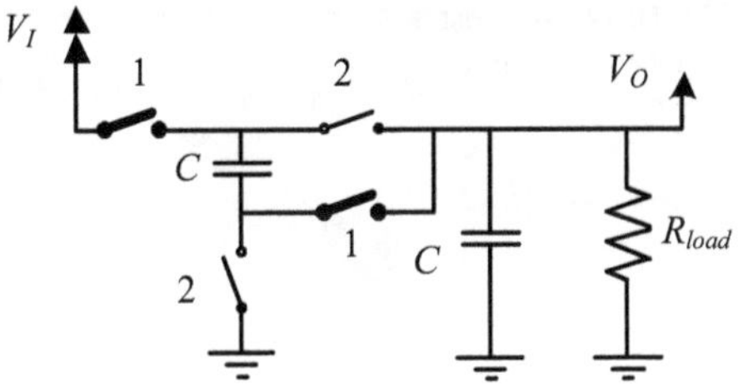

Fig. 10.11 A buck switched-capacitor converter with ideal switches

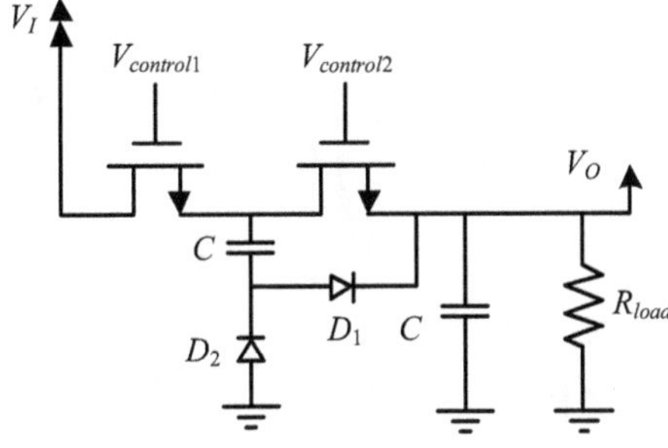

Fig. 10.12 A buck switched-capacitor converter

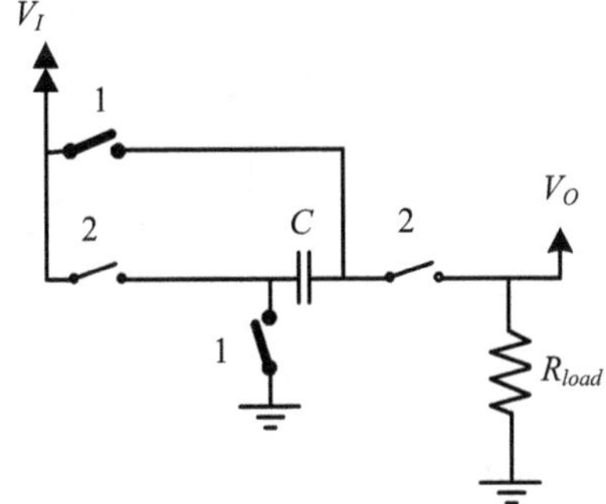

Fig. 10.13 A boost switched-capacitor converter with ideal switches

in parallel. The resulting output voltage is half that at the input since the capacitors are equal.

A simple implementation using transistors and diodes as switches is shown in Fig. 10.12.

10.2.3.2 Boost Switched-Capacitor Converters

A Boost switched-capacitor converter gives a DC output higher than its DC input, resulting in a step-up behavior. Conceptually, this can be done by taking a capacitor and charging it by connecting it in parallel with the source. Then, the capacitor is reconnected in series with the source resulting in an output which is double the input as can be seen in Fig. 10.13.

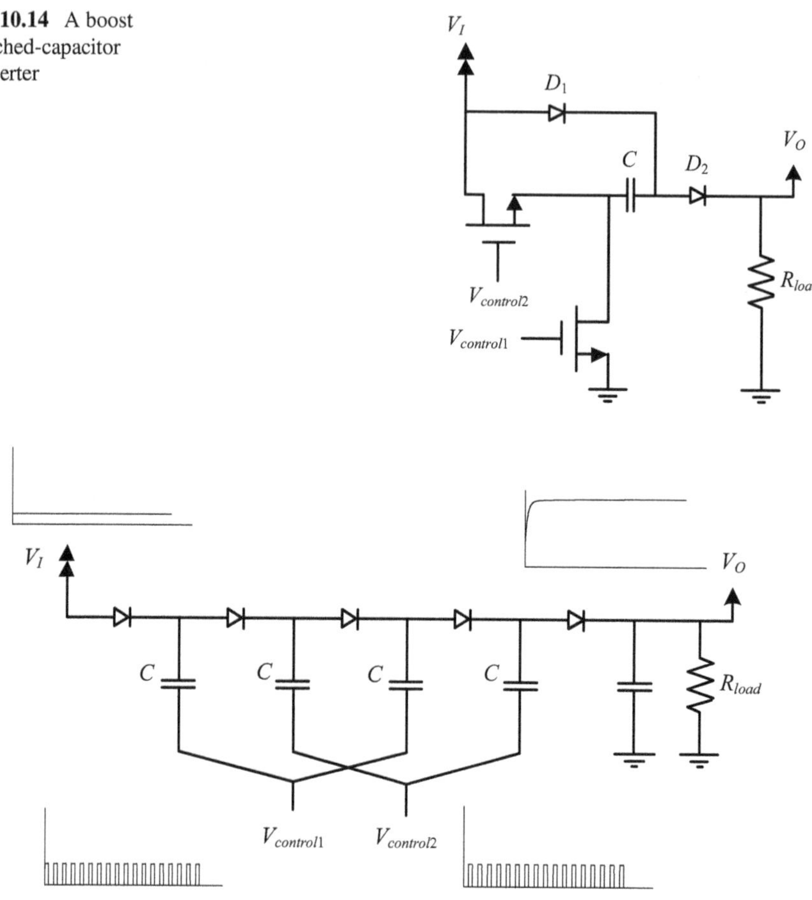

Fig. 10.14 A boost switched-capacitor converter

Fig. 10.15 A Dickson charge pump

Switch sets 1 and 2 are never closed at the same time. When switch set 1 is closed and switch set 2 is open, the capacitor is connected in parallel with the source and charged to V_I. When switch set 1 is open and switch set 2 is closed, the capacitor will be connected in series with the source, with a polarity reversal. The resulting output voltage is double that at the input.

A simple implementation using transistors and diodes as switches is shown in Fig. 10.14.

A Dickson charge pump is another type of boost switched-capacitor converter. In addition to the DC input, the Dickson charge pump requires also two control pulse trains that do not overlap.

A four-stage Dickson charge pump is shown in Fig. 10.15.

The signals in the Dickson charge pump are shown in Fig. 10.16.

If N is the number of stages in the Dickson charge pump, $V_{control}$ is the control voltage, which is assumed to be the same for both, and V_D is the voltage drop across

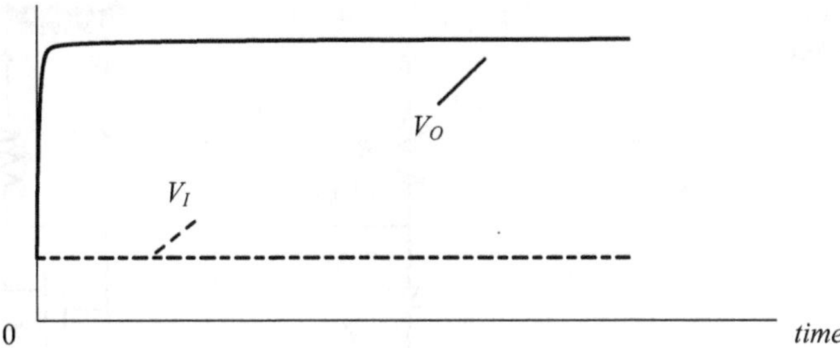

Fig. 10.16 Signals in a Dickson charge pump

the diode when it is ON, then the relation between V_I and V_O is approximately as follows:

$$V_O \approx V_I + N(V_{control} - V_D) \tag{10.10}$$

10.3 Reference Circuits

As shown in Fig. 10.1, reference circuits provide a DC signal, typically a voltage, with a particular level, which is fixed. These circuits can range from being quite simple to being quite sophisticated, for example, by providing a temperature-independent value. Their level of sophistication depends on the requirements of the subsequent circuits.

10.3.1 Simple Voltage Reference

The simplest voltage reference circuit is shown in Fig. 10.17a. It is based on the idea that a diode has a steep I-V characteristic, meaning that the voltage across its terminals changes very little even with a large change in the current passing through it. This current varies based on the load connected at V_{REF}.

Its counterpart in integrated circuit design consists of replacing the diode with a diode-connected MOSFET in this case as in Fig. 10.17b. Even though the I-V curve is quadratic instead of being exponential as in the previous case, the stability of V_{REF} still applies. In cases where V_{REF} is connected to an open circuit, such as a gate terminal, for example, there will be no variations in the current passing through the transistor due to the load, and the circuit will still be used to stabilize the variations due to noise.

In this case, the governing equations are as follows:

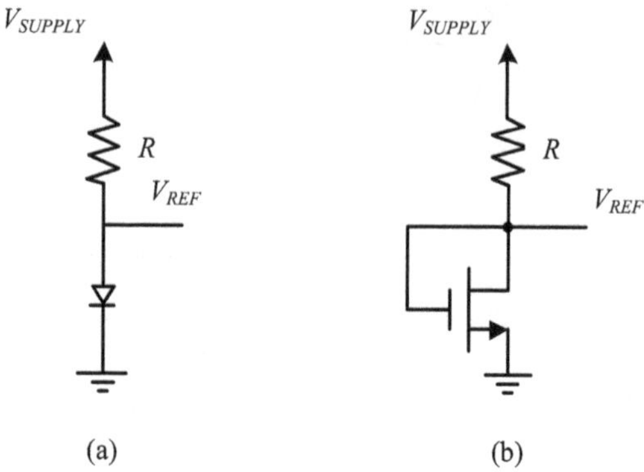

Fig. 10.17 A simple voltage reference circuit

$$I = \frac{V_{SUPPLY} - V_{REF}}{R} = \frac{1}{2}k'\left(\frac{W}{L}\right)\left(V_{REF} - V_{th}\right)^2 \tag{10.11}$$

Solving for V_{REF}, we get

$$V_{REF} = V_{th} + \frac{1}{k'\left(\frac{W}{L}\right)R}\left(\sqrt{1 + 2(V_{SUPPLY} - V_{th})k'\left(\frac{W}{L}\right)R} - 1\right) \tag{10.12}$$

A drawback of this circuit is that V_{REF} is a function of V_{th}, which itself is a function of temperature. A better solution would be to use a Bandgap voltage reference.

10.3.2   Bandgap Voltage Reference

The main idea of a Bandgap voltage reference is to use two structures with opposite temperature dependencies and adding their outputs in order to get a V_{REF} which is independent of temperature. The two structures are:

- A complementary-to-absolute-temperature (CTAT) circuit. This circuit has a negative temperature dependence, which means that its output voltage decreases as the temperature increases.
- A proportional-to-absolute-temperature (PTAT) circuit. This circuit has a positive temperature dependence, which means that its output voltage increases as the temperature increases.

In principle, only a CTAT circuit is needed. The PTAT circuit can be implemented out of two CTAT circuits. Therefore, it all starts with two slightly different CTAT circuits generating V_{CTAT1} and V_{CTAT2} and proceeds as in Fig. 10.18.

V_{CTAT1} has a stronger temperature dependency than V_{CTAT2}, resulting in a higher slope, that is, from Fig. 10.18, $slope_1$ is greater than $slope_2$. Therefore, their

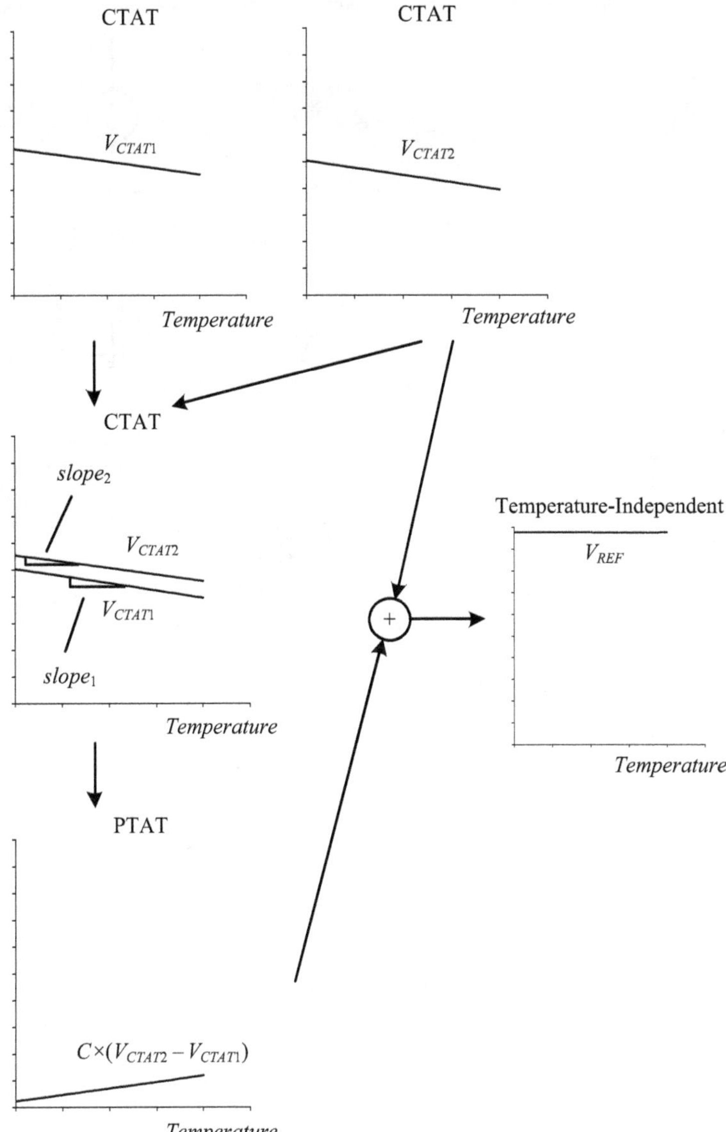

Fig. 10.18 Bandgap voltage reference generation concept

difference results in a PTAT, which is scaled by an appropriate constant C so that when it is added to V_{CTAT2}, the result is the temperature-independent V_{REF}.

This temperature-independent V_{REF} can be expressed as follows:

Fig. 10.19 A simple CTAT circuit

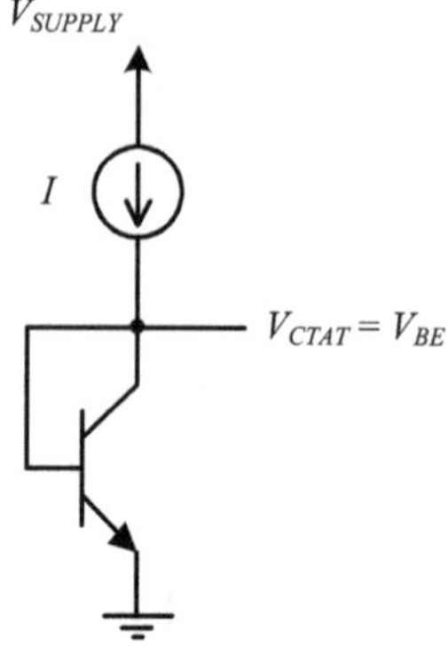

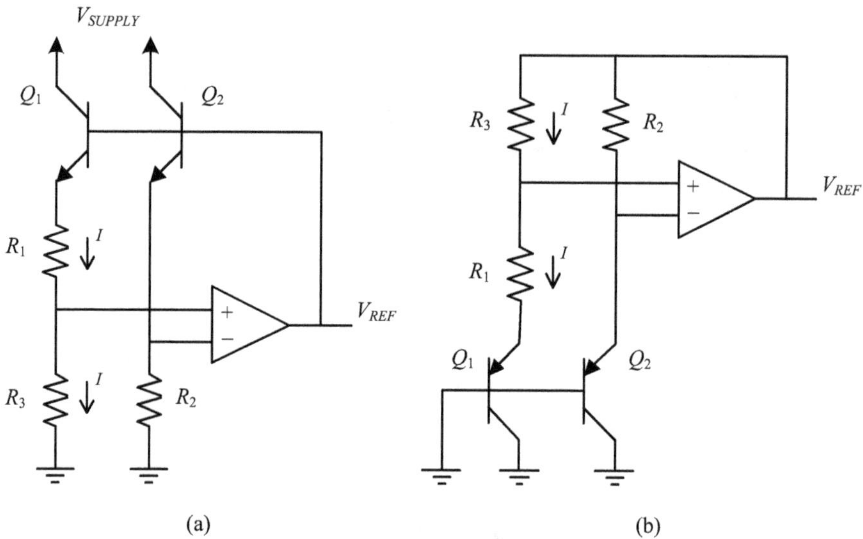

(a)   (b)

Fig. 10.20 Sample bandgap reference circuits using NPN BJTs (**a**) and PNP BJTs (**b**)

$$V_{REF} = V_{CTAT2} + C \times (V_{CTAT2} - V_{CTAT1})$$ (10.13)

The simplest way to implement a CTAT is using a pn junction as shown in Fig. 10.19.

For the same transistor, different currents will result in a different V_{CTAT}. In this case, a smaller current results in V_{CTAT1} as in Fig. 10.18, and a larger one results in V_{CTAT2}.

In order to create V_{REF}, we need a circuit that implements (10.13) using the idea in Fig. 10.19. Two popular implementations are shown in Fig. 10.20. The one in Fig. 10.20a makes use of NPN BJT transistors and the one in Fig. 10.20b makes use of PNP BJT transistors.

We will assume that the op-amps in Fig. 10.20 are ideal. For both circuits, we have

$$V_{REF} = |V_{BE2}| + V_{R2} \tag{10.14}$$

where V_{R2} is the voltage across resistor R_2.

Since after settling, the voltages at the inputs of the op-amps are equal, then we have

$$V_{R2} = V_{R3} \tag{10.15}$$

As a result

$$V_{REF} = |V_{BE2}| + V_{R3} = |V_{BE2}| + IR_3 \tag{10.16}$$

As for the current I, it can be expressed as follows

$$I = \frac{V_{R1}}{R_1} \tag{10.17}$$

and V_{R1} can be expressed in terms of $|V_{BE1}|$ and $|V_{BE2}|$ as follows

$$V_{R1} = \frac{|V_{BE2}| - |V_{BE1}|}{R_1} \tag{10.18}$$

Combining (10.16) with (10.17) and (10.18) we get

$$V_{REF} = |V_{BE2}| + \frac{R_3}{R_1}\left(|V_{BE2}| - |V_{BE1}|\right) \tag{10.19}$$

which is in the same form as (10.13) with

$$C = \frac{R_3}{R_1} \tag{10.20}$$

10.4   Bias Circuits

In integrated implementations, bias circuits primarily provide currents for the core circuits. Sometimes voltage is also needed, but that was covered in the previous sections. As a result, the main aim of a bias circuit is to implement a current source with a predictable current and a high output resistance. This is what we will tackle in this section.

10.4.1   Simple Current Source

The main idea is to start with a fixed voltage using any of the methods above and then transform that into a current using the bias circuit as shown in Fig. 10.21.

Consequently, the bias circuit is fundamentally a transconductor, which is a voltage-controlled current source. The simplest implementation would be a single transistor. Therefore, combining this idea with the simplest reference circuit as in Fig. 10.17b, we get the simple current source as in Fig. 10.22. Two variations are presented. Figure 10.22a shows the NMOS implementation, where the current source is in fact a current sink from the core circuit into ground, whereas Fig. 10.22b shows the PMOS implementation, where the current is being sourced from the supply into the core circuit. The choice between the two implementations depends on the requirements of the core circuit.

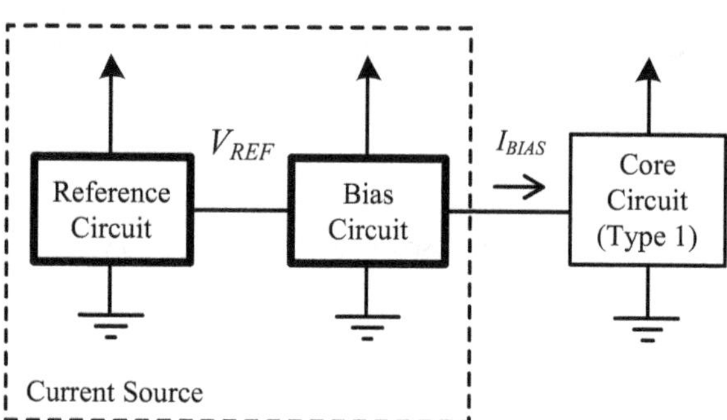

Fig. 10.21 Bias circuit concept

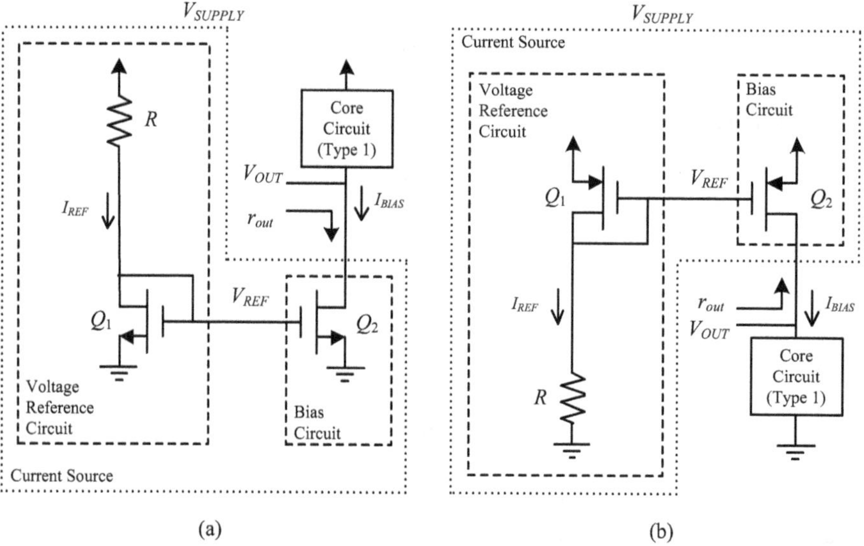

(a)   (b)

Fig. 10.22 Simple current sources of an NMOS type (**a**) and a PMOS type (**b**)

V_{OUT} has to stay within the allowable range. We will analyze the range for the circuit in Fig. 10.22a. A similar analysis can be done for the circuit in Fig. 10.22b.

V_{OUT_max} is limited by the minimum voltage drop allowed across the core circuit. If we call this voltage drop V_{CORE_min}, then

$$V_{OUT\_max} = V_{SUPPLY} - V_{CORE\_min} \qquad (10.21)$$

V_{D2_min} is limited by the fact that Q_2 has to be in the saturation region. This essentially means that

$$V_{DS2} > V_{GS2} - V_{th}$$
$$V_{OUT} > V_{GS2} - V_{th} \qquad (10.22)$$

As a result

$$V_{OUT\_min} = V_{GS} - V_{th} \qquad (10.23)$$

After doing that, we will now find the relation between I_{REF} and I_{BIAS}.

$$I_{REF} = \frac{1}{2}\mu C_{ox}\left(\frac{W}{L}\right)_1 (|V_{GS1}| - |V_{th}|)^2(1 + \lambda|V_{DS1}|)$$
$$I_{BIAS} = \frac{1}{2}\mu C_{ox}\left(\frac{W}{L}\right)_2 (|V_{GS2}| - |V_{th}|)^2(1 + \lambda|V_{DS2}|) \qquad (10.24)$$

The two transistors share the same V_{REF} which is their $|V_G|$ and in both cases, they have the same $|V_S|$. As a result, they have the same $|V_{GS}|$. Taking the ratio between the currents, we get

$$\frac{I_{BIAS}}{I_{REF}} = \frac{\left(\frac{W}{L}\right)_2(1 + \lambda|V_{DS2}|)}{\left(\frac{W}{L}\right)_1(1 + \lambda|V_{DS1}|)} \qquad (10.25)$$

If we design the circuit such that

$$|V_{DS1}| = |V_{DS2}| \qquad (10.26)$$

then the ratio will be

$$\frac{I_{BIAS}}{I_{REF}} = \frac{\left(\frac{W}{L}\right)_2}{\left(\frac{W}{L}\right)_1} \qquad (10.27)$$

We say in this case that I_{REF} is mirrored to I_{BIAS} with the mirroring factor equal to the ratio of aspect ratios. This is why this structure is also called a current mirror.

As a result, we can get accurate mirroring factors by simply choosing the appropriate ratios between the transistor sizes.

Ideally, the output resistance of a current source is infinite. However, in this case, the output resistance as indicated in Fig. 10.22 is the resistance seen looking through the drain of Q_2 leading to

$$r_{out} = r_{o2} \approx \frac{|V_A|}{|I_{BIAS}|} \tag{10.28}$$

This output resistance is not very high since V_A for modern technologies is not very high. Therefore, an important way to improve the situation is by coming up with other current sources that have a higher output resistance.

10.4.2   Improved Current Sources

The output resistance of the simple current source can be increased by employing the cascoding method. The NMOS version is shown in Fig. 10.23. It is expected that the output resistance will increase compared to that of a simple current source, but also the range of V_{OUT} will decrease.

Similarly to the simple current source, V_{OUT_max} is limited by the minimum voltage drop allowed across the core circuit. If we call this voltage drop V_{CORE_min}, then

$$V_{OUT\_max} = V_{SUPPLY} - V_{CORE\_min} \tag{10.29}$$

This result is the same as that for a simple current source.

V_{OUT_min} is limited by the fact that Q_4 has to be in the saturation region, along of course with Q_2. Knowing that

$$\begin{aligned} V_{DS4} &> V_{GS4} - V_{th} \\ V_{D4} &> V_{G4} - V_{th} \\ V_{OUT} &> V_{G4} - V_{th} \end{aligned} \tag{10.30}$$

As for V_{REF2}, it can be written as

$$V_{G4} = V_{GS3} + V_{GS1} = 2V_{GS} \tag{10.31}$$

assuming that the V_{GS} of transistors Q_1 and Q_3 are equal.

Combining (10.30) with (10.31) we get

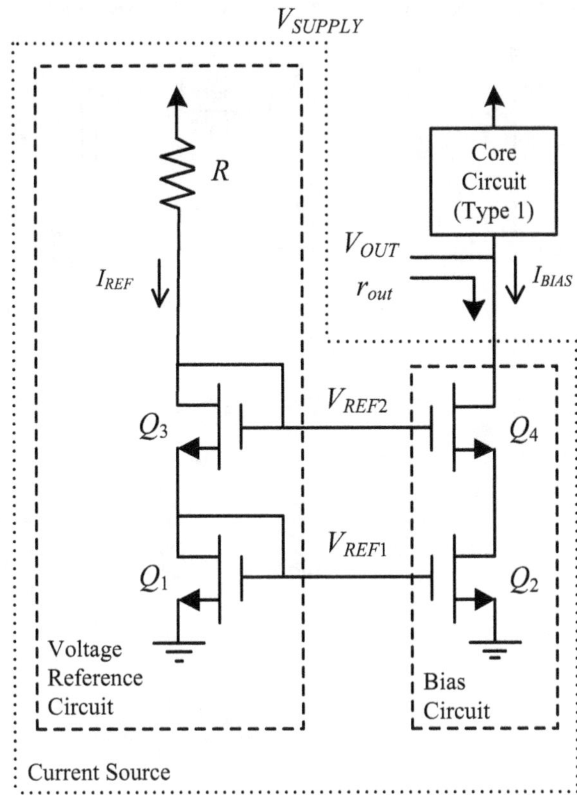

Fig. 10.23 Cascode NMOS current source

$$V_{OUT} > 2V_{GS} - V_{th} \tag{10.32}$$

which leads to

$$V_{OUT\_min} = 2V_{GS} - V_{th} \tag{10.33}$$

This result is higher than that of a simple current source by V_{GS}.

Therefore a cascode current source has a smaller allowable maximum output swing compared to a simple currrent source.

The relation between I_{REF} and I_{BIAS} for a cascode current source is the same as that for a simple current source as in (10.25), (10.27).

As for the output resistance, it is the same as that of a cascode amplifier, namely

$$r_{out} \approx g_{m4} r_{o4} r_{o1} \tag{10.34}$$

which is much higher, and thus better, than that of a simple current source.

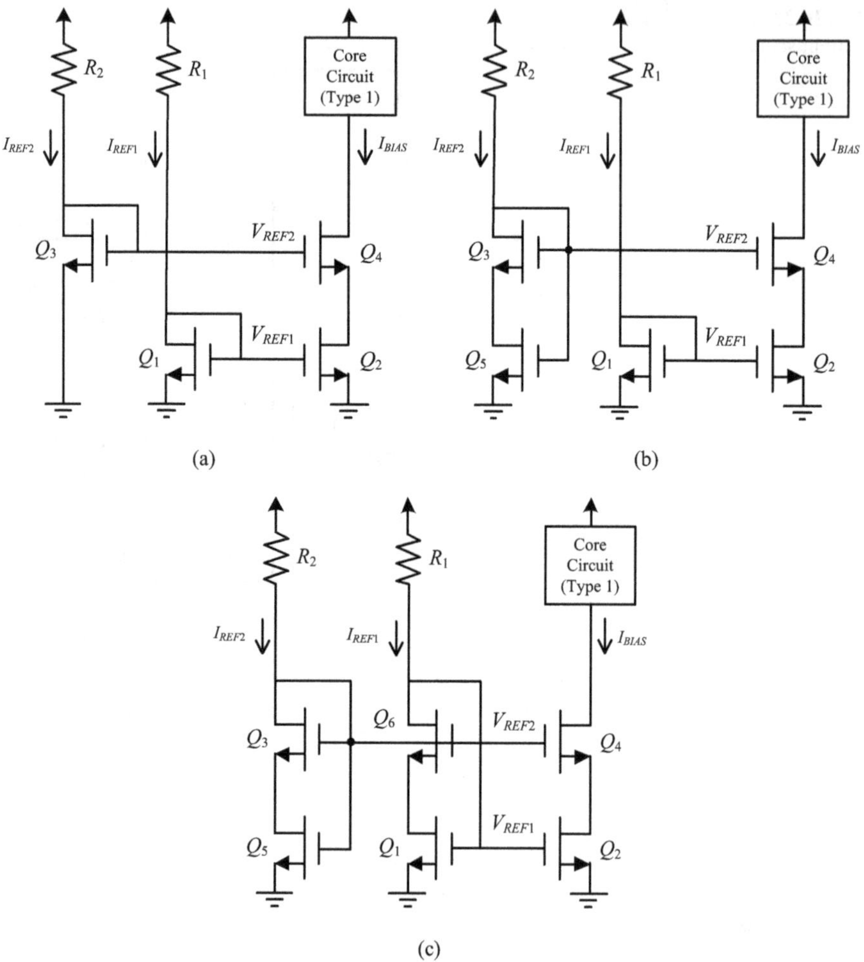

Fig. 10.24 Improved Cascode NMOS current source

A drawback of this circuit is that V_{DS1} might not be equal to V_{DS2}, leading to an error in the mirroring factor due to the presence of λ (channel-length modulation effect) as seen in (10.25).

To improve this cascode circuit let us first generate V_{REF1} and V_{REF2} independently as shown in Fig. 10.24a. The output bias circuit to the right has three stacked stages consisting of Q_2, Q_4, and the core circuit. On the other hand, the voltage reference circuit generating V_{REF1} only has two stacked stages, namely Q_1 and R_1, whereas the voltage reference circuit generating V_{REF2} only has two stacked stages, namely Q_3 and R_2. This makes the circuit unbalanced, which is why it would be difficult for V_{DS1}, for example, to match V_{DS2}. The general solution would be to add stages to the voltage reference circuits generating V_{REF1} and V_{REF2} in order to match the number of stages in the output bias circuit.

To remedy the voltage reference circuit generating V_{REF2}, we can add a transistor Q_5 below Q_3. This transistor can be either diode-connected or its gate can be connected to that of Q_3 as in Fig. 10.24b. With this connection, V_{GS5} is now divided between V_{GS3} and V_{DS5}. Therefore, in case the needed V_{GS5} is high, V_{DS5} does not need to be as high.

As for the voltage reference circuit generating V_{REF1}, one solution would be to add a transistor Q_6 on top of transistor Q_1 as in Fig. 10.24c. Q1 has its gate connected to the drain of Q_6. The benefit in this is that, unlike in Fig. 10.24b, V_{REF1} now is divided between V_{DS6} and V_{DS1}. Therefore, in case the needed V_{REF1} is high, V_{DS1} does not need to be as high, and it will be easier for V_{DS1} to match V_{DS2}. Also, in this case, V_{G6} can be taken directly from V_{G3} in the previous voltage reference circuit.

Usually, we prefer to use transistors instead of resistors. Therefore, the resistors at the top of the voltage reference circuits in Fig. 10.24 can themselves be replaced with current bias circuits, as will be discussed in the following section.

10.4.3 Biasing a Complete Integrated Circuit

Different blocks in an integrated circuit typically require different currents with some of them being NMOS-based while the others being PMOS-based. One way to achieve this is to have a separate current source for each block.

However, there are two problems with this. The first is that the implementation will take a lot of space owing to the generation of a reference voltage for every bias circuit. The second is that with process variations and the absence of a link between the bias circuits, there is a tendency for the currents to vary in an uncorrelated manner reducing the possibility of the circuit to operate correctly.

Alternatively, we can have a single reference voltage circuit whose output is shared by several bias circuits as shown in Fig. 10.25. This is called a voltage-mode bias distribution scheme. The advantage of this scheme is that it does not occupy a large area. Its drawback is that the shared voltage V_{REF} has the tendency to vary and pick up noise especially if there is a large number of bias circuits.

A sample implementation of such a scheme is shown in Fig. 10.26.

All the currents are related to I_{REF} approximately as follows:

$$\frac{I_{BIAS\_II}}{I_{BIAS\_I}} = \frac{\left(\frac{W}{L}\right)_3}{\left(\frac{W}{L}\right)_2} \qquad \frac{I_{BIAS\_I}}{I_{REF}} = \frac{\left(\frac{W}{L}\right)_2}{\left(\frac{W}{L}\right)_1} \tag{10.35}$$

An alternative method would be to use the output current and use this to generate reference voltages for the other current bias circuits. This is called the current-mode bias distribution scheme and is shown in Fig. 10.27. Since the current is being shared, it is less prone to voltage noise, thus suitable for large and/or sensitive circuit. Its drawback is that it occupies a larger area than the voltage-mode scheme.

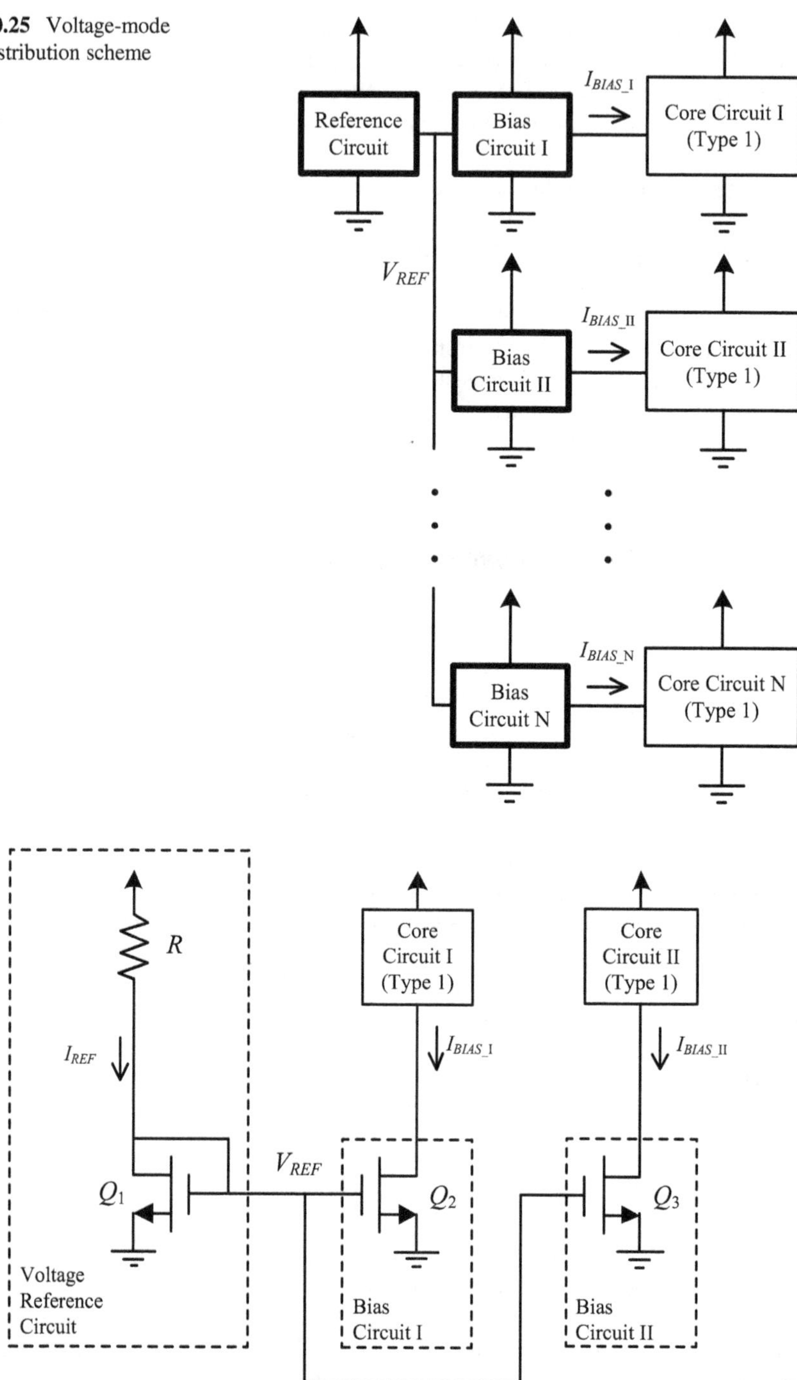

Fig. 10.25 Voltage-mode bias distribution scheme

Fig. 10.26 Sample implementation of a voltage-mode bias distribution scheme

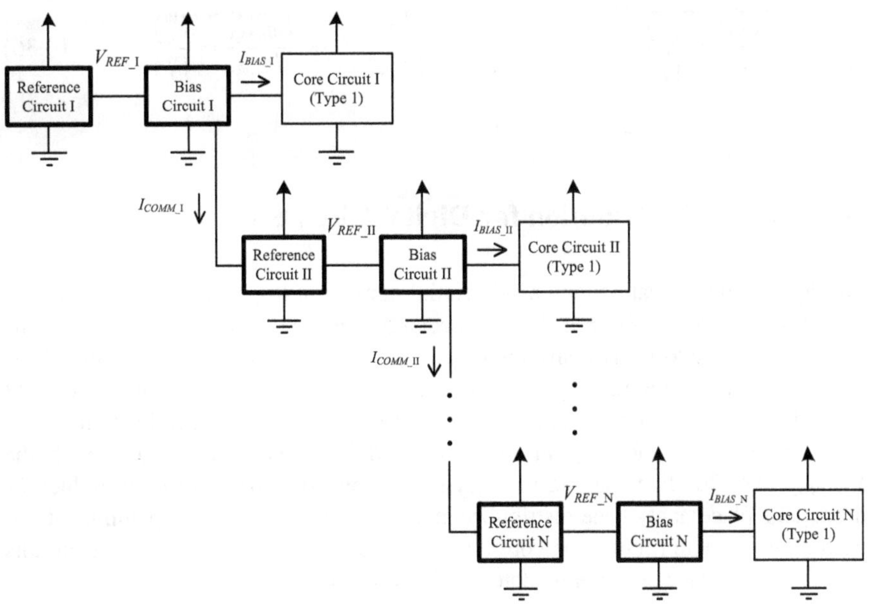

Fig. 10.27 Current-mode bias distribution scheme

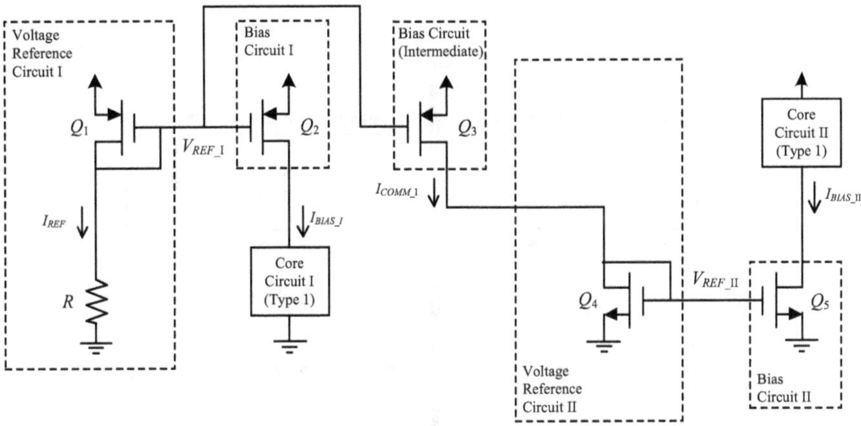

Fig. 10.28 Sample implementation of a current-mode bias distribution scheme

A sample implementation of such a scheme is shown in Fig. 10.28. Note the presence of the intermediate bias circuit using Q_3, necessary to generate I_{COMM_I}.

All the currents are related to I_{REF} approximately as follows:

$$\frac{I_{BIAS\_II}}{I_{COMM\_I}} = \frac{\left(\frac{W}{L}\right)_5}{\left(\frac{W}{L}\right)_4} \qquad \frac{I_{COMM\_I}}{I_{REF}} = \frac{\left(\frac{W}{L}\right)_3}{\left(\frac{W}{L}\right)_1} \qquad \frac{I_{BIAS\_I}}{I_{REF}} = \frac{\left(\frac{W}{L}\right)_2}{\left(\frac{W}{L}\right)_1} \qquad (10.36)$$

10.5   Power Distribution for Digital Circuits

For digital circuits that contain a substantial number of logic gates, effectively core circuits of the type 3 as in Fig. 10.1, a special power distribution method needs to be adopted in order to fit as many logic gates as possible within a certain area. This should be done while making sure that the voltage drop, also called the *IR* (current multiplied by resistance) drop, across the power lines is within specifications.

An example of such a distribution method is shown in Fig. 10.29 with the V_{SUPPLY} and *Gnd* lines having more than one point of entry in order to reduce *IR* drop. This placement of the distribution network is done at the floorplanning stage, before placing and routing the core circuits. Observe that some of the core circuits are flipped in order to accommodate for the position of V_{SUPPLY} and *Gnd*.

The power lines, both V_{SUPPLY} and *Gnd*, can be modeled as a distributed *RC* network as shown in Fig. 10.30.

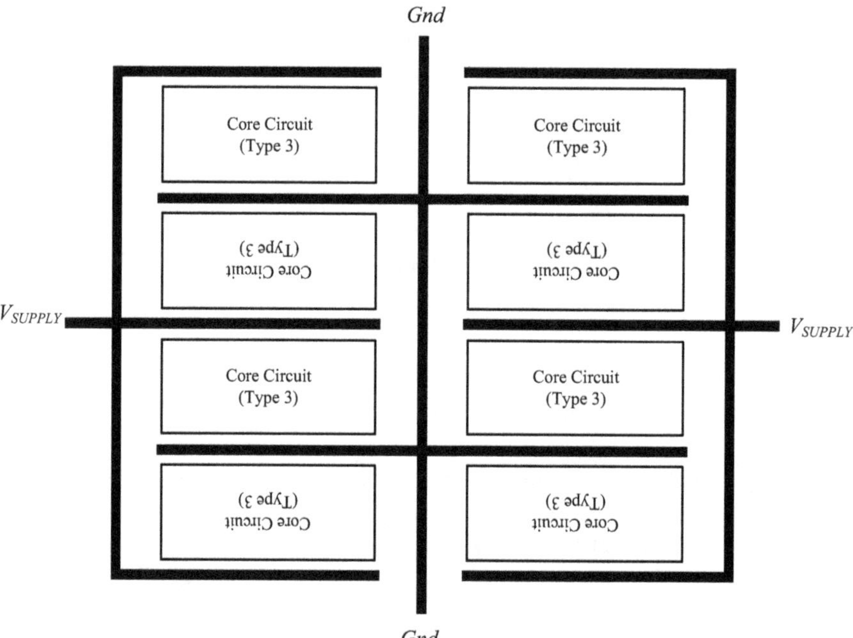

Fig. 10.29 Power distribution method for digital circuits

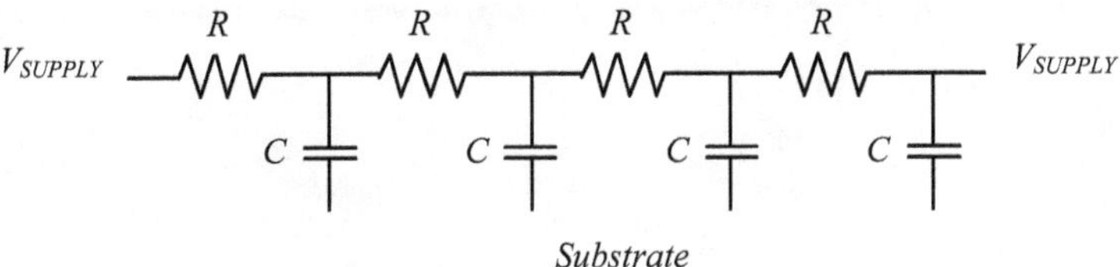

Fig. 10.30 Distributed RC model of the V_{SUPPLY} line

In order to have a small IR drop, It is desirable to reduce the resistance of these lines. This can be achieved by making them wide, which increases their cross-sectional area.

Also, by making them wide, their area on top of the substrate increases, which increases their capacitance. This helps in stabilizing V_{SUPPLY} and in shunting high-frequency noise into the substrate.

Chapter 11
Output Stages

This chapter focuses on output stages designed to drive loads with small input impedances. These structures are different from voltage amplifiers in that they take a large input voltage with a large input impedance and provide a large output voltage but with small output impedance, thus acting as current amplifiers. Therefore, small-signal analysis does not apply in these cases and a different kind of analysis aiming at the relevant specifications will be presented.

11.1 Introduction

Towards the end of a signal processing chain, when the data is in the voltage domain, which is typically the case, the voltage signal swing is large. As for the load impedance, it is typically small, in the range of a few Ohms or tens of Ohms. As a result, an output stage is needed to drive the load, effectively ensuring that the voltage swing across the load stays large despite the load having a low input impedance. In order to achieve this, the output stage needs to sink a large amount of current into the load without taking so much current from the previous stage. This makes the output stage a current amplifier rather than a voltage amplifier.

A general overview of the setup is shown in Fig. 11.1.

In general, we would be dealing with input and output impedances, but here we will restrict the discussion to resistances.

11.2 Performance Parameters

Table 11.1 lists the important performance parameters on which we will focus.

J. G. Atallah, M. Ismail, *Integrated Electronic Circuits*,
https://doi.org/10.1007/978-3-031-62707-1_11

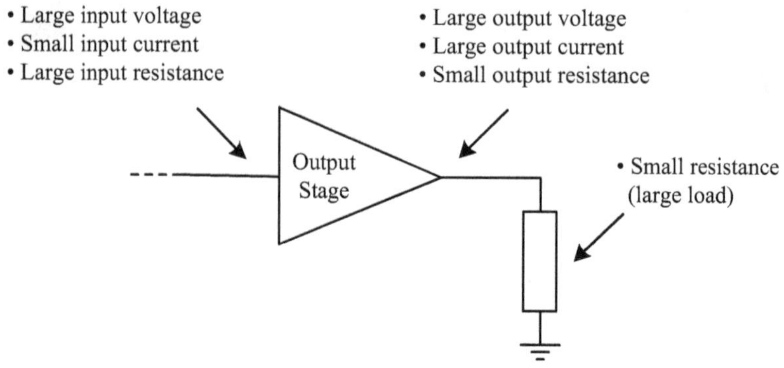

Fig. 11.1 Overview of the setup

Table 11.1 Output stage performance parameters

Parameter
Output voltage range
Input voltage range
Linearity and total harmonic distortion
Power consumption
Power dissipation
Efficiency

Owing to their position at the end of the signal processing chain, output stages take a large voltage signal at their input and output also a large voltage signal. Therefore, it is important for these signal ranges to be as large as possible.

This large input and output range has its drawbacks. Since transistors are inherently nonlinear components, having a wide swing at their terminals results in nonlinear behavior, which creates frequencies at the output that did not exist at the input. Take, for example, a sinusoidal input with a fundamental frequency f_{in} expressed as

$$v_{in} = \cos\left(2\pi f_{in}t\right) \tag{11.1}$$

Let this signal pass through a nonlinear block, for example, one whose transfer characteristic is quadratic with constants A and B as follows

$$v_{out} = A v_{in}^2 + B v_{in} \tag{11.2}$$

The resulting output is

$$
\begin{aligned}
v_{out} &= A\left[\cos\left(2\pi f_{in}t\right)\right]^2 + B\cos\left(2\pi f_{in}t\right) \\
&= A\left[\frac{1}{2} + \frac{1}{2}\cos\left(4\pi f_{in}t\right)\right] + B\cos\left(2\pi f_{in}t\right) \\
&= B\cos\left(2\pi f_{in}t\right) + \frac{A}{2}\cos\left(4\pi f_{in}t\right) + \frac{A}{2}
\end{aligned} \tag{11.3}
$$

As can be seen, the first term in the output contains the fundamental frequency f_{in}, but the second term contains double this frequency, which did not exist in the input, and the third term is just a DC component. The second term is called the second harmonic component. As a result, and owing to the presence of the harmonic components, the output is a distorted version of the input.

One way to characterize this distortion is to take the ratio of the sum of RMS values of all the harmonic components, excluding DC, to the RMS value of the fundamental component, usually expressed as a percentage called the total harmonic distortion (THD):

$$\text{THD}\% = \frac{\sqrt{V_2^2 + V_3^2 + V_4^2 \cdots}}{V_1} \times 100\% \tag{11.4}$$

where V_1 is the RMS value of the fundamental component and V_n is the RMS value of the nth harmonic component.

Applying this to our example in Eq. (11.3), we get

$$\text{THD}\% = \frac{\sqrt{\left(\frac{|A_2|}{\sqrt{2}}\right)^2}}{\frac{|B|}{\sqrt{2}}} \times 100\% = \left|\frac{A}{2B}\right| \times 100\% \tag{11.5}$$

Obviously, the lower the THD, the better is the output quality.

Given that output stages are essentially current buffers, they tend to take a lot of current from the power supply and send it to the output. This leads to high power consumption. Additionally, not all the consumed power reaches the load. Part of it is dissipated in the form of heat. As a result, if we call the consumed power taken from the supply $P_{CONSUMED}$, the power sent to the load P_{LOAD}, and the power dissipated as heat $P_{DISSIPATED}$, then

$$P_{DISSIPATED} = P_{CONSUMED} - P_{LOAD} \tag{11.6}$$

Also, we can define the efficiency η, typically expressed as a percentage:

$$\eta = \frac{P_{LOAD}}{P_{CONSUMED}} \times 100\% \tag{11.7}$$

Note that the dissipated power and the efficiency cannot predict each other. In other words a system can have a high dissipated power and a high efficiency or a low dissipated power and a low efficiency. Of course, by design, we aim to reduce the dissipated power and increase the efficiency.

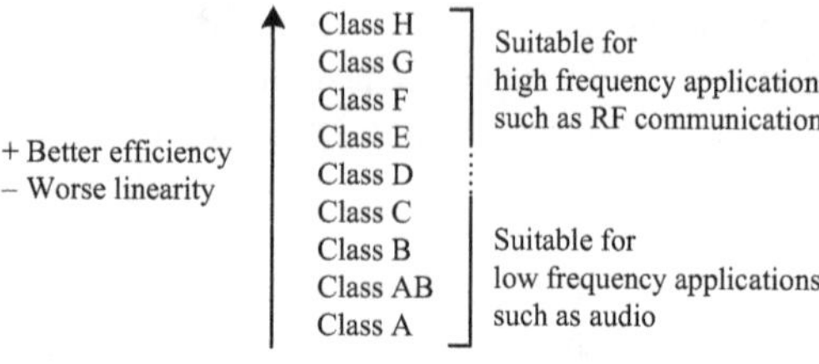

Fig. 11.2 Classes of output stages and their tradeoffs

11.3   Classes: Introduction

Output stages are divided into classes based on their linearity and efficiency. The division starts with Class A which is the most linear, while also being the least efficient with the most power dissipation. As we go higher in classes, linearity becomes worse, while efficiency improves and power dissipation decreases. Classes G and H take a lower class such as Class AB and add methods that modify the power supply in order to improve the performance. Figure 11.2 shows the tradeoff between the different classes of output stages.

The difference in the power consumption and efficiency between the classes is best-seen by plotting the current passing through the amplifying transistor(s) versus time for a sinusoidal input. This will be done when describing every class. The DC component of this current will contribute to the stand-by power consumption, which is when the input only contains a DC component (no data), also known as the quiescent state. Additionally, the AC component will contribute to the dynamic power consumption. Therefore, to reduce the overall power consumption, both current components have to be reduced.

In this chapter, we will assume the usage of two power supplies, a positive one and a negative one, with equal magnitudes:

$$V_{SUP+} = -V_{SUP-} = V_{SUP} \qquad (11.8)$$

As a result, the quiescent output voltage (initial condition, with the circuit powered ON, but with no input), being in the middle of the complete voltage range, will have a value of zero.

11.4   Class A

The current through a Class A output stage has a significant DC component. The voltage at the input and the current through the stage are shown in Fig. 11.3. The high DC current component denoted as I_D happens to be the same as the quiescent current I_Q. This high quiescent current is the reason for the high stand-by power consumption of a Class A output stage.

Fig. 11.3 Class A input voltage (**a**) and transistor current (**b**)

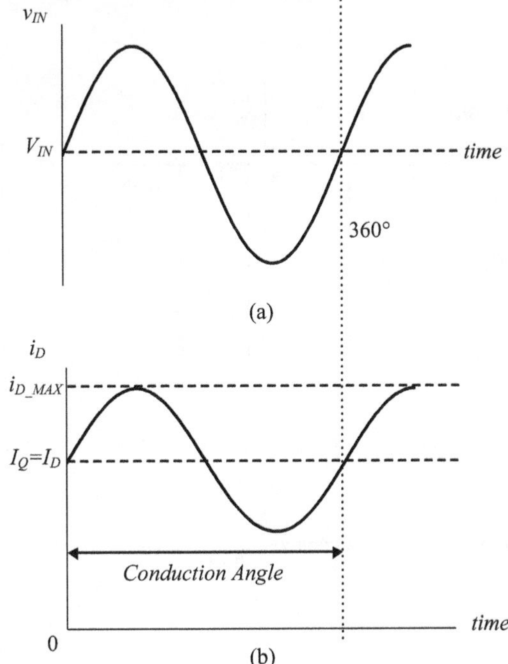

One transistor is needed to conduct all the current for the complete input cycle, which means that the conduction angle is 360 degrees.

As an example of a Class A output stage, Fig. 11.4 shows a common-drain amplifier. The circuits surrounding the transistor are modeled as follows:

- The previous stage (signal source) is modeled using its Thevenin equivalent model. This signal source contains a DC component and an AC component. Also, the source has an output resistance R_{SRC}.
- The next stage (load) is modeled using its input resistance, indicated as R_{LD}.

Note that v_{SRC} and v_{IN} are the same since there is no current through the gate. Therefore v_{SRC} and v_{IN} can be used interchangeably.

Typical v_{IN} and v_{LD} waveforms of a Class A configuration as in Fig. 11.4 are shown in Fig. 11.5. Note that in this example, i_D is the same as i_{LD}.

11.4.1 Class A Output and Input Voltage Range

When it comes to the output voltage range, looking at the NMOS-based implementation in Fig. 11.4a, we can see that the maximum value the output voltage can take is governed by what is happening above the output node. Therefore:

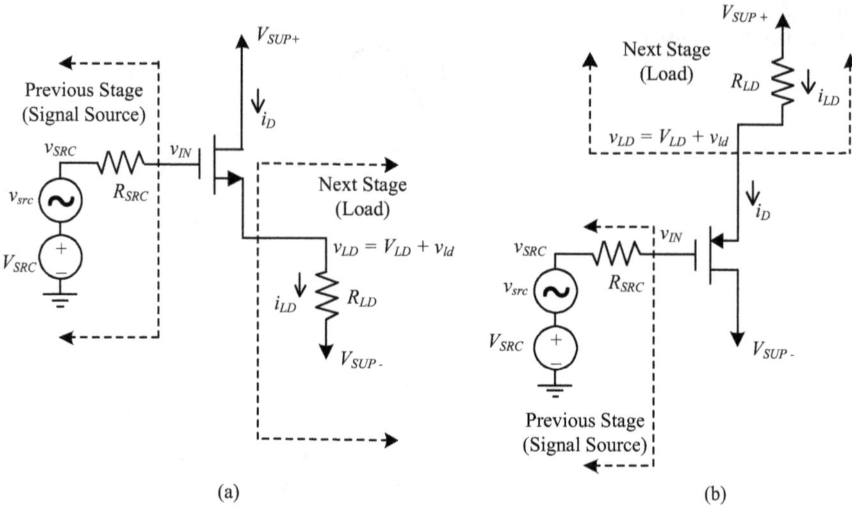

(a)   (b)

Fig. 11.4 Class A common-drain example configuration: (**a**) NMOS-based and (**b**) PMOS-based

Fig. 11.5 Typical (**a**) v_{IN} and (**b**) v_{LD} waveforms of a Class A output stage

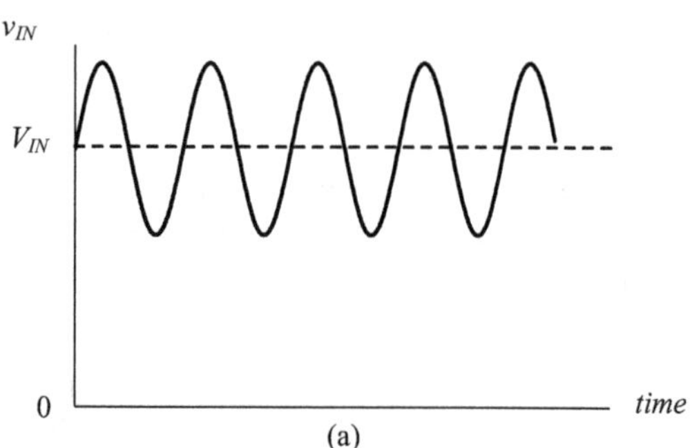

(a)

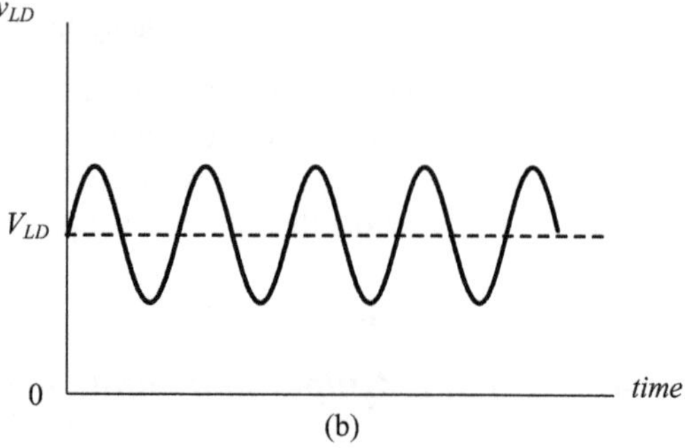

(b)

$$v_{LD\_MAX} = V_{SUP+} - V_{OV\_REQ} = V_{SUP+} - v_{GS\_MAX} - V_{th} \qquad (11.9)$$

where V_{OV_REQ} is the required overdrive voltage of the NMOS transistor. Note that this voltage will coincide with the maximum v_{GS} since this is when v_{LD} is maximum. If v_{LD} goes above this value, the transistor will get out of the saturation region and into the triode region.

The minimum value of the output voltage is governed by what is happening below the output node. Since R_{LD} is a passive component, it does not set a limit on the output voltage. Therefore:

$$v_{LD\_MIN} = V_{SUP-} \qquad (11.10)$$

Consequently, the output voltage range is

$$V_{SUP-} \leq v_{LD} \leq V_{SUP+} - V_{OV\_REQ} \qquad (11.11)$$

As a result, in order to get the maximum swing at the output, it is sensible to have the DC point at the output (V_{LD}) in the middle between v_{LD_MIN} and v_{LD_MAX}.

Similar reasoning can be applied to the PMOS-based circuit in Fig. 11.4b.

As for the input voltage range, we can see from Fig. 11.4a that since the input resistance is infinite, the maximum input voltage is

$$v_{IN\_MAX} = V_{SUP+} \qquad (11.12)$$

Regarding the minimum input voltage, it is limited by what is happening at the output. In this case, the input and the output are in phase. Therefore, when the input is at its minimum, so is the output. Additionally, the minimum V_{GS} is the threshold voltage in order to keep the transistor ON. Therefore,

$$v_{IN\_MIN} = V_{GS\_MIN} + v_{LD\_MIN} = V_{th} + V_{SUP-} \qquad (11.13)$$

11.4.2   Class A Transfer Characteristic and Linearity

Since a single transistor is conducting the complete input cycle, a Class A amplifier has a quite linear transfer characteristic. The transfer characteristic consists of the output voltage amplitude versus the input voltage amplitude excluding any DC shift. The transfer characteristic for the circuit shown in Fig. 11.4a is v_{ld} plotted versus v_{in} as shown in Fig. 11.6a. The slope of the transfer characteristic is shown in Fig. 11.6b.

As a result, two observations can be made.

Fig. 11.6 Typical (**a**) transfer characteristic and (**b**) transfer characteristic slope of a Class A output stage

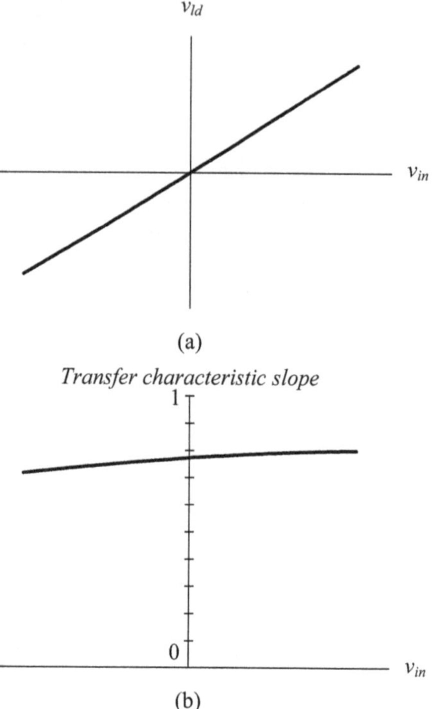

The first observation is that the slope is almost constant, confirming our expectation of good linearity. The slope is not perfectly constant since, looking at Fig. 11.4a, we see that

$$v_{SRC} = v_{IN} = v_{GS} + v_{LD} \tag{11.14}$$

Therefore, as the input increases, v_{GS} will also increase to enable more current into R_{LD} so that v_{LD} increases too. Since v_{GS} is a nonlinear function of the current, this results in a slightly nonlinear behavior.

The second observation is that the transfer characteristic slope, which is the gain, is less than unity, an expected behavior from the common-drain amplifier.

11.4.3   Class A Power Consumption, Power Dissipation, and Efficiency

In a Class A output stage, the power supply has only a DC voltage component and delivers a DC current of I_D as in Fig. 11.3. This current will always be delivered even in a stand-by situation, thus contributing to power consumption in the presence as well as the absence of an input. The output stage is connected to two power supplies as in Fig. 11.4. Therefore, the average power consumption P_{CONS} is

$$
\begin{aligned}
P_{CONS} &= V_{SUP+} \times I_D - V_{SUP-} \times I_D \\
&= (V_{SUP+} - V_{SUP-}) \times I_D
\end{aligned}
\tag{11.15}
$$

In general, the average power delivered to the load P_{LD} consists of the DC power as well as the AC power as follows

$$
\begin{aligned}
P_{LD} &= P_{LD\_DC} + P_{LD\_AC} \\
&= V_{LD} \times I_D + v_{LD\_RMS} \times i_{LD\_RMS}
\end{aligned}
\tag{11.16}
$$

where v_{LD_RMS} is the RMS value of the AC component of the output voltage.

The average power dissipated as heat is calculated using (11.6) as

$$
\begin{aligned}
P_{DISSIP} &= P_{CONS} - P_{LD} \\
&= (V_{SUP+} - V_{SUP-} - V_{LD}) \times I_D - v_{LD\_RMS} \times i_{LD\_RMS}
\end{aligned}
\tag{11.17}
$$

The average efficiency can be calculated using (11.7) as

$$
\begin{aligned}
\eta &= \frac{P_{LD}}{P_{CONS}} \times 100\% \\
&= \frac{V_{LD} \times I_D + v_{LD\_RMS} \times i_{LD\_RMS}}{(V_{SUP+} - V_{SUP-}) \times I_D} \times 100\%
\end{aligned}
\tag{11.18}
$$

To get an idea of the efficiency range for a Class A, and knowing that the positive and negative supply voltages have equal magnitudes as in (11.8), let us allow the output voltage swing to cover its maximum range. According to (11.11), and neglecting V_{OV_REQ},

$$
V_{SUP-} \leq v_{LD} \leq V_{SUP+}
\tag{11.19}
$$

In this case the output voltage and current will be as in Fig. 11.7.

As a consequence, V_{LD} is zero and

$$
v_{LD\_RMS} = \frac{V_{SUP}}{\sqrt{2}}
\tag{11.20}
$$

As for the current, the DC part is

$$
I_D = \frac{V_{SUP}}{R_{LD}}
\tag{11.21}
$$

and the current RMS AC part is

Fig. 11.7 Class A output voltage waveform (**a**) and current waveform (**b**) for the efficiency example

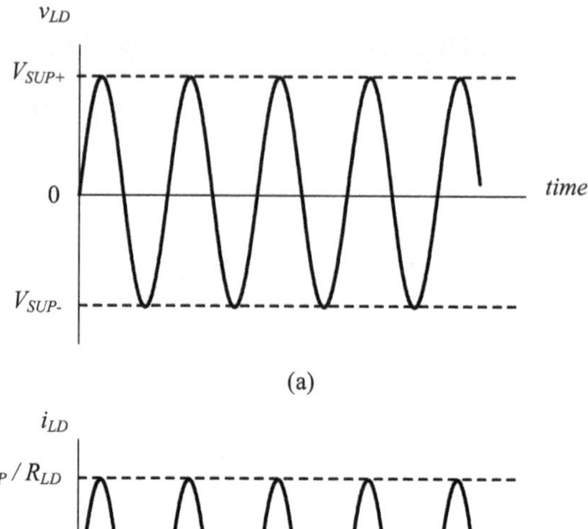

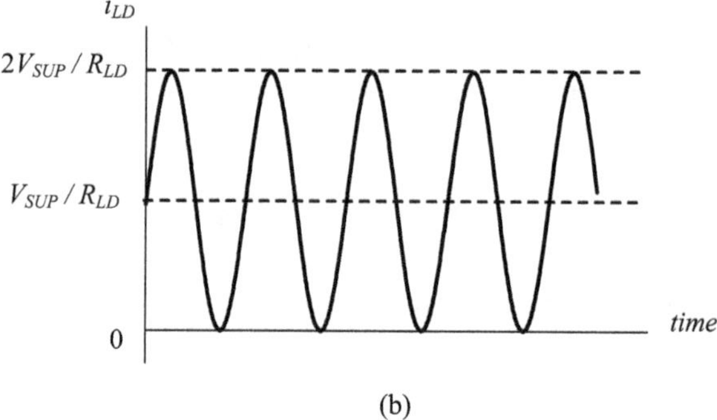

$$i_{LD\_RMS} = \frac{V_{SUP}}{\sqrt{2}R_{LD}} \tag{11.22}$$

Substituting (11.8), (11.20), (11.21), and (11.22) in (11.18) we get the efficiency in this case:

$$\eta = \frac{0 \times \frac{V_{SUP}}{R_{LD}} + \frac{V_{SUP}}{\sqrt{2}} \times \frac{V_{SUP}}{\sqrt{2}R_{LD}}}{2V_{SUP} \times \frac{V_{SUP}}{R_{LD}}} \times 100\% = 25\% \tag{11.23}$$

This maximum efficiency is quite low as will be seen when compared with the higher classes.

11.5   Class B

The current through a Class B output stage has a zero DC component. The voltage at the input and the current through the stage are shown in Fig. 11.8. As can be seen, in the quiescent state, I_Q is zero so no current is conducted. This absence of stand-by power consumption will enormously reduce the total power consumed compared to a Class A.

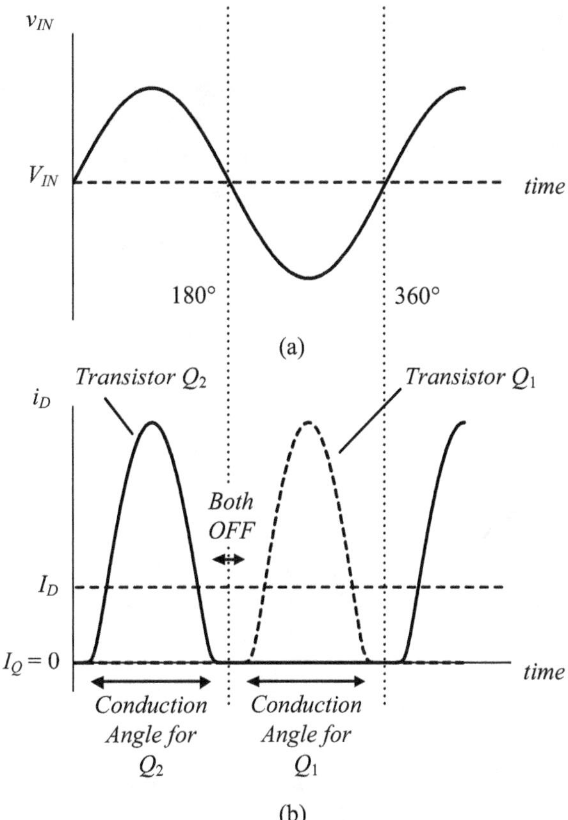

Fig. 11.8 Class B input voltage (**a**) and transistor current (**b**)

Two transistors are needed to conduct all the current for the complete input cycle. The conduction angle for each transistor is slightly less than 180 degrees.

Figure 11.9 shows an example of a Class B output stage. As can be seen, two transistors are needed, Q_1 and Q_2. Transistor Q_1 conducts current during the positive part of the input cycle and Q_2 conducts during the negative part.

The circuits surrounding the transistor are modeled as follows:

- The previous stage (signal source) is modeled using its Thevenin equivalent model. This signal source contains a DC component and an AC component. Also, the source has an output resistance R_{SRC}.
- The next stage (load) is modeled using its input resistance, indicated as R_{LD}.

Note that v_{SRC} and v_{IN} are the same since there is no current through the gates of Q_1 and Q_2. Therefore v_{SRC} and v_{IN} can be used interchangeably.

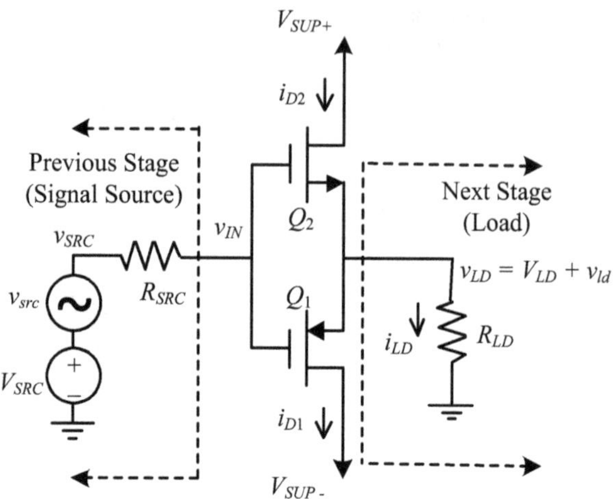

Fig. 11.9 Class B example configuration

Typical v_{IN} and v_{LD} waveforms of a Class B configuration as in Fig. 11.9 are shown in Fig. 11.10. As can be seen, the output is subject to crossover distortion as it crosses zero. This is due to the nonlinearity in the transfer characteristic as will be discussed shortly.

11.5.1   Class B Output and Input Voltage Range

When it comes to the output voltage range, looking at the circuit in Fig. 11.9, we can see that the maximum value the output voltage can take is governed by what is happening above the output node. Therefore:

$$v_{LD\_MAX} = V_{SUP+} - V_{OV2\_REQ} = V_{SUP+} - v_{GS2\_MAX} - V_{th} \tag{11.24}$$

where V_{OV2_REQ} is the required overdrive voltage of Q_2. Note that this voltage will coincide with the maximum v_{GS} since this is when v_{LD} is maximum. If v_{LD} goes above this value, the transistor will get out of the saturation region and into the triode region.

Similarly, the minimum value of the output voltage is governed by what is happening below the output node. Since R_{LD} is a passive component, it does not set a limit on the output voltage. Therefore:

$$v_{LD\_MIN} = V_{SUP-} + \left| V_{OV1\_REQ} \right| = V_{SUP-} + \left| v_{GS1\_MAX} \right| + \left| V_{th} \right| \tag{11.25}$$

Fig. 11.10 Typical (**a**) v_{IN} and (**b**) v_{LD} waveforms of a Class B output stage

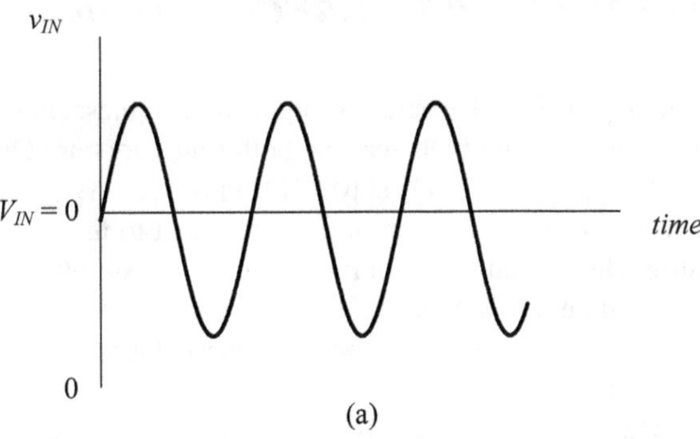

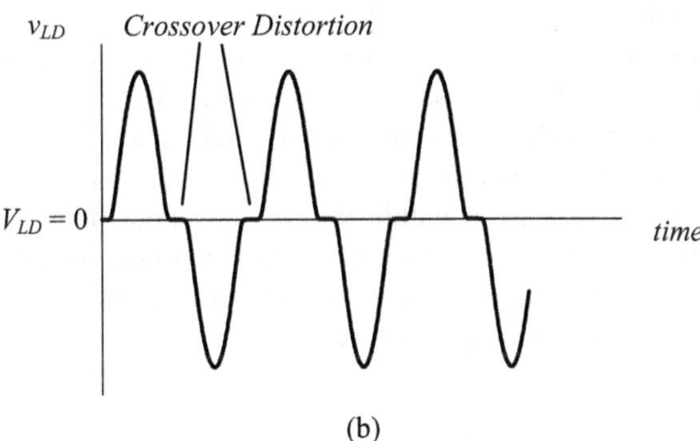

where V_{OV1_REQ} is the required overdrive voltage of Q_1. This voltage will coincide with the maximum $|v_{GS}|$ since this is when v_{LD} is minimum.

As a result, the output voltage range is

$$V_{SUP-} + \left| V_{OV1\_REQ} \right| \leq v_{LD} \leq V_{SUP+} - V_{OV2\_REQ} \tag{11.26}$$

In order to get the maximum swing at the output, it is sensible to have the DC point at the output (V_{LD}) in the middle between v_{LD_MIN} and v_{LD_MAX}.

As for the input voltage range, we can see from Fig. 11.9 that since the input resistance is infinite, the maximum input voltage is

$$v_{IN\_MAX} = V_{SUP+} \tag{11.27}$$

Similarly, the minimum input voltage is

$$v_{IN\_MIN} = V_{SUP-} \tag{11.28}$$

11.5.2   Class B Transfer Characteristic and Linearity

Looking at Fig. 11.9, and starting from the quiescent state when v_{IN} and v_{LD} are zero, v_{GS1} and v_{GS2} are both zero, so both transistors are OFF.

As v_{IN} increases, Q_2 stays OFF until $v_{IN} - v_{LD} = v_{GS2}$ exceeds V_{th}. During that duration of time, the output does not react to the input. When v_{GS2} exceeds V_{th}, Q_2 turns ON, turning the configuration into a simple common-drain one. During this whole time, Q_1 is always OFF.

As v_{IN} decreases, the opposite scenario takes place, this time Q_2 stays OFF and Q_1 turns on.

The transfer characteristic consists of the output voltage amplitude versus the input voltage amplitude. With the circuit in Fig. 11.9, there is no DC shift anyway, so the AC voltages are the same as composite ones. The transfer characteristic is shown in Fig. 11.11a. The slope of the transfer characteristic is shown in Fig. 11.11b.

As a result, three observations can be made.

The first observation is that the transfer characteristic is not linear. It exhibits a dead zone region, within which the output does not react to the input. This is undesirable since it distorts the output signal as seen in Fig. 11.10b. This makes the Class B output stage unsuitable for any signal whose information is in the phase such as a clock signal, for example.

Fig. 11.11 Typical (**a**) transfer characteristic and (**b**) transfer characteristic slope of a Class B output stage

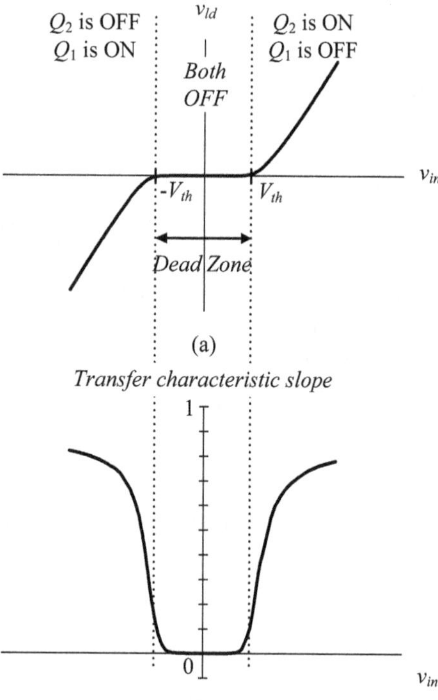

The second observation is that this dead zone can be easily seen in the large variations around the origin (quiescent point) of the transfer characteristic slope as in Fig. 11.11b.

The third observation is that the transfer characteristic slope, which is the gain, does not reach unity at its extreme value, an expected behavior from the common-drain amplifier.

11.5.3   Class B Power Consumption, Power Dissipation, and Efficiency

During half of the input cycle, the current passes through Q_1, while it passes through Q_2 during the second half as in Fig. 11.8. Therefore, the average power consumption is

$$P_{CONS} = V_{SUP+} \times I_D - V_{SUP-} \times I_D$$
$$= (V_{SUP+} - V_{SUP-}) \times I_D$$

(11.29)

where I_D is the average value of the current through every transistor.

The current through every transistor is shown in Fig. 11.12.

Neglecting the dead zone, the average value of this current is

$$I_D = \frac{1}{2\pi} \int_0^\pi i_D d\phi = \frac{1}{2\pi} \int_0^\pi i_{D\_MAX} \sin(\phi) d\phi$$
$$= \frac{i_{D\_MAX}}{2\pi} \left(-\cos(\phi) \right)\Big|_{\phi=0}^{\phi=\pi} = \frac{i_{D\_MAX}}{\pi}$$

(11.30)

Replacing (11.30) in (11.29), we get

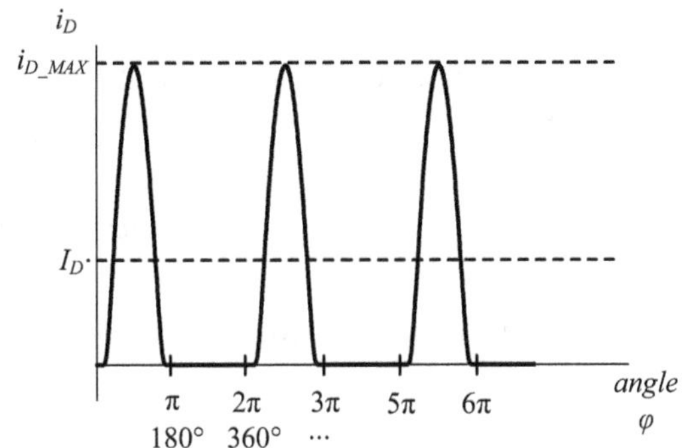

Fig. 11.12 Current through every transistor in a Class B output stage

$$P_{CONS} = (V_{SUP+} - V_{SUP-}) \times \frac{i_{D\_MAX}}{\pi} \tag{11.31}$$

In general, the average power delivered to the load consists of the DC power as well as the AC power as follows. However, in this case, based on Fig. 11.10b, V_{LD} is zero. As a result, we have

$$\begin{aligned} P_{LD} &= P_{LD\_DC} + P_{LD\_AC} \\ &= V_{LD} \times I_{LD} + v_{LD\_RMS} \times i_{LD\_RMS} \\ &= v_{LD\_RMS} \times i_{LD\_RMS} \end{aligned} \tag{11.32}$$

where v_{LD_RMS} and i_{LD_RMS} are the RMS values of the output voltage and current, respectively.

The average power dissipated as heat is calculated using (11.6) as

$$\begin{aligned} P_{DISSIP} &= P_{CONS} - P_{LD} \\ &= (V_{SUP+} - V_{SUP-}) \times \frac{i_{D\_MAX}}{\pi} - v_{LD\_RMS} \times i_{LD\_RMS} \end{aligned} \tag{11.33}$$

The average efficiency can be calculated using (11.7) as

$$\begin{aligned} \eta &= \frac{P_{LD}}{P_{CONS}} \times 100\% \\ &= \frac{v_{LD\_RMS} \times i_{LD\_RMS}}{(V_{SUP+} - V_{SUP-}) \times \dfrac{i_{D\_MAX}}{\pi}} \times 100\% \end{aligned} \tag{11.34}$$

To get an idea of the efficiency range for a Class B, and knowing that the positive and negative supply voltages have equal magnitudes as in (11.8), let us allow the output voltage swing to cover its maximum range. According to (11.26), and neglecting V_{OV_REQ} for both transistors,

$$V_{SUP-} \leq v_{LD} \leq V_{SUP+} \tag{11.35}$$

In this case the output voltage and current will be as in Fig. 11.13.
As an approximation, let us assume that the dead zone does not exist.
For the voltage, we have

$$v_{LD\_RMS} = \frac{V_{SUP}}{\sqrt{2}} \tag{11.36}$$

As for the current, we have

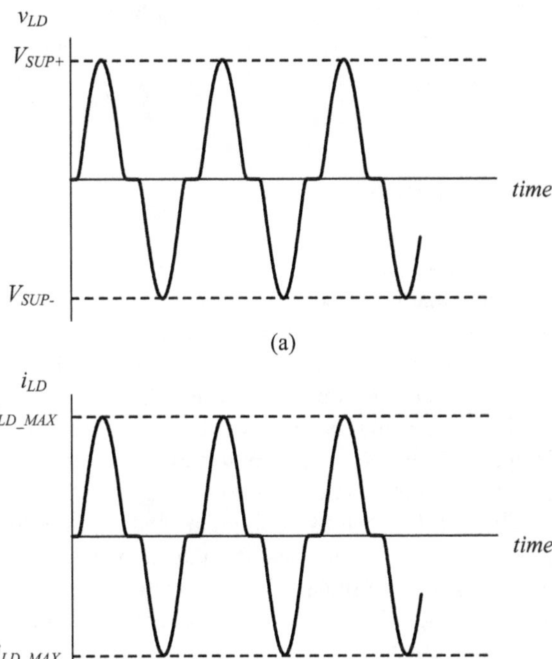

Fig. 11.13 Class B output voltage waveform (**a**) and current waveform (**b**) for the efficiency example

$$i_{LD\_RMS} = \frac{i_{LD\_MAX}}{\sqrt{2}} \tag{11.37}$$

where

$$i_{LD\_MAX} = \frac{V_{SUP}}{R_{LD}} \tag{11.38}$$

The maximum current passing through any of the two transistors is the same as the maximum current through the load, thus

$$i_{D\_MAX} = i_{LD\_MAX} \tag{11.39}$$

Substituting (11.36), (11.37), (11.38), and (11.39) into (11.34), we get the efficiency in this case:

$$\eta = \frac{\frac{V_{SUP}}{\sqrt{2}} \times \frac{V_{SUP}}{\sqrt{2}R_{LD}}}{2V_{SUP} \times \frac{V_{SUP}}{\pi R_{LD}}} \times 100\% = \frac{\pi}{4} \times 100\% = 78.5\% \tag{11.40}$$

This maximum efficiency is much higher than that of a Class A output stage. However, the crossover distortion is not acceptable for many applications. As a result, a compromise between a Class A and a Class B exists, and that is Class AB.

11.6   Class AB

The Class AB output stage aims to mitigate the main drawback of a Class B output stage and that is crossover distortion.

The source of the problem in the Class B lies in the fact that the transistors are allowed to turn OFF while the signal is crossing its middle point, which is zero in this case. In order to turn these transistors ON again (i.e. wake them up again) a certain minimum voltage is needed, which is V_{th} in this case. One way to mitigate this effect is not to allow the transistors to turn OFF, and that is by keeping them a bit ON (i.e. by keeping them drowsing).

This can be done by applying a small and constant DC voltage V_{GG} across the gate terminals of the transistors on top of which our signal is applied. In this manner, even if there is no signal, V_{GG} will keep the transistors a bit ON. Subsequently, any signal on top of V_{GG} will make the transistors react instantly, thus mitigating the dead zone effect.

The current through a Class AB output stage has a nonzero DC component. The voltage at the input and the current through the stage are shown in Fig. 11.14. As can be seen, in the quiescent state, I_Q is small, but not zero due to the presence of V_{GG}. This results in a small stand-by power consumption, but one which is much lower than that of a Class A.

Since it is based on a Class B output stage, a Class AB output stage requires two transistors to conduct all the current for the complete input cycle. The conduction angle for each transistor is slightly greater than 180 degrees.

Figure 11.15 shows an example of a Class AB output stage. As can be seen, two transistors are needed, Q_1 and Q_2. As for V_{GG}, it is supplied by the V_{GG} generator, which consists simply of two diode-connected transistors Q_3 and Q_4 effectively acting as a voltage regulator.

The circuits surrounding the transistor are modeled as follows:

- The previous stage (signal source) is modeled using its Thevenin equivalent model. This signal source contains a DC component and an AC component. Also, the source has an output resistance R_{SRC}.
- The next stage (load) is modeled using its input resistance, indicated as R_{LD}.

Note that v_{SRC} and v_{IN} are not the same in this circuit since the current through Q_3 creates a voltage drop across R_{SRC}.

Typical v_{IN} and v_{LD} waveforms of a Class AB configuration as in Fig. 11.15 are shown in Fig. 11.16. As can be seen, the output is subject to a crossover distortion that is much smaller than that of a Class B. This is due to a more linear transfer characteristic compared to that of a Class B, as will be discussed shortly.

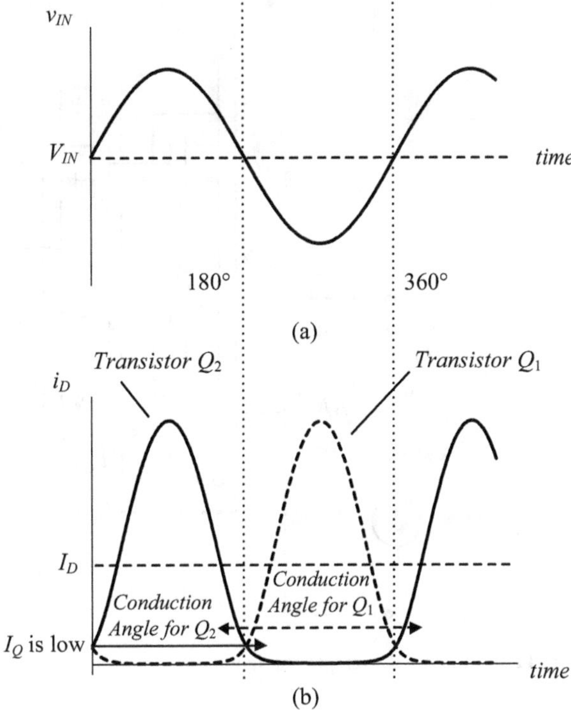

Fig. 11.14 Class AB input voltage (**a**) and transistor current (**b**)

11.6.1 Class AB Output and Input Voltage Range

When it comes to the output voltage range, looking at the circuit in Fig. 11.15, we can see that the maximum value the output voltage can take is governed by what is happening above the output node. Therefore:

$$v_{LD\_MAX} = V_{SUP+} - V_{OV2\_REQ} = V_{SUP+} - v_{GS2\_MAX} - V_{th} \tag{11.41}$$

where V_{OV2_REQ} is the overdrive voltage of Q_2.

For the minimum value, looking at the circuit in Fig. 11.15, we can see that this value is governed by what is happening below the output node. Therefore:

$$v_{LD\_MIN} = V_{SUP-} + \left| V_{OV1\_REQ} \right| = V_{SUP-} + \left| v_{GS1\_MAX} \right| + \left| V_{th} \right| \tag{11.42}$$

where V_{OV1_REQ} is the required overdrive voltage of Q_1.

As a result, the output voltage range is

$$V_{SUP-} + \left| V_{OV1\_REQ} \right| \leq v_{LD} \leq V_{SUP+} - V_{OV2\_REQ} \tag{11.43}$$

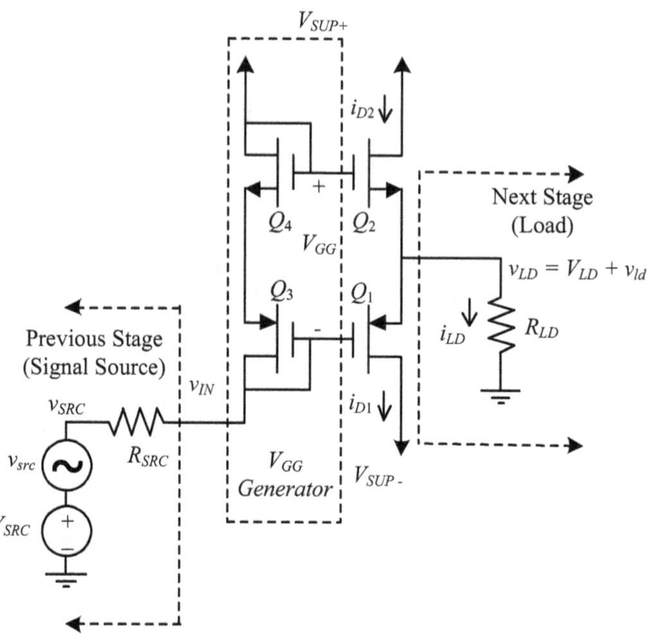

Fig. 11.15 Class AB example configuration

In order to get the maximum swing at the output, it is sensible to have the DC point at the output (V_{LD}) in the middle between v_{LD_MIN} and v_{LD_MAX}.

As for the input voltage range, we can see from Fig. 11.15 that since the input resistance is infinite, the absolute maximum input voltage is

$$v_{IN\_MAX} = V_{SUP+} - V_{GG} \tag{11.44}$$

On the other hand, the absolute minimum input voltage is

$$v_{IN\_MIN} = V_{SUP-} \tag{11.45}$$

11.6.2 Class AB Transfer Characteristic and Linearity

Looking at Fig. 11.15, and starting from the quiescent state when v_{in} and v_{LD} are zero, the voltage drop across the gate of Q_1 and Q_2 is V_{GG}. Note that the DC V_{IN} in the quiescent state here is not zero. Since usually the power supplies have equal magnitudes as in (11.8), then

Fig. 11.16 Typical (**a**) v_{IN} and (**b**) v_{LD} waveforms of a Class AB output stage

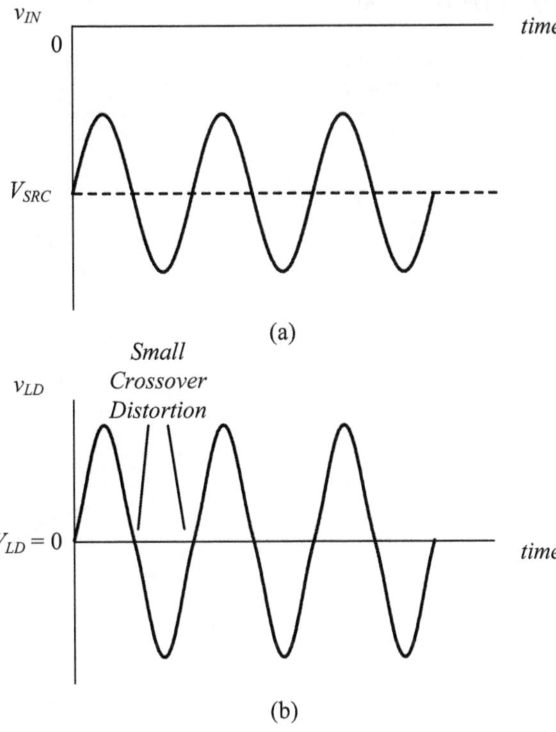

$$|V_{GS1}| = V_{GS2} = \frac{V_{GG}}{2} \tag{11.46}$$

As a result, if Q_3 and Q_4 and V_{IN} are chosen such that $V_{GG}/2$ is slightly greater than V_{th}, then a small DC current I_Q will pass through the transistors in the quiescent state, effectively keeping both transistors ON at the same time. This is the main reason why crossover distortion is very small.

As v_{IN} increases, Q_2 sources more current and Q_1 sinks less current. Similarly, as v_{IN} decreases, Q_2 sources less current and Q_1 sinks more current.

The transfer characteristic consists of the output voltage amplitude versus the input voltage amplitude. The transfer characteristic is shown in Fig. 11.17a. The slope of the transfer characteristic is shown in Fig. 11.17b.

As a result, three observations can be made.

The first observation is that the transfer characteristic is more linear than that of a Class B, but less linear than that of a Class A. This makes the Class AB output stage more suitable for audio applications than a Class B from a distortion point-of-view.

The second observation is that the small nonlinearity around the origin can be easily seen in the small variations around the origin (quiescent point) of the transfer characteristic slope as in Fig. 11.17b.

Fig. 11.17 Typical (**a**)
transfer characteristic and
(**b**) transfer characteristic
slope of a Class AB output
stage

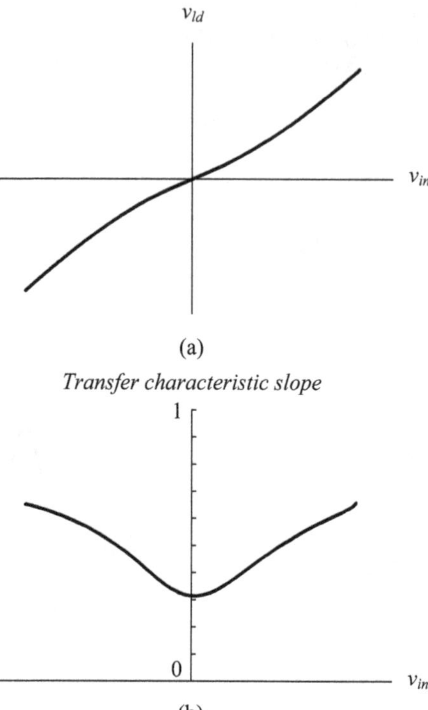

The third observation is that the transfer characteristic slope, which is the gain,
does not reach unity at its extreme value, an expected behavior from the common-
drain amplifier.

11.6.3   Class AB Power Consumption, Power Dissipation, and Efficiency

Power and efficiency analysis of a Class AB stage is very similar to that of a Class B
stage.

During half of the input cycle, the current passes primarily through Q_1, while it
passes primarily through Q_2 during the second half as in Fig. 11.14. Therefore, the
average power consumption is

$$P_{CONS} = V_{SUP+} \times I_D - V_{SUP-} \times I_D$$
$$= (V_{SUP+} - V_{SUP-}) \times I_D$$

(11.47)

where I_D is the average value of the current through every transistor.

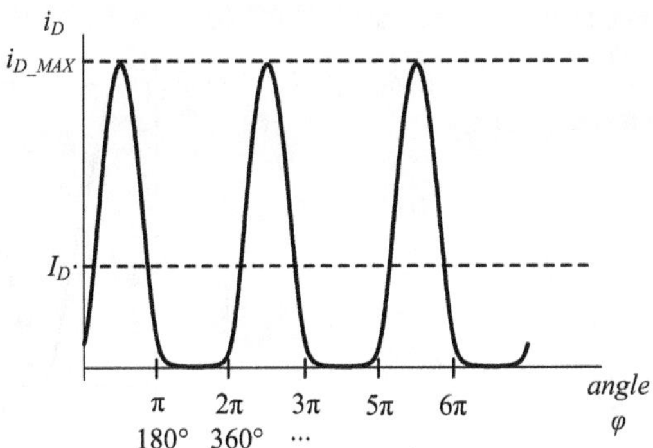

Fig. 11.18 Current through every transistor in a Class AB output stage

The current through every transistor is shown in Fig. 11.18.
The average value of this current is approximately

$$I_D \approx \frac{1}{2\pi} \int_0^\pi i_D d\phi = \frac{1}{2\pi} \int_0^\pi i_{D\_MAX} \sin(\phi) d\phi$$

$$= \frac{i_{D\_MAX}}{2\pi} \left(-\cos(\phi) \big|_{\phi=0}^{\phi=\pi} \right) = \frac{i_{D\_MAX}}{\pi} \tag{11.48}$$

Replacing (11.48) in (11.47), we get

$$P_{CONS} = (V_{SUP+} - V_{SUP-}) \times \frac{i_{D\_MAX}}{\pi} \tag{11.49}$$

In general, the average power delivered to the load consists of the DC power as well as the AC power as follows. However, in this case, based on Fig. 11.16b, V_{LD} is zero. As a result, we have

$$P_{LD} = P_{LD\_DC} + P_{LD\_AC}$$

$$= V_{LD} \times I_{LD} + v_{LD\_RMS} \times i_{LD\_RMS} \tag{11.50}$$

$$= v_{LD\_RMS} \times i_{LD\_RMS}$$

where v_{LD_RMS} and i_{LD_RMS} are the RMS values of the output voltage and current respectively.

The average power dissipated as heat is calculated using (11.6) as

$$P_{DISSIP} = P_{CONS} - P_{LD}$$

$$= (V_{SUP+} - V_{SUP-}) \times \frac{i_{D\_MAX}}{\pi} - v_{LD\_RMS} \times i_{LD\_RMS} \tag{11.51}$$

The average efficiency can be calculated using (11.7) as

Fig. 11.19 Class AB output voltage waveform (**a**) and current waveform (**b**) for the efficiency example

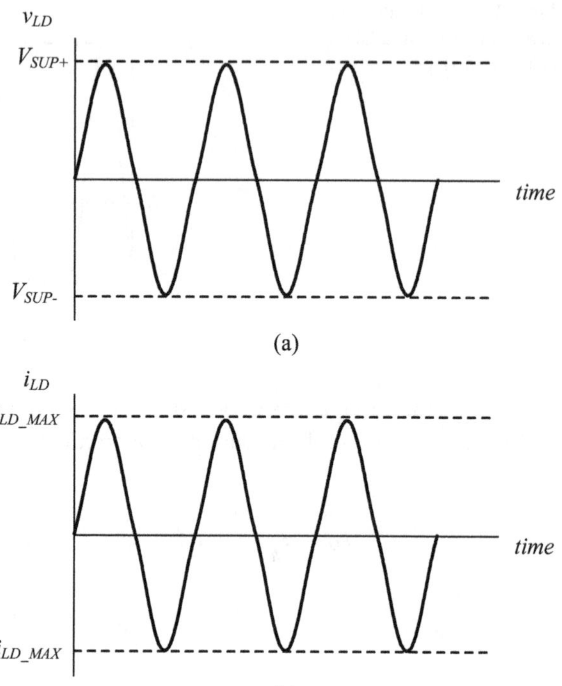

$$\eta = \frac{P_{LD}}{P_{CONS}} \times 100\%$$

$$= \frac{v_{LD\_RMS} \times i_{LD\_RMS}}{(V_{SUP+} - V_{SUP-}) \times \dfrac{i_{D\_MAX}}{\pi}} \times 100\% \tag{11.52}$$

To get an idea of the efficiency range for a Class AB, and knowing that the positive and negative supply voltages have equal magnitudes as in (11.8), let us allow the output voltage swing to cover its maximum range. According to (11.43), and neglecting V_{OV_REQ} for both transistors,

$$V_{SUP-} \leq v_{LD} \leq V_{SUP+} \tag{11.53}$$

In this case the output voltage and current will be as in Fig. 11.19.
For the voltage, we have

$$v_{LD\_RMS} = \frac{V_{SUP}}{\sqrt{2}} \tag{11.54}$$

As for the current, we have

$$i_{LD\_RMS} = \frac{i_{LD\_MAX}}{\sqrt{2}} \tag{11.55}$$

where

$$i_{LD\_MAX} = \frac{V_{SUP}}{R_{LD}} \tag{11.56}$$

The maximum current passing through any of the two transistors is the same as the maximum current through the load, thus

$$i_{D\_MAX} = i_{LD\_MAX} \tag{11.57}$$

Substituting (11.54), (11.55), (11.56), and (11.57) into (11.52), we get the efficiency in this case:

$$\eta = \frac{\frac{V_{SUP}}{\sqrt{2}} \times \frac{V_{SUP}}{\sqrt{2}R_{LD}}}{2V_{SUP} \times \frac{V_{SUP}}{\pi R_{LD}}} \times 100\% = \frac{\pi}{4} \times 100\% = 78.5\% \tag{11.58}$$

With the approximations that were done, this maximum efficiency is the same as that of a Class B. In practice, and due to the nonzero quiescent current, the efficiency is lower.

Chapter 12
Mixed-Signal Circuits

Many applications require circuits that deal with signals that are neither purely analog nor purely digital. These circuits are called mixed-signal circuits. In this chapter, we will introduce the concept of mixed-signal circuits and some of their applications. The added complexity allows these circuits to achieve some performance levels and functionalities that cannot be attained otherwise. This chapter looks into comparators, switched-capacitor circuits, sample-and-hold circuits, digital-to-analog converters, analog-to-digital converters, as well as oscillators and phase-locked loops.

12.1 Comparators

A comparator takes two signals at its input, compares them by subtracting them from each other, and outputs either a high signal or low signal depending on whether the result is positive or negative. The input signals are typically continuous, and since the output ideally is either high or low, then it is discrete. Figure 12.1a shows a general comparator and Fig. 12.1b shows a typical input–output transfer characteristic.

As can be seen from Fig. 12.1a, the comparator can take either two varying inputs *Input+* and *Input−*, or one varying input *Input+* and a fixed input *Reference*. In our context, *Input−* and *Reference* will be used interchangeably. The relations used are as follows:

$$Input = (Input+) - (Input-) \tag{12.1}$$

Notice from Fig. 12.1b that a comparator does not switch in a vertical manner. There is rather a range of inputs within which the comparator output transitions from one extreme to the other.

J. G. Atallah, M. Ismail, *Integrated Electronic Circuits*,
https://doi.org/10.1007/978-3-031-62707-1_12

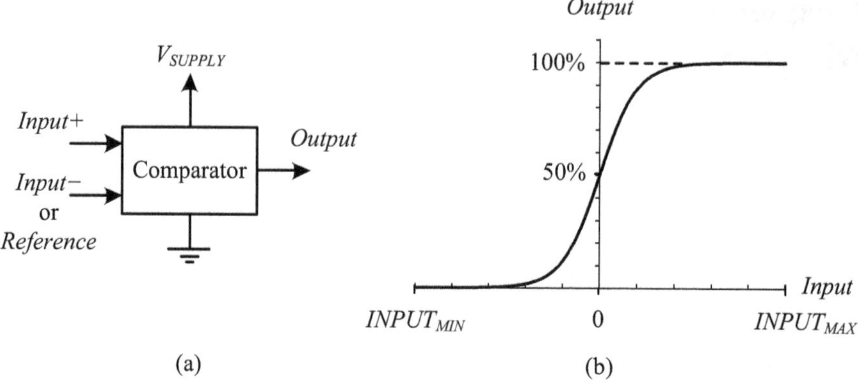

(a)   (b)

Fig. 12.1 A general comparator (**a**) and a typical comparator transfer characteristic (**b**)

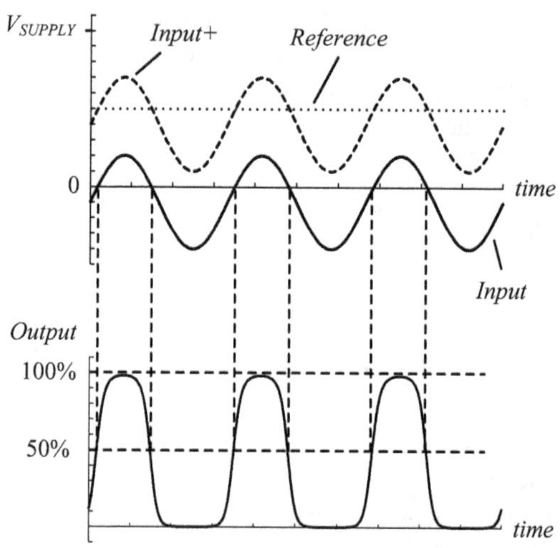

Fig. 12.2 Time-domain behavior of a general comparator

A sample time-domain behavior of the general comparator is shown in Fig. 12.2. *Input+* and *Reference* have to be within the power supply voltage range of the comparator. Their difference, *Input*, can have any value. The comparator simply outputs a high signal if *Input* is positive and a low signal if it is negative.

Ideally, the switching point, which is when the output reaches 50%, coincides with the input being exactly equal to zero as in Fig. 12.1b. In this case, the comparator is said to have no input offset.

On the other hand, Fig. 12.3 shows the transfer characteristic of a comparator with an input offset called IN_{OFFSET}. As a result, the input at which the output is at 50% will be offset from zero by IN_{OFFSET} either to the right or to the left. Figure 12.3 shows the situation when the offset is to the right.

The simplest way to implement a comparator is by using an op-amp as in Fig. 12.4.

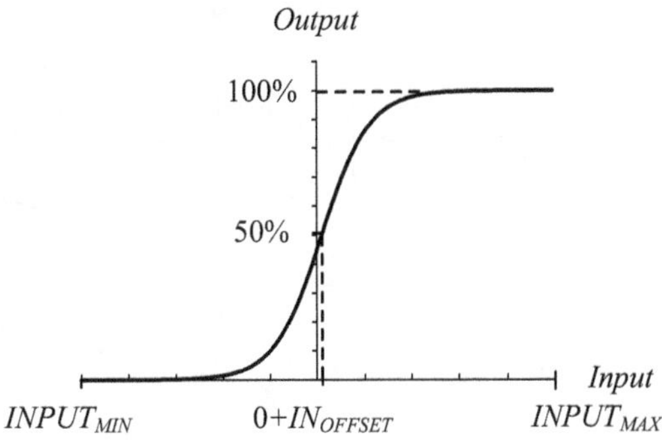

Fig. 12.3 Transfer characteristic of a typical comparator with an input offset

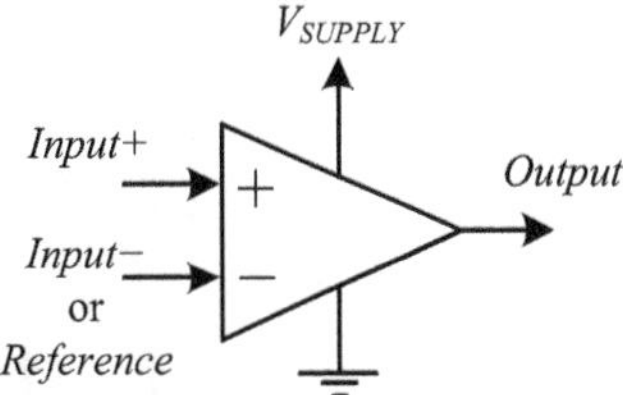

Fig. 12.4 Comparator implemented using an op-amp

This implementation has several drawbacks, namely:

1. The op-amp is continuously comparing the inputs. This creates glitches in the output if the inputs did not settle to their final values.
2. The op-amp has a finite gain and slew rate which will limit its switching gain and speed.
3. The op-amp is continuously drawing power from the supply resulting in high overall power consumption.
4. The output is single-ended, which makes it sensitive to noise.

A better implementation attempts to remedy all the above issues as follows:

1. We will be using a clock signal, *CLK*, in order to compare the inputs during specific durations of time.
2. Positive feedback will be employed to increase the gain and the switching speed.
3. The *CLK* signal itself will be employed in such a way as to stop the power consumption when the comparison is done.
4. A differential output will be used.

All of the above requirements are met in the StrongARM comparator. The general architecture is shown in Fig. 12.5.

Capacitors C_{apA} and C_{apB} model the output capacitances of this stage combined with the input capacitances of the next stage. Switch Sw_C is closed when *CLK* is high and open when *CLK* is low. Switches Sw_A and Sw_B work in the opposite manner. The whole differential structure is assumed to be perfectly matched. Note how the inverters are connected in a positive feedback manner creating a saturated and stable output.

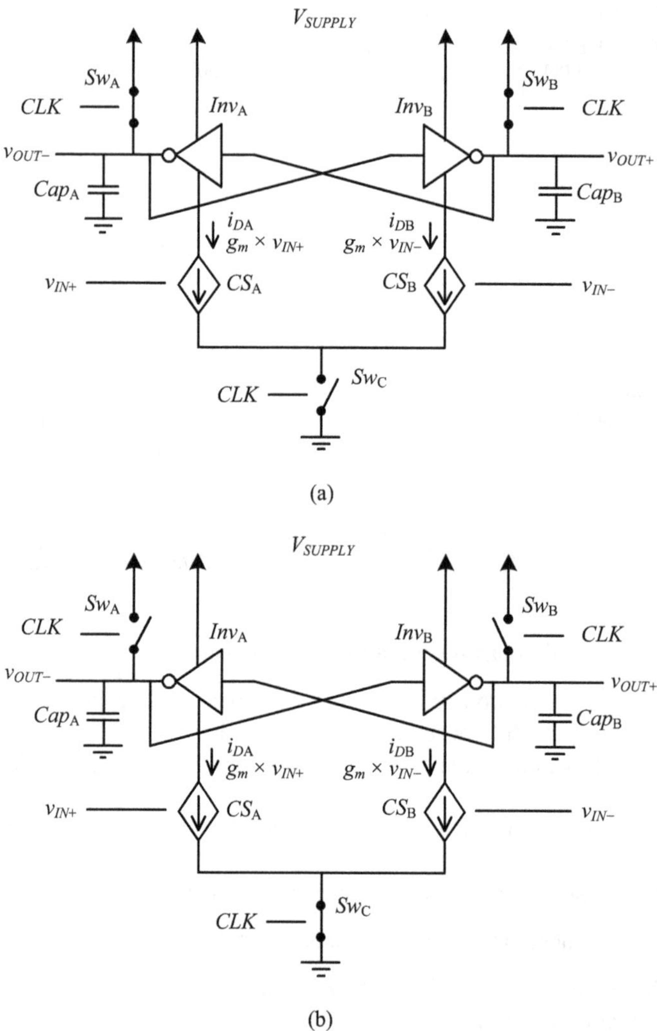

Fig. 12.5 StrongARM comparator architecture when *CLK* is low (**a**) and when *CLK* is high (**b**)

The StrongARM comparator has two main phases:

1. During the first phase, called the pre-charge phase, as in Fig. 12.5a, *CLK* is low, switches Sw_A and SW_B are closed and switch Sw_C is open. Capacitors Cap_A and Cap_B are now charged to V_{SUPPLY} and the input signals v_{IN+} and v_{IN-} are not read since Sw_C is open. Note that after the capacitors are charged, ideally no power is consumed.

2. During the second phase, called the comparison phase, as in Fig. 12.5b, *CLK* is high, switches Sw_A and Sw_B are open, and switch Sw_C is closed. Now, the inputs will create corresponding currents i_{DA} and i_{DB} depending on their strengths. If v_{IN}

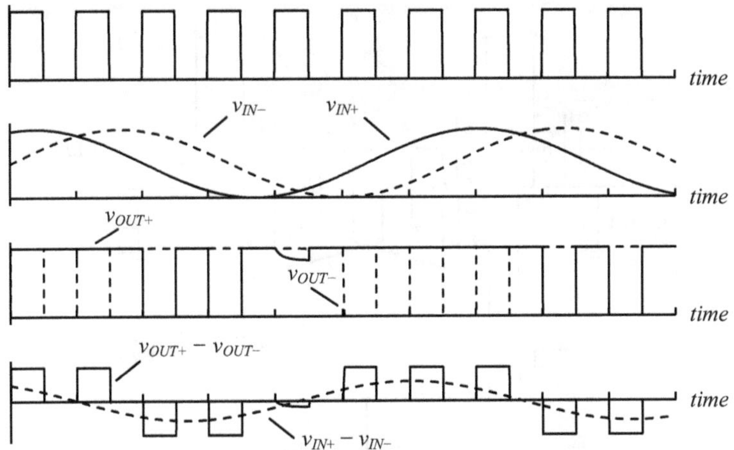

Fig. 12.6 Sample time-domain waveforms of the StrongARM comparator signals

$_+$ is higher than v_{IN-}, i_{DA} will be higher than i_{DB}, so inverter Inv_A will be stronger than Inv_B, thus pulling v_{OUT-} down. This will pull the input of Inv_B down which will keep v_{OUT+} high, which itself is the input of Inv_A, accelerating the decrease in v_{OUT-}. This creates a stable situation where v_{OUT+} is high and v_{OUT-} is low, effectively acting as a latch. Note that when the output is stable, ideally no power is consumed.

Sample time-domain waveforms of the above signals are shown in Fig. 12.6.

A straightforward transistor-level implementation of the StrongARM comparator is shown in Fig. 12.7. Note how switches Sw_A and Sw_B are implemented using PMOS transistors and switch Sw_C is implemented using an NMOS transistor in order to operate in the opposite manner. Also note that the inverters Inv_A and Inv_B are implemented using simple CMOS logic.

12.2 Switched-Capacitor Circuits

Many circuits, such as filters, rely on a particular value of resistance and a particular value of capacitance, resulting in a targeted time constant. The challenge in this case lies in the fact that resistors and capacitors have physically very different natures and process variations will inevitably result in a time constant different from the targeted one.

However, if we can create a resistor out of a capacitor, particularly one that is inversely proportional to a capacitor with α being a constant of proportionality, then the target time constant can be expressed as a ratio of capacitors as follows:

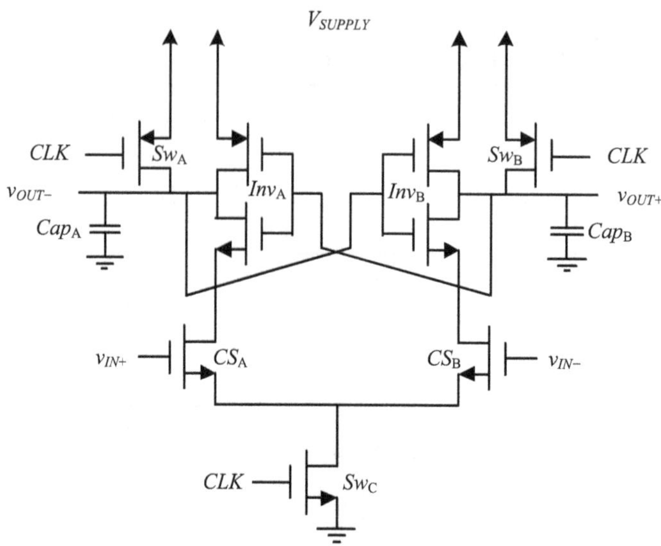

Fig. 12.7 Transistor-level implementation of the StrongARM comparator

Fig. 12.8 A resistor (**a**) can be implemented using a switched-capacitor circuit (**b**)

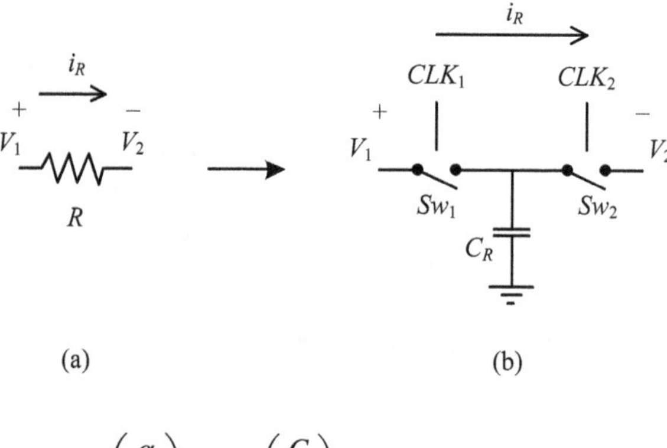

$$\tau = RC = \left(\frac{\alpha}{C_R}\right) C = \alpha\left(\frac{C}{C_R}\right) \tag{12.2}$$

Owing to the nature of process variations, it is much easier to achieve a particular ratio of capacitors, for example, by choosing a capacitor C and then combining several of these Cs in parallel in order to form C_R, than by trying to achieve a particular value of R and a particular value of C.

A switched-capacitor circuit achieves exactly this. By combining a capacitor with some switches, a resistive effect can be created as in Fig. 12.8. Getting the same resistance from these circuits means that for the same voltage $V_1 - V_2$ applied to both circuits, the same average current is drawn.

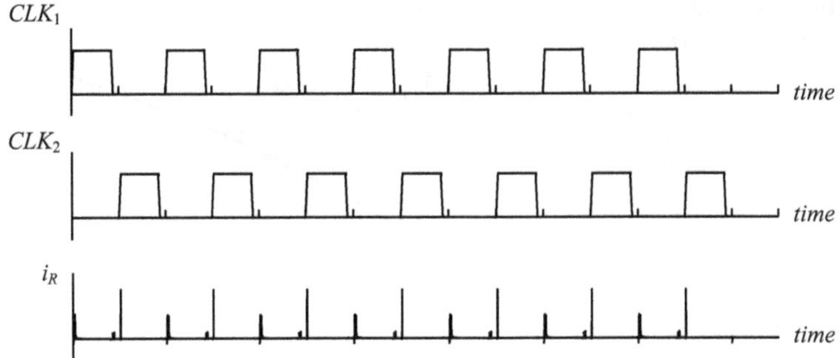

Fig. 12.9 Sample signals for a constant voltage drop across a switched-capacitor circuit

CLK_1 and CLK_2 are two clock signals having the same frequency but opposite phases. Also, they are nonoverlapping, so when one is ON, the other is completely OFF.

After applying a positive constant voltage between V_1 and V_2 to the switched-capacitor circuit in Fig. 12.8b, we get the sample signals for CLK_1, CLK_2, and i_R as shown in Fig. 12.9.

As can be seen, current i_R is not constant. However, by dividing the applied constant voltage by the average of this current, we get the average equivalent resistance R:

$$R = \frac{V_1 - V_2}{i_{R\_avg}} \tag{12.3}$$

The reason this equivalence works is because, based on Fig. 12.8b, when Sw_1 is closed, the capacitor is charged to V_1 and when Sw_2 is closed, the capacitor is charged to V_2. This is repeated every clock period. As a result, the change in the capacitor charge is

$$\Delta Q = C_R(V_1 - V_2) \tag{12.4}$$

This change in capacitor charge can be written as

$$\Delta Q = i_{R\_avg} \times T \tag{12.5}$$

where i_{R_avg} is the average current and T is the clock period.

Combining (12.4) and (12.5) we get

$$i_{R\_avg} = \frac{C_R(V_1 - V_2)}{T} = \frac{V_1 - V_2}{R} \tag{12.6}$$

As a result of that,

Fig. 12.10 Change of
equivalent resistance versus
clock period

Fig. 12.11 Transistor-level
implementation of a
switched-capacitor circuit

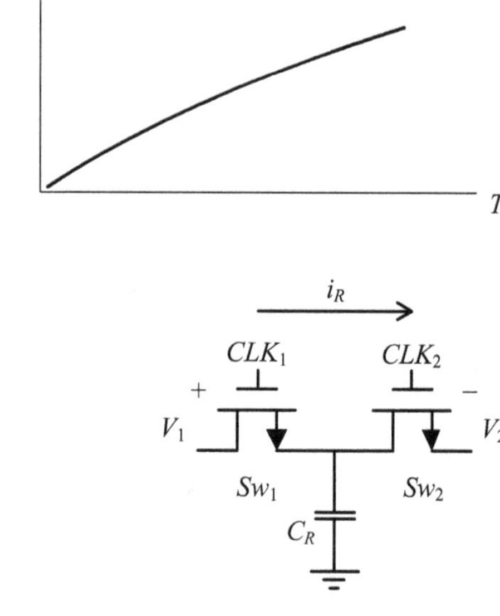

$$R = \frac{T}{C_R} \tag{12.7}$$

Therefore, α mentioned in (12.2) is actually T, the clock period.

The interesting thing about this circuit is that we can change the value of the equivalent resistance R simply by changing the clock period T. Figure 12.10 shows the change in resistance R versus the clock period T. Ideally, the relation is supposed to be linear; however, non-idealities in the implementation result in a slightly curved relation.

The major non-idealities stem from the switches themselves. For example, a simple transistor-level implementation is shown in Fig. 12.11 where the internal parasitics of the NMOS transistors used, especially the internal capacitances, interact with the capacitor C_R.

The input voltage in the discussion above was taken to be constant. In reality, this input voltage varies. If the variations are very slow compared to the clock period, the above circuit works as expected. Otherwise, either a discrete-time analysis is needed, or a sample-and-hold circuit is used at the input in order to keep the input voltage constant for a while.

12.3   Sample-and-Hold Circuits

A sample-and-hold circuit is employed to take a sample of the incoming signal and hold its value long enough for the next stage to process it. This makes the design of the next stage much easier since now it only needs to deal momentarily with a constant signal rather than a varying one.

The simplest ideal implementation of a sample-and-hold circuit is shown in Fig. 12.12.

Sample waveforms of the ideal implementations are shown in Fig. 12.13.

The circuit operates in two phases:

1. During the first phase, the switch is closed and V_{out} tracks V_{in} while the value is stored on the capacitor. It is because of this behavior that this circuit is also called ta track-and-hold circuit and this phase is called a track phase, indicated as 'T' in Fig. 12.13.
2. During the second phase, the switch is open and the capacitor retains its charge, thus V_{out} keeps the last sampled value of V_{in}. Ideally, during this phase, the sampled value should stay constant since the input resistance r_{next} to the next stage should be infinite. This can be achieved by interposing a circuit with a high input resistance (a buffer) between this stage and the next one. This phase is indicated as 'H' in Fig. 12.13.

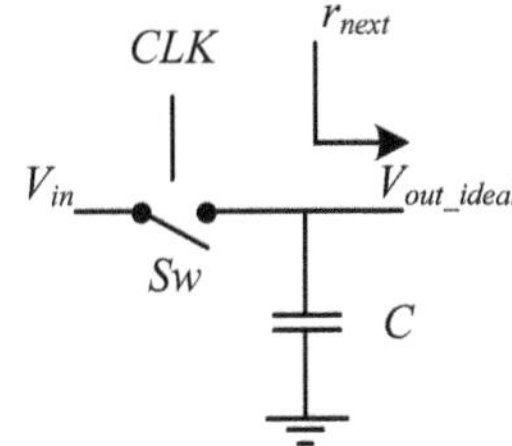

Fig. 12.12 Simple ideal implementation of a sample-and-hold circuit

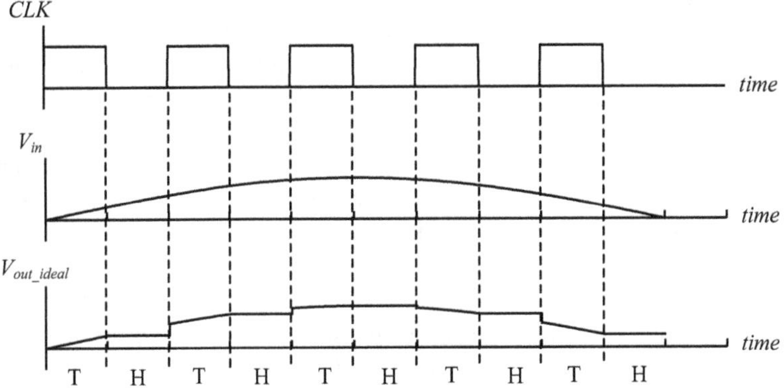

Fig. 12.13 Sample waveforms of the ideal sample-and-hold implementation

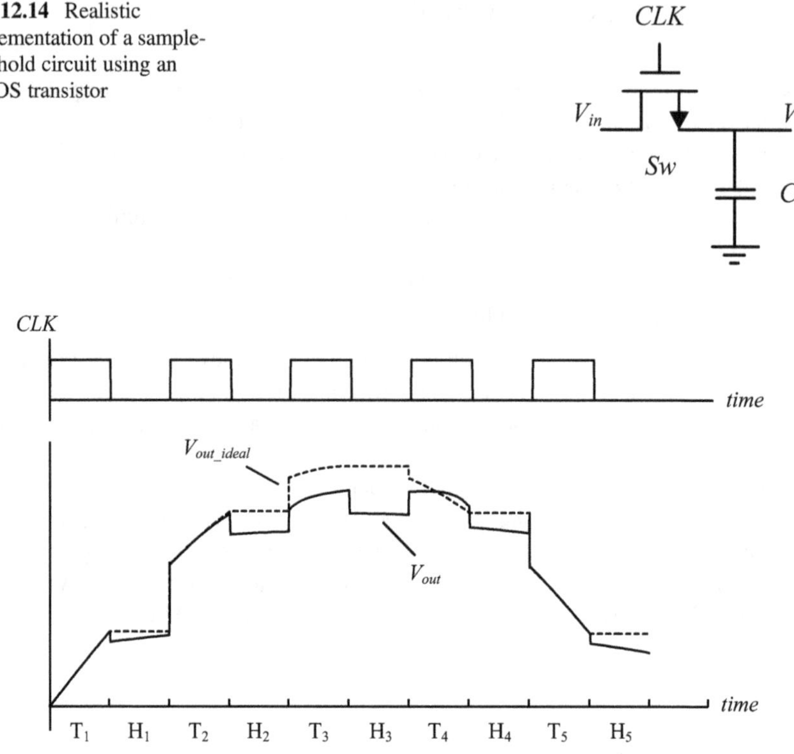

Fig. 12.14 Realistic implementation of a sample-and-hold circuit using an NMOS transistor

Fig. 12.15 Sample waveforms of a realistic versus ideal sample-and-hold implementation

In practice, several non-idealities are introduced due to a realistic switch. A sample circuit with the switched implemented using an NMOS transistor is shown in Fig. 12.14.

Sample waveforms of this realistic implementation versus an ideal implementation are shown in Fig. 12.15.

In Fig. 12.15, we can compare the ideal output signal V_{out_ideal} to the realistic one V_{out}. As can be seen, some non-idealities show up during the track phase, while other during the hold phase.

During the track phase such as T_3, we can see that the output cannot reach its intended value. This is because the NMOS transistor requires a voltage drop across its terminals when it is ON, effectively creating a ceiling for the output. This can be prevented by using a larger transistor that requires a smaller V_{DS} while providing the same current.

During all the hold phases, we can see that when the transistor turns OFF, the output voltage drops instantly. This is due to the internal capacitances of the NMOS transistor that create a path from its gate to its other terminals resulting in what is called clock feedthrough. This can be alleviated by reducing the high-frequency components of the clock, for example, by increasing its fall time in this case. It can

Fig. 12.16 A sample-and-hold circuit using NMOS and PMOS transistors as a switch

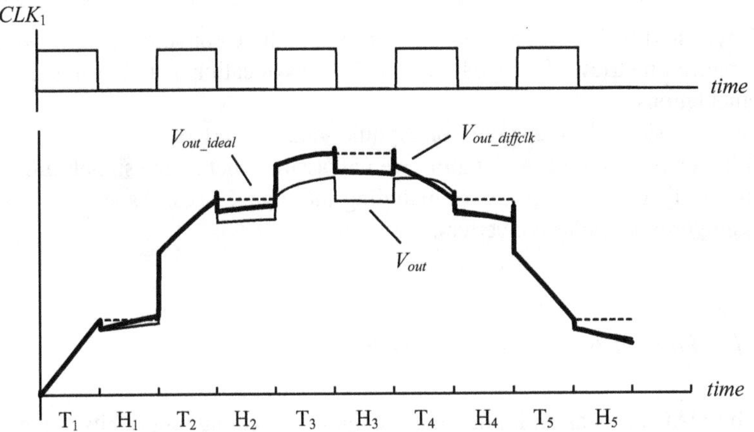

Fig. 12.17 Sample waveforms comparing the output of all three implementations

also be alleviated by using another switching technique where the clock signal is farther away from the output node, or by using a differential structure. Another problem during the hold phases is that the value of the output voltage is not held perfectly constant. This is can be either due to clock feedthrough or due to the finite time constant created by the capacitor and the finite resistance of the NMOS switch.

One way to alleviate these problems is to use a PMOS in parallel with the NMOS in order to form a complementary switch. This requires two clock phases, which adds to the complexity. A sample circuit is shown in Fig. 12.16.

Sample waveforms comparing the ideal implementation with the single-transistor switch with the complementary switch implementation are shown in Fig. 12.17.

As can be seen from Fig. 12.17, the complementary switch implementation waveform $V_{out\_diffclk}$ is closer to the ideal implementation waveform V_{out_ideal}. In particular, we can see that the output can take a higher value compared to the NMOS implementation as in the T_3 phase. Also, the drop in the output is less severe as during the H_2 phase.

12.4   Data Converters

Data converters sit at the interfaces between analog circuits and digital circuits. These data converters are indispensable whenever a circuit needs to interact with external signals, which are predominantly analog.

There are two types of data converters:

1. Digital-to-analog converters (DACs): convert digital signals to analog signals.
2. Analog-to-digital converters (ADCs): convert analog signals to digital signals.

We will start with the system level overview of these converters, along with their performance metrics, followed by their classifications and sample circuit implementations.

The discussion will focus only on Nyquist-rate converters.

Without any loss of generalization, we will assume that all the signals are voltage signals and that they are positive, including the digital ones. As such, we will be discussing only unipolar converters.

12.4.1   Digital-to-Analog Converters

An N-bit DAC takes as an input a digital signal B_{IN} containing N bits along with a reference signal V_{REF}, combines them, and outputs an analog signal V_{OUT} as shown in Fig. 12.18. In practice, these systems use a clock signal that clocks in the B_{IN} in a serial manner and clocks out v_{OUT}.

Regarding B_{IN}, b_0 is the least significant bit and $b_{N\text{-}1}$ is the most significant bit. As a result, B_{IN} can be written as

$$B_{IN} = \left(b_{N-1} \times 2^{N-1}\right) + \left(b_{N-2} \times 2^{N-2}\right) + \ldots + \left(b_1 \times 2^1\right) + \left(b_0 \times 2^0\right) \qquad (12.8)$$

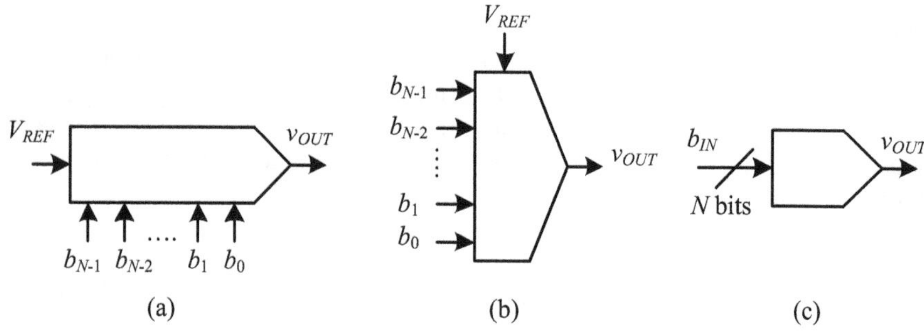

Fig. 12.18 Three possible symbols for an N-bit DAC

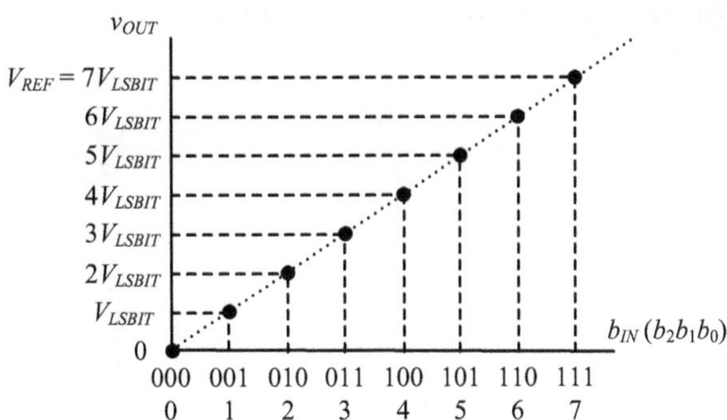

Fig. 12.19 Ideal input–output characteristic of a 3-bit DAC

Table 12.1 Classification of DACs

DAC Class	Advantages	Drawbacks
Decoder-based	Simple to implement	Relies on accurate resistor values
Binary-weighted	Can make use of binary-weighted current sources, resistors, capacitors, etc.	Subject to nonlinearities
Thermometer-code	Can make use of current sources, resistors, capacitors, etc.	Better performance than binary-weighted DACs
Hybrid	Combination of the above	Combination of the above

The ideal transfer characteristic of a 3-bit DAC, where $N = 3$, is shown in Fig. 12.19. V_{REF} is the maximum output voltage. V_{REF} is divided by the resolution 2^N, to give V_{LSBIT} which is the voltage of the least-significant bit. Therefore, V_{LSBIT} can be written as

$$V_{LSBIT} = \frac{V_{REF}}{2^N} \tag{12.9}$$

The output voltage v_{OUT} can be written as

$$v_{OUT} = b_{IN} \times V_{LSBIT} \tag{12.10}$$

In other words, the relation between v_{OUT} and b_{IN} is a linear one and V_{LSBIT} is the slope of the line formed by all the possible outputs in Fig. 12.19.

There are several types of DAC implementations with different advantages and drawbacks. Generally, they can be classified as shown in Table 12.1.

Figure 12.20 shows a decoder-based circuit-level implementation of a 3-bit DAC. Several reference voltages are generated using matched resistors in the resistor string at the right. The digital input bits at the left are decoded, and only one of the switching transistors is turned ON, passing the corresponding resistor string voltage to the output.

Fig. 12.20 Decoder-based circuit-level implementation of a 3-bit DAC

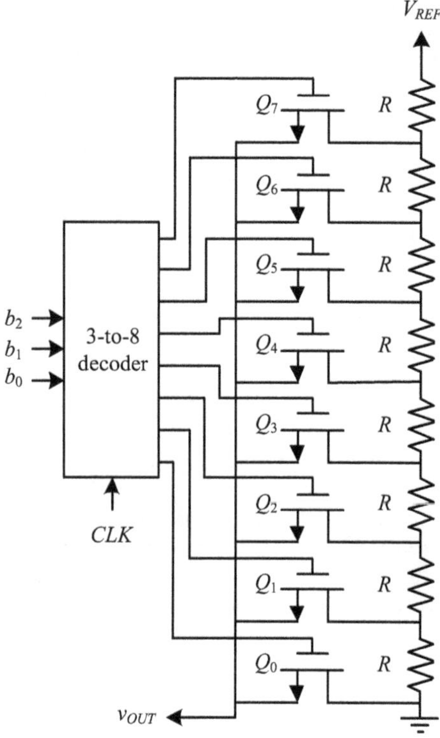

This implementation uses matched resistors, which makes it a challenge to implement considering process variations. This limits the number of resistors used and thus the number of bits of the converter.

Note that the decoder makes use of a clock signal originating from a clock generator that will be discussed later.

12.4.2 Analog-to-Digital Converters

A N-bit ADC takes as an input an analog signal v_{IN} along with a reference signal V_{REF}, combines them, and outputs a digital signal b_{OUT} containing N bits as shown in Fig. 12.21. In practice, these systems use a clock signal that clocks in v_{IN} and clocks out b_{OUT} in a serial manner.

Regarding b_{OUT}, b_0 is the least significant bit and $b_{N\text{-}1}$ is the most significant bit. As a result, b_{OUT} can be written as

$$b_{OUT} = \left(b_{N-1} \times 2^{N-1}\right) + \left(b_{N-2} \times 2^{N-2}\right) + \ldots + \left(b_1 \times 2^1\right) + \left(b_0 \times 2^0\right) \quad (12.11)$$

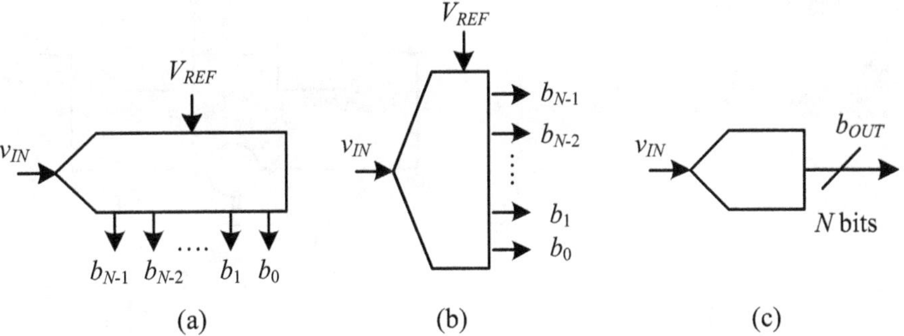

Fig. 12.21 Three possible symbols for an N-bit ADC

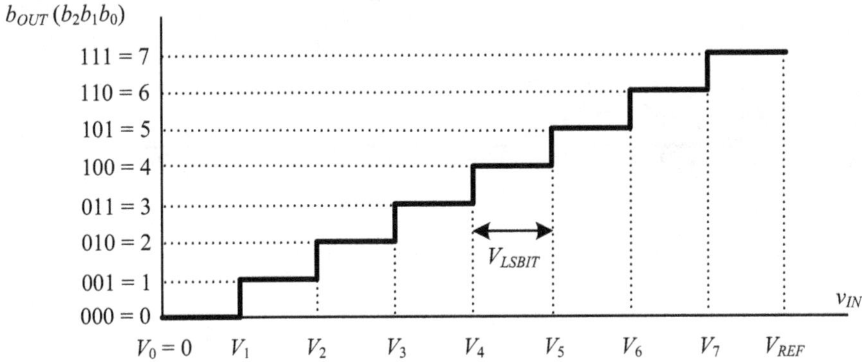

Fig. 12.22 Ideal input–output characteristic of a 3-bit ADC

The ideal transfer characteristic of a 3-bit ADC, where $N = 3$, is shown in Fig. 12.22.

V_{REF} is divided by 2^N, to give V_{LSBIT} which is the tread size. Therefore, V_{LSBIT} can be written as

$$V_{LSBIT} = \frac{V_{REF}}{2^N} \tag{12.12}$$

As can be seen, unlike in an ideal DAC, an ideal ADC does not have a linear transfer characteristic, which means that errors are inherent to its operation. The errors due to this step-like characteristic are called quantization errors.

Quantization errors are a natural result of ADCs owing to the digital, thus finite, values that they can output. In other words, they are the result of the discrete y-axis in Fig. 12.22.

To visualize and, eventually characterize, the quantization errors, we can follow the ADC with an ideal DAC as in Fig. 12.23.

As can be seen, the quantization error signal v_{QUAN} can be expressed as

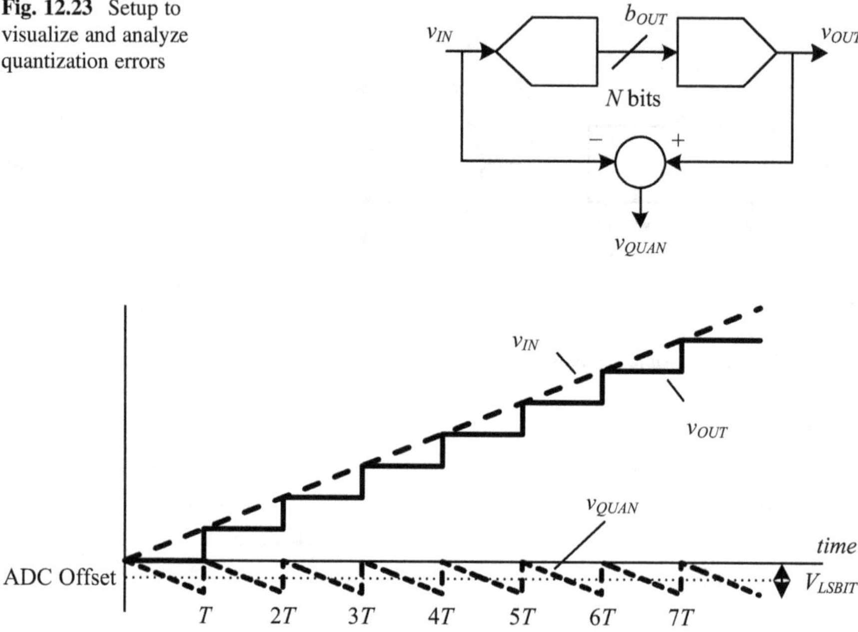

Fig. 12.23 Setup to visualize and analyze quantization errors

Fig. 12.24 Input, output, and quantization error signals versus time

$$v_{QUAN} = v_{OUT} - v_{IN} \tag{12.13}$$

Note that in this case, the ADC and the DAC are assumed to be ideal, with no delay, for example.

Plotting the above signals versus time, we get Fig. 12.24.

The quantization signal has a maximum excursion of V_{LSBIT}. Additionally, it has a nonzero DC component, also called the offset of the ADC. This offset can be easily handled when processing the digital signal, depending on the application requirements.

In order to quantify the root-mean-square value of the quantization error signal v_{QUAN} we will first remove the offset. Also, let the time taken for v_{IN} to cover V_{LSBIT} be called T. Additionally, to ease the mathematics, center the first period of v_{QUAN} at zero. As a result, the first period will look as in Fig. 12.25.

Consequently, from $-T/2$ to $+T/2$, v_{QUAN} can be expressed as

$$v_{QUAN} = \left(\frac{-t}{T}\right) V_{LSBIT} \tag{12.14}$$

Therefore, the root-mean-square value of the quantization error signal is

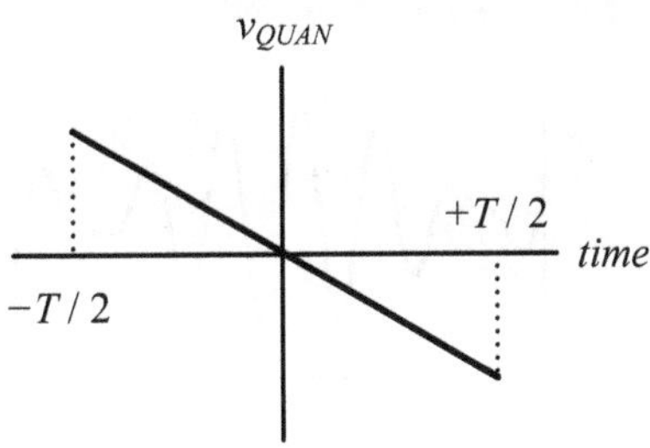

Fig. 12.25 One period of the quantization error signal used to calculate its root-mean-square value

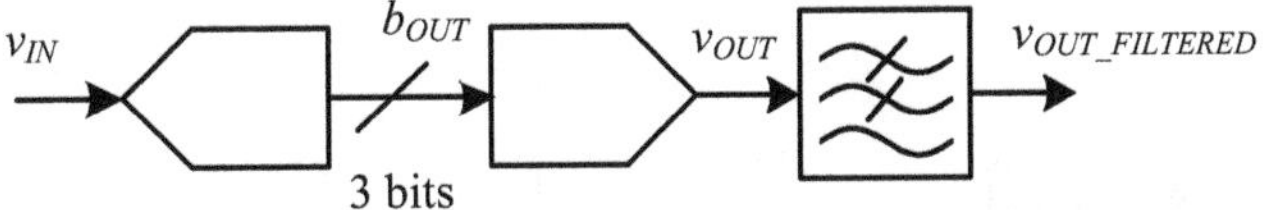

Fig. 12.26 Filtering the output of a 3-bit DAC to reduce the quantization errors

$$V_{QUAN\_RMS} = \sqrt{\frac{1}{T} \int_{-T/2}^{+T/2} \left(v_{QUAN}^2 \right) dt} = \sqrt{\frac{1}{T} \int_{-T/2}^{+T/2} \left(\left(\frac{-t}{T} \right) V_{LSBIT} \right)^2 dt}$$

$$= \sqrt{\frac{V_{LSBIT}^2}{T^3} \int_{-T/2}^{+T/2} (-t)^2 dt} = \sqrt{\frac{V_{LSBIT}^2}{T^3} \left(\frac{t^3}{3} \Big|_{-T/2}^{+T/2} \right)} \qquad (12.15)$$

$$= \sqrt{\frac{V_{LSBIT}^2}{12}} = \frac{V_{LSBIT}}{\sqrt{12}}$$

As can be seen, this value is proportional to V_{LSBIT}, which decreases as N increases for a given V_{REF}. As a result, increasing the number bits in an ADC reduces the quantization errors.

Quantization errors, though inevitable, can be dramatically reduced by filtering v_{OUT} as in Fig. 12.26. In this case, 3-bit converters are used, followed by a low-pass filter.

The corresponding signals are seen in Fig. 12.27.

Even a simple filter can restore the signal back to what it was, albeit with some distortion. The amount of distortion can be seen clearly in the spectra in Fig. 12.27.

There are several types of ADC implementations with different advantages and drawbacks. Generally, they can be classified based on their speed and accuracy as shown in Table 12.2.

Figure 12.28 shows an implementation of a 3-bit flash ADC. It works in the opposite direction compared to the decoder-based DAC shown in Fig. 12.20.

When the positive input of every comparator exceeds its negative input coming from the reference-generating resistive string, the comparator outputs a high value. The priority encoder generates a binary value based on its highest-ranking active input, ignoring all the others.

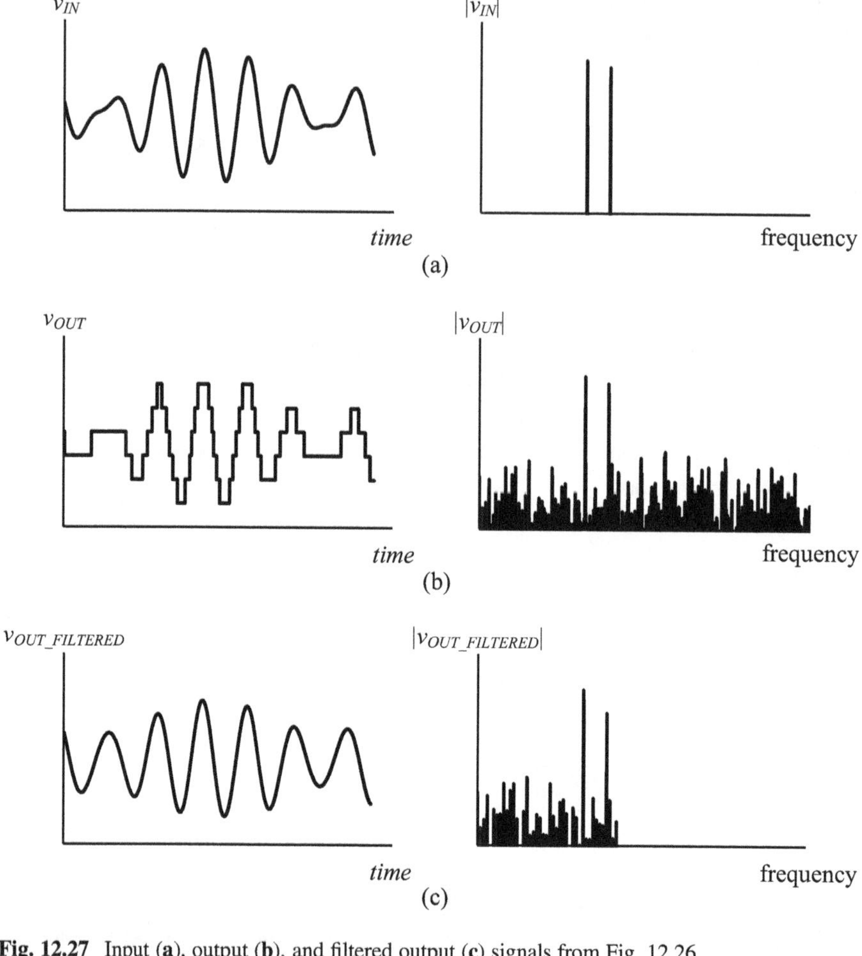

Fig. 12.27 Input (**a**), output (**b**), and filtered output (**c**) signals from Fig. 12.26

Table 12.2 Classification of ADCs

Low-to-medium speed High accuracy	Medium speed Medium accuracy	High speed Low-to-medium accuracy
Integrating	Algorithmic Successive-Approximation	Flash Folding Interpolating Pipelined Time-Interleaved Two-Step

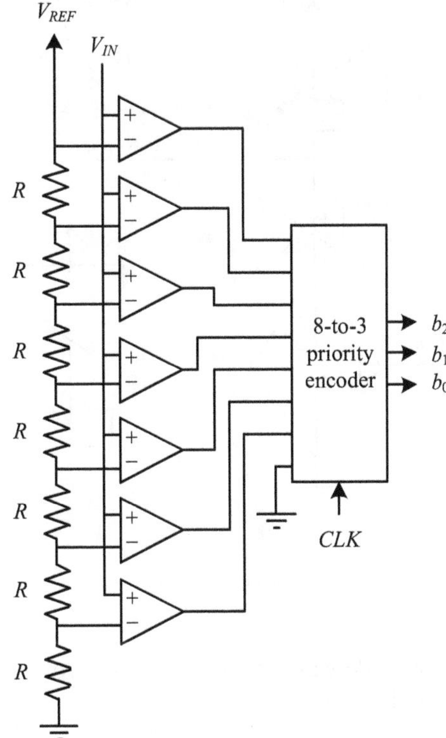

Fig. 12.28 Flash implementation of a 3-bit ADC

Similarly to the decoder-based DAC, this implementation uses matched resistors, which makes it a challenge to implement considering process variations. This limits the number of resistors used and thus the number of bits of the converter. On the other hand, it can support high-speed applications.

Note that the decoder makes use of a clock signal originating from a clock generator that will be discussed later.

12.5 Oscillators and Phase-Locked Loops

Many systems need a periodic signal for synchronization purposes. For example, a clock is needed in sequential logic implementations and a stable frequency is needed in communication systems. All this can be achieved using a programmable oscillator, which is usually placed within a Phase-Locked Loop structure (PLL). An alternative way to implement this at low frequencies is to make use of purely digital implementations, such as direct digital synthesis, but we will not cover this in this section.

An oscillator is a structure that issues a periodic output signal. The frequency of this output signal is controlled using a current or voltage control input signal. The

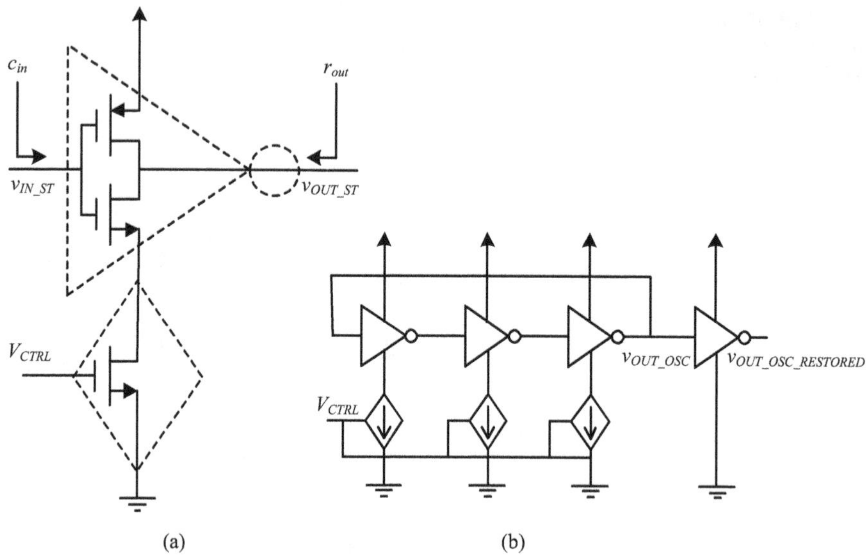

Fig. 12.29 Current-starved inverter stage (**a**) used to implement a 3-stage VCO (**b**)

most widely used control signal is a voltage signal resulting in a Voltage-Controlled Oscillator (VCO).

There are many ways to implement a VCO, but one of the simplest is based on the ring oscillator as shown in Fig. 12.29.

A ring oscillator is implemented using an odd number of inverting stages as in Fig. 12.29a connected in a ring fashion as in Fig. 12.29b. In this case, three stages are used. The delay in every stage dictates the frequency of the oscillator output signal. The longer the delay, the lower is the frequency.

In order to control the frequency at the output, current-starved CMOS inverters are used. The current is limited using a voltage-controlled current source at the power feed of the inverter. The less the current, the more time it takes for every stage to charge and discharge the input capacitance of the subsequent stage, thus the lower is the output signal frequency.

Sample VCO output signals are shown in Fig. 12.30. The signal in Fig. 12.30a is taken directly from the VCO. As can be seen, it contains some distortions particularly related to its ability to reach its maximum value. To remedy this, the signal is passed through a buffer at the output of the VCO, which might be a simple inverter, and the intended levels are restored as shown in Fig. 12.30b. Another advantage of this output buffer is that it isolates the VCO from the next stage, thus reducing its effect on the VCO signal.

The frequency of the VCO output signal is controlled by a DC voltage V_{CTRL}. Therefore, every output frequency corresponds to a value of V_{CTRL}. It is customary to plot the output frequency versus V_{CTRL} resulting in a graph called the tuning curve as in Fig. 12.31.

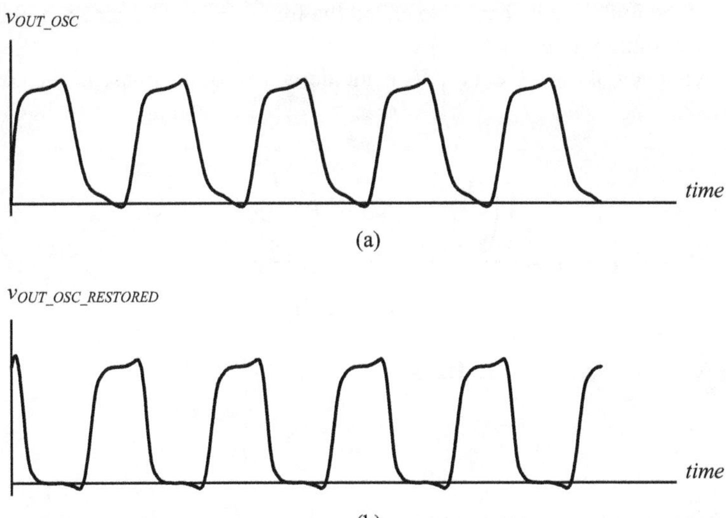

Fig. 12.30   VCO output signal (**a**) and restored output signal (**b**)

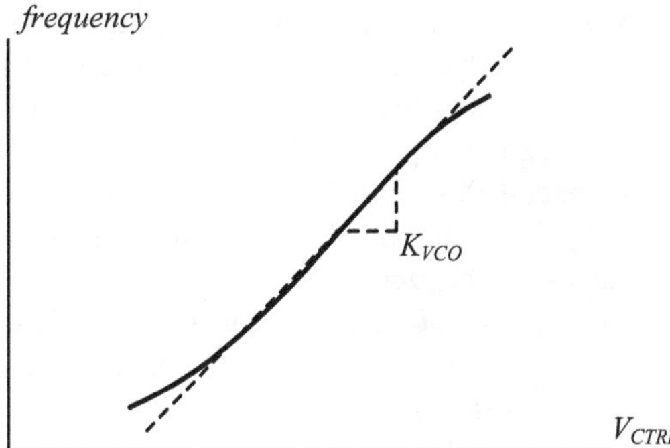

Fig. 12.31   VCO tuning curve

It is desirable, from a controllability point-of-view, to have a linear tuning curve. However, in practice, tuning curves are not perfectly linear, but can be approximated as a straight line such as the dotted line in Fig. 12.31. The slope of this line is called the sensitivity of the VCO, usually denoted as K_{VCO}. If the frequency is in rad/s, then K_{VCO} will be in rad/s•V.

An ideal VCO output is given as follows

$$v_{OUT\_OSC}(t) = A\cos\left(2\pi f_c t + K_{VCO}\int_0^t v_{CTRL}(t)\,dt + \varphi\right) \tag{12.16}$$

where f_c is the initial frequency, also called the free-running frequency, of the VCO and ϕ is the initial phase.

The frequency of the VCO in rad/s is the derivative of the phase of the cosine in (12.16):

$$\omega_{OUT\_OSC}(t) = \frac{d\left(2\pi f_c t + K_{VCO}\int_0^t v_{CTRL}(t)\,dt + \varphi\right)}{dt} \tag{12.17}$$
$$= 2\pi f_c + K_{VCO}v_{CTRL}(t)$$

As a result, the frequency in Hz is

$$f_{OUT\_OSC}(t) = f_c + \frac{K_{VCO}}{2\pi}v_{CTRL}(t) \tag{12.18}$$

which is ideally a linear function of the control voltage.

An important drawback of using only a VCO is that V_{CTRL} controls the output frequency and not its phase. In other words, even though we can know the output signal frequency of the VCO, we cannot know its phase since there is nothing that fixes it. As a result, the phase might change from time to time due to external or internal perturbations such as interference and noise. Most applications cannot tolerate this phase change, so we need to stabilize this phase and this is where a Phase-Locked Loop (PLL) is used.

A general PLL with its associated signal waveforms are shown in Fig. 12.32. Since we are using this PLL to synthesize a signal, then this is called a PLL-based frequency synthesizer.

The input is the reference signal sig_{REF} that comes from a phase-stable source such as a crystal oscillator. The output is sig_{VCO} that can have either the same frequency as sig_{VCO} or a multiple of its frequency, depending on the divider. Either way, sig_{VCO} is phase-locked to sig_{REF}, which means that whenever there is an edge occurrence in sig_{REF}, an edge occurrence will also happen in sig_{VCO}.

A short description of every block is as follows:

- The phase-frequency detector (PFD) takes as an input sig_{REF} and sig_{DIV} and outputs a pulse-width-modulated signal sig_{PFD}, whose average is proportional to the phase (and frequency) difference between sig_{REF} and sig_{DIV}.
- The low-pass filter (LPF) takes the average of sig_{PFD} and outputs a signal sig_{LPF} to control the frequency of the VCO. Changing the frequency of the VCO also changes its phase.
- The divider takes sig_{VCO} and divides it by a certain factor N to generate sig_{DIV}. The divider can be thought of as a counter that issues one edge for every N edges of sig_{VCO}.

The loop will converge when sig_{DIV} has the same frequency and phase as sig_{REF}. After that, in steady state, sig_{LPF} will be constant.

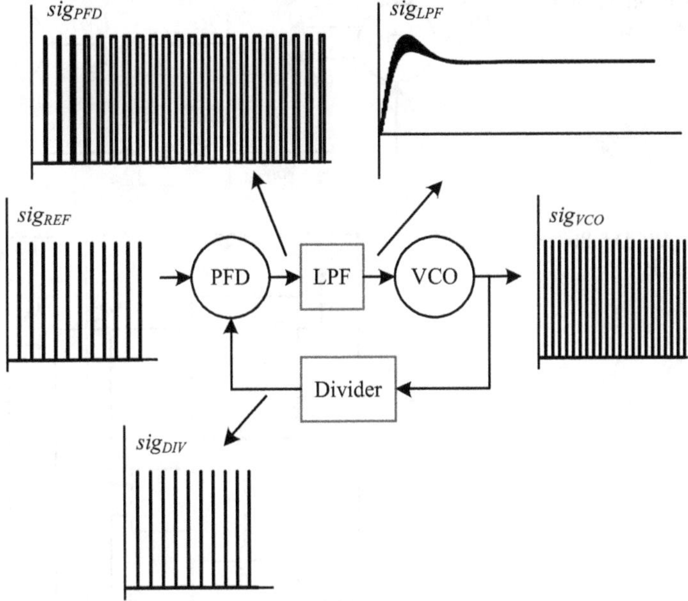

Fig. 12.32 General PLL with associated signal waveforms

If any perturbation happens at sig_{VCO}, the PLL will automatically adjust its phase, thus keeping it locked to sig_{REF}. This is the main advantage of a PLL compared to a stand-alone VCO. The time needed for this adjustment is called the settling time.

With the above signals being in the voltage domain, there is a difficulty implementing the above architecture if the supply voltage is low, which is the case with modern technologies. Therefore, it is desirable to have some of these signals in the current domain, where the level can be higher, while keeping the others in the voltage domain. This is done by splitting the PFD in Fig. 12.32 into two parts:

- The first is a PFD that issues a high signal v_{UP} or a low signal v_{DN} defending on whether the phase of the reference signal precedes or follows the phase of divider output.
- The second is a charge pump (CP) that acts as a transconductor, converting the output of the PFD from the voltage domain to the current domain.

Since the VCO takes a voltage as a control signal, the loop filter will have to be modified so that it takes a current signal at its input and outputs a voltage signal.

The resulting architecture is a charge-pump-based PLL as shown in Fig. 12.33.

An implementation of a PFD and a simple CP is shown in Fig. 12.34. The PFD consists of two D-type Flip Flops (D-FF) with a NAND gate connected in the feedback path. As for the CP, it consists of two transistors acting as transconductors.

Sample PFD-CP signal waveforms are shown in Fig. 12.35.

Fig. 12.33 Charge pump-based PLL

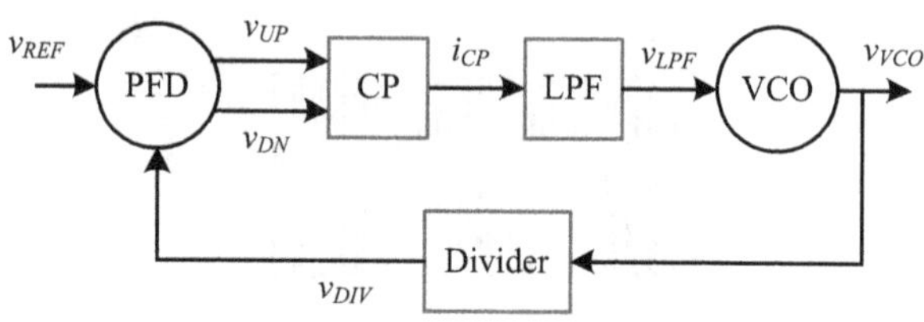

Fig. 12.34 A PFD followed by a simple CP

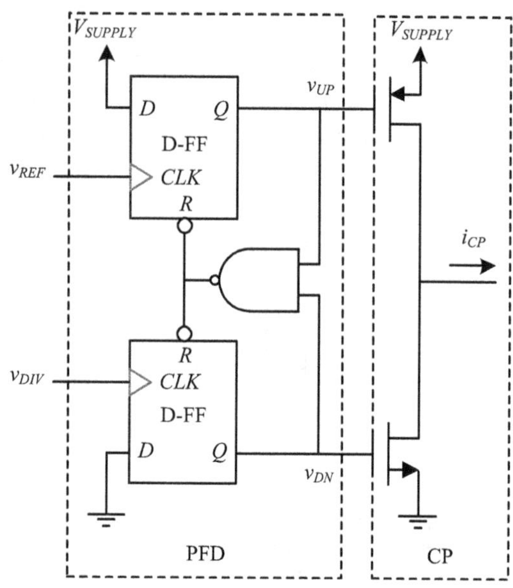

The system reads the phases based on the rising edges of the signals since the D-FFs are positive-edge-triggered. Signal v_{REF} leads signal v_{DIV}, therefore, the PFD generates a wide signal v_{UP}, which, when subtracted from the narrow v_{DN}, results in a positive i_{CP} from the CP. Note that the highest level of this current is I_{CP}. This tells the VCO in Fig. 12.33 to increase its frequency so that the phase of v_{DIV} gets closer to that of v_{REF}.

The transfer function of the PFD-CP combination in Fig. 12.34 consists of the output current i_{CP} versus the difference between the phases of the v_{REF} signal and the v_{DIV} signal. The ideal transfer function is shown in Fig. 12.36. As can be seen, the ideal transfer function is linear with a slope of $K_{PFD\text{-}CP}$.

The LPF takes a current signal at its input and outputs a voltage signal. As an example, the LPF shown in Fig. 12.37 will be used.

The transfer function of this LPF is

$$Z(s) = \frac{sRC_1 + 1}{s(sRC_1C_2 + C_1 + C_2)} \tag{12.19}$$

The order of s in the denominator is two, so this is a second-order filter.

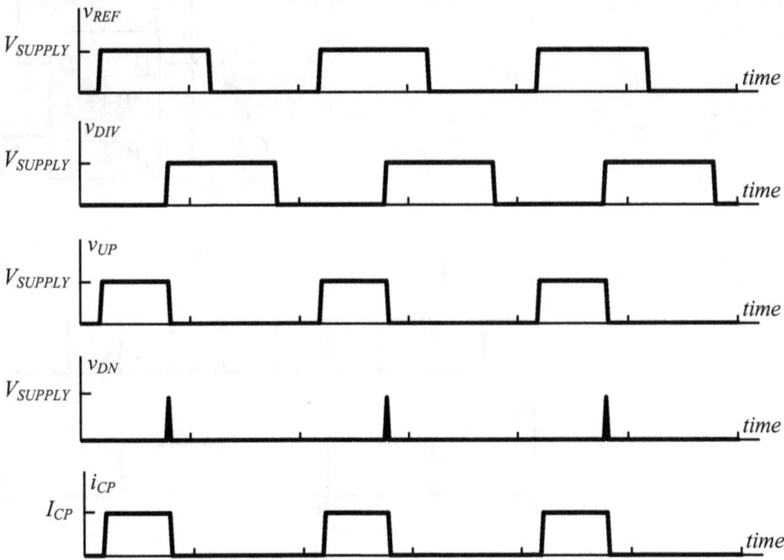

Fig. 12.35 Sample PFD-CP signal waveforms

Fig. 12.36 Ideal PFD-CP transfer function

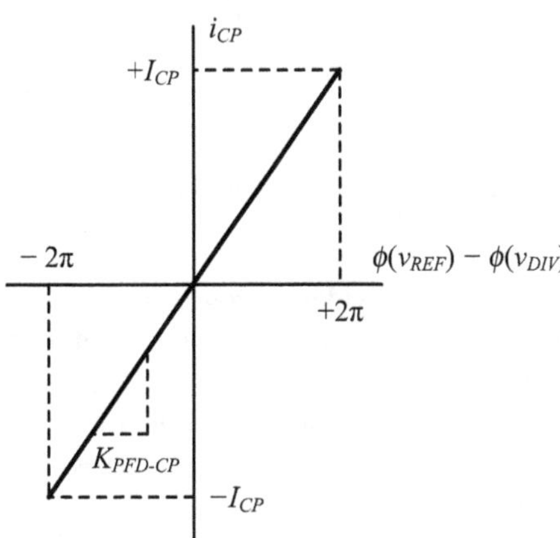

Fig. 12.37 Second-order current-to-voltage LPF

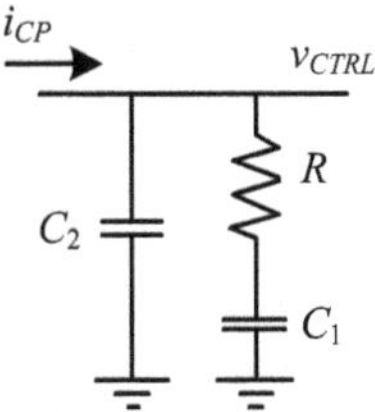

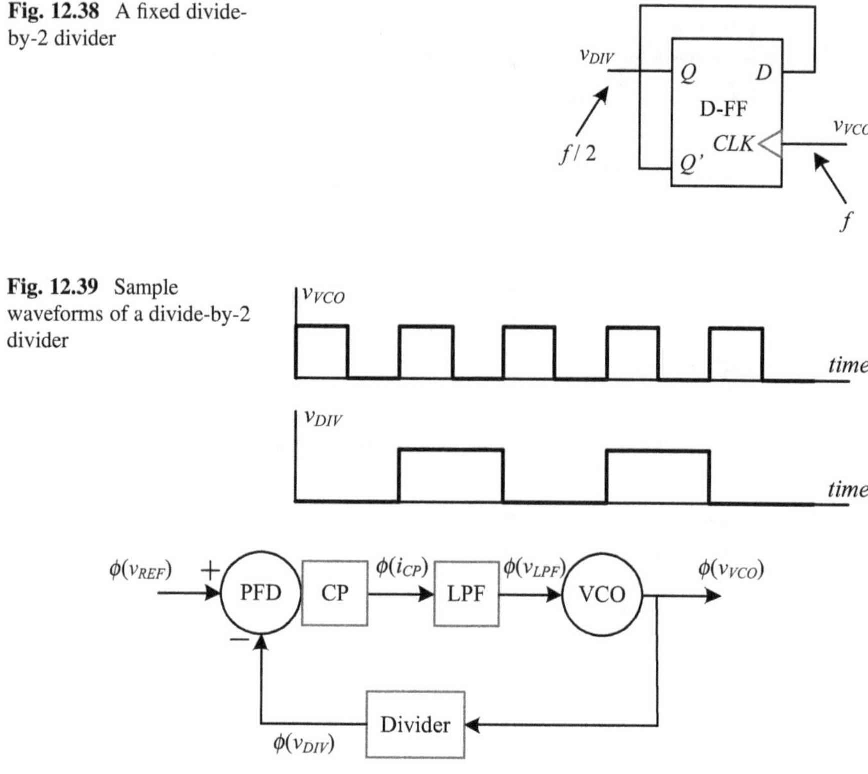

Fig. 12.38 A fixed divide-by-2 divider

Fig. 12.39 Sample waveforms of a divide-by-2 divider

Fig. 12.40 Linear phase-domain model of a CP-based PLL

As for the divider, a programmable one is usually used. In the simplest case where a fixed frequency is needed at the output of the VCO, a counter with fixed N can be used. If N is a power of 2, a cascade of the D-FF configuration shown in Fig. 12.38 can be used, with the input being from the right side as placed in the PLL. One stage will give a division by 2, two stages will give a division by 4, three stages will give a division by 8, and so on.

Sample waveforms from this divider are shown in Fig. 12.39. Notice how the change in the output happens at the rising edge of the input since the D-FF is positive-edge-triggered.

One way to model the CP-based PLL in Fig. 12.33 in order to analyze and design it is to linearize its behavior and model it in the phase domain. The result is a feedback Single-Input-Single-Output (SISO) Linear-Time-Invariant (LTI) model as shown in Fig. 12.40.

Note how the variable is now a phase instead of being a voltage or a current. Also, the PFD is combined with the CP for convenience reasons, and note the sequence in which the PFD-CP subtracts its inputs.

As such, the linear model of each block is shown in Table 12.3.

Table 12.3 Linear models of the blocks in a CP-based PLL

Block	Model
PFD-CP	$K_{PFD-CP} = \frac{I_p}{2\pi}$
LPF	$Z(s) = \frac{sRC_1+1}{s(sRC_1C_2+C_1+C_2)}$
VCO	$\frac{K_{VCO}}{s}$
Divider	$\frac{1}{N}$

The model of the PFD-CP comes directly from Fig. 12.36. The model of the VCO comes from (12.17) while realizing that the phase is the integration of the frequency, which in the s-domain translates to a division by s. Consequently, given this linear phase-frequency relation in the s-domain, the divider divides the frequency by N, which will also apply to the phase.

Following the convention used in feedback controls, the feed-forward transfer function $G(s)$ consists of the combined models of the PFD-CP, LPF, and VCO, whereas $H(s)$ consists of the model of the divider. As a result, the open-loop transfer function is

$$G(s)H(s) = \frac{K_{PFD-CP}Z(s)K_{VCO}}{sN} = \frac{K_{VCO}I_D}{s2\pi N}\frac{sRC_1+1}{s(sRC_1C_2+C_1+C_2)}$$

$$= \frac{K_{VCO}I_D}{s^2(C_1+C_2)2\pi N}\left(\frac{\frac{s}{\frac{1}{RC_1}}+1}{\frac{s}{\frac{C_1+C_2}{RC_1C_2}}+1}\right) \tag{12.20}$$

Looking at (12.20), we can see that the open-loop transfer function has a zero ω_z and a pole ω_p as follows:

$$\omega_z = \frac{1}{RC_1} \qquad \omega_p = \frac{C_1+C_2}{RC_1C_2} = \frac{1+C_2/C_1}{RC_2} \approx \frac{1}{RC_2} \tag{12.21}$$

This last approximation is due to the fact that in an optimized system, as we will see later, $C_2 << C_1$.

As a result of (12.20), the closed-loop transfer function is

$$\frac{G(s)}{1+G(s)H(s)}$$

$$= \frac{K_{VCO}K_{PFD-CP}N(sRC_1+1)}{s^3NRC_1C_2 + s^2N(C_1+C_2) + sRC_1K_{VCO}K_{PFD-CP} + K_{VCO}K_{PFD-CP}} \tag{12.22}$$

From the maximum power of s in the denominator, we can see that this is a third-order system, always one more than the order of the filter in (12.19).

To analyze the behavior of this PLL, we will approximate it as a second-order system whose analysis is much simpler.

In a second-order system, we will optimize the settling time of the PLL by relating it to the damping ratio ζ, which will then be related to the phase margin γ.

A second-order system has a standard open-loop transfer function

$$G_2(s)H_2(s) = \frac{\omega_n^2}{s(s + 2\omega_n\zeta)} \tag{12.23}$$

and a closed-loop transfer function

$$\frac{G_2(s)}{1 + G_2(s)H_2(s)} = \frac{\omega_n^2}{s^2 + 2\omega_n\zeta s + \omega_n^2} \tag{12.24}$$

where ω_n is the natural frequency.

The phase margin γ in degrees is defined as

$$\gamma = 180 + \angle(G_2(j\omega_c)H_2(j\omega_c)) \tag{12.25}$$

where ω_c is the gain cross-over frequency defined as

$$|G_2(j\omega_c)H_2(j\omega_c)| = 1 \tag{12.26}$$

ω_c consists of the bandwidth of the LPF and it approximates the bandwidth of the whole PLL.

Combining (12.23) with (12.26), we find that

$$\omega_c = \omega_n\sqrt{\sqrt{1 + 4\zeta^4} - 2\zeta^2} \tag{12.27}$$

Also, combining (12.23) with (12.25) and (12.27) gives the phase margin

$$\gamma = \tan^{-1}\left(\frac{2\zeta}{\sqrt{\sqrt{1 + 4\zeta^4} - 2\zeta^2}}\right) \tag{12.28}$$

which is independent of ω_n.

Now, we will optimize the PLL to have a small settling time. We choose the damping ratio slightly higher than 0.5, thus $\zeta \approx 0.51$, which, from (12.28), results in a phase margin $\gamma \approx 53°$.

Combining the closed-loop transfer function in (12.20) with the definition of the gain cross-over frequency in (12.26) we get the CP current

$$I_D = \sqrt{\frac{4\pi^2 N^2 \left(\omega_c{}^4 (C_1 + C_2)^2 + \omega_c{}^6 R^2 C_1{}^2 C_2{}^2\right)}{K_{VCO}{}^2 \left(1 + \omega_c{}^2 R^2 C_1{}^2\right)}} \qquad (12.29)$$

Since $\omega_c{}^6 R^2 C_1{}^2 C_2{}^2 < < \omega_c{}^4 (C_1 + C_2)^2$ and $1 < < \omega_c{}^2 R^2 C_1{}^2$, then

$$I_D \approx \frac{2\pi N \omega_c (C_1 + C_2)}{K_{VCO} R C_1} \Leftrightarrow \omega_c \approx \frac{C_1}{C_1 + C_2} \frac{I_D K_{VCO} R}{2\pi N} \qquad (12.30)$$

As we will see, $C_2 < < C_1$, in which case

$$\omega_c \approx \frac{I_D K_{VCO} R}{2\pi N} \Leftrightarrow R \approx \frac{2\pi N}{I_D K_{VCO}} \omega_c \qquad (12.31)$$

Placing ω_c in the middle between ω_z and ω_p on a logarithmic scale in order to take advantage of the filter, ω_c will be the geometric mean of ω_z and ω_p:

$$\omega_c = \sqrt{\omega_z \omega_p} \Rightarrow \frac{\omega_c}{\omega_z} = \frac{\omega_p}{\omega_c} \qquad (12.32)$$

These two ratios are important, so we give them the following symbols:

$$\alpha = \frac{\omega_c}{\omega_z} \approx \omega_c R C_1 \qquad (12.33)$$

$$\beta = \frac{\omega_p}{\omega_c} \approx \frac{1}{\omega_c R C_2} \qquad (12.34)$$

Combining (12.31) with (12.33) and (12.34) by substituting for R, we get

$$C_1 = \frac{I_D K_{VCO}}{2\pi N} \frac{\alpha}{\omega_c{}^2} \qquad (12.35)$$

$$C_2 = \frac{I_D K_{VCO}}{2\pi N} \frac{1}{\omega_c{}^2 \beta} \qquad (12.36)$$

Solving for the phase margin in (12.20), we get

$$\gamma = \tan^{-1}(\omega_c R C_1) - \tan^{-1}\left(\frac{\omega_c R C_1 C_2}{C_1 + C_2}\right) \approx \tan^{-1}\left(\frac{\omega_c}{\omega_z}\right) - \tan^{-1}\left(\frac{\omega_c}{\omega_p}\right) \qquad (12.37)$$

and

Table 12.4 Equations used to design the PLL LPF

Parameter	Equation
R	$\dfrac{2\pi N}{I_D K_{VCO}}\,\omega_c$
C_1	$\dfrac{I_D K_{VCO}}{2\pi N}\,\dfrac{3}{\omega_c^2} = \dfrac{3}{R\omega_c}$
C_2	$\dfrac{I_D K_{VCO}}{2\pi N}\,\dfrac{1}{3\omega_c^2} = \dfrac{1}{3R\omega_c}$

$$\gamma \approx \tan^{-1}(\alpha) - \tan^{-1}\left(\frac{1}{\beta}\right) \tag{12.38}$$

This value will be set close to 53°. Solving for α and β, we get $\alpha = \beta = 3$. The result of this is that

$$\alpha\beta = 9 \Rightarrow \frac{\omega_p}{\omega_z} = 9 \Rightarrow \frac{C_1}{C_2} = 9 \tag{12.39}$$

This justifies the assumption made earlier that $C_2 < \, < C_1$.

Resulting from this, and as a summary, given a certain system with a certain K_{VCO}, N, PLL bandwidth ω_c, and I_D, we can use the equations in Table 12.4 to get the parameters R, C_1, and C_2 of the LPF.

Index

© The Editor(s) (if applicable) and The Author(s), under exclusive license to
Springer Nature Switzerland AG 2024
J. G. Atallah, M. Ismail, *Integrated Electronic Circuits*,
https://doi.org/10.1007/978-3-031-62707-1

9 783031 627064